Hyperspherical Harmonics and Their Physical Applications

Hypersphereical Harmonics and Their Physical Applications

James Emil Avery
John Scales Avery

Niels Bohr Institute and H.C. Ørsted Institute,
University of Copenhagen, Denmark

World Scientific

NEW JERSEY • LONDON • SINGAPORE • BEIJING • SHANGHAI • HONG KONG • TAIPEI • CHENNAI • TOKYO

Published by

World Scientific Publishing Co. Pte. Ltd.

5 Toh Tuck Link, Singapore 596224

USA office: 27 Warren Street, Suite 401-402, Hackensack, NJ 07601

UK office: 57 Shelton Street, Covent Garden, London WC2H 9HE

Library of Congress Cataloging-in-Publication Data

Names: Avery, James Emil, author. | Avery, John, 1933– author.

Title: Hyperspherical harmonics and their physical applications / James Emil Avery (University of Copenhagen, Denmark), John Scales Avery (University of Copenhagen, Denmark).

Description: New Jersey : World Scientific, 2017. | Includes bibliographical references and index.

Identifiers: LCCN 2017026745 | ISBN 9789813229297 (hardcover : alk. paper)

Subjects: LCSH: Spherical harmonics--Problems, exercises, etc. | Scattering (Physics)--Problems, exercises, etc.

Classification: LCC QC20.7.S645 A94 2017 | DDC 515/.785--dc23

LC record available at https://lccn.loc.gov/2017026745

British Library Cataloguing-in-Publication Data

A catalogue record for this book is available from the British Library.

Cover image: The real part of the Coulomb Sturmian function $\chi_{12,6,4}(x, y, z)$ in a contour plot taken through the plane $z = 0$. Positive values are coloured blue, while negative values are set in orange. Through Fock's transformation of momentum space onto the sphere in four dimensions, this function becomes the hyperspherical harmonic $Y_{12,6,4}(\hat{\mathbf{u}})$.

For any available supplementary material, please visit
http://www.worldscientific.com/worldscibooks/10.1142/10690#t=suppl

Printed in Singapore

For Birthe

Contents

Preface xiii

1. HARMONIC FUNCTIONS 1

1.1 Definitions 1
1.2 The canonical decomposition of a homogeneous polynomial 3
1.3 More general canonical decompositions 8
1.4 Exercises 9

2. GENERALIZED ANGULAR MOMENTUM 11

2.1 Eigenfunctions of generalized angular momentum 11
2.2 Hyperspherical harmonics 12
2.3 Generalized solid angle 13
2.4 Hyperangular integration 14
2.5 A general theorem for hyperangular integration 15
2.6 Chains of subgroups 17
2.7 Degeneracy of hyperspherical harmonics 18
2.8 Representations of $\Lambda_{s,t}$ 19
2.9 Exercises 22

3. GEGENBAUER POLYNOMIALS 25

3.1 The generating function for Gegenbauer polynomials . . . 25
3.2 Properties of Gegenbauer polynomials 27
3.3 Determination of the constant K_λ 29
3.4 The generalization of $P_l^m(\mathbf{u} \cdot \mathbf{u}')$ 30
3.5 The standard tree 33
3.6 Normalization 35

3.7 Nonstandard trees 40
3.8 Exercises 44

4. FOURIER TRANSFORMS IN d DIMENSIONS 45

4.1 Notation and basic properties 45
4.2 Expansions of a d-dimensional plane wave 46
4.3 The Green's function of Δ 48
4.4 Hyperspherical Bessel transforms 49
4.5 An alternative expansion of a plane wave 50
4.6 The Fourier convolution theorem 52
4.7 Harmonic analysis for non-Euclidean spaces 53
4.8 Exercises 54

5. FOCK'S TREATMENT OF HYDROGENLIKE ATOMS AND ITS GENERALIZATION 55

5.1 Fock's original treatment 55
5.2 Generalization of Fock's treatment 60
5.3 Normalization 64
5.4 Exercises 65

6. D-DIMENSIONAL HYDROGENLIKE ORBITALS IN DIRECT SPACE 67

6.1 Generalization of the 3-dimensional solution 67
6.2 Orthonormality 70
6.3 d-dimensional Coulomb Sturmians 71
6.4 Potential-weighted orthonormality 74
6.5 Fourier transforms of Coulomb Sturmians 76
6.6 Use of d-dimensional Coulomb Sturmians as a basis 76
6.7 Cases where the potential may have angular dependence . 78
6.8 Exercises 80

7. GENERALIZED STURMIANS 81

7.1 Generalized Sturmians and many-particle bound state problems 81
7.2 Sturmians and the many-particle Schrödinger equation . . 83
7.3 Momentum-space orthonormality of generalized Sturmian basis sets 84
7.4 Sturmian expansions of d-dimensional plane waves 85

7.5 Iteration of the Schrödinger equation 86
7.6 Molecular spectra 89
7.7 Goscinskian configurations in atomic physics 91
7.8 Derivation of the secular equations 93
7.9 The Large-Z approximation 96
7.10 Exercises 102

8. CHOOSING APPROPRIATE HYPERSPHERICAL REPRESENTATIONS 103

8.1 The alternative Coulomb Sturmians of Aquilanti and Coletti 103
8.2 Transformations between matrix representations 107
8.3 An example: Alternative Coulomb Sturmians corresponding to real spherical harmonics 108
8.4 The d-dimensional case 115
8.5 The d-dimensional Schrödinger equation in momentum space 119
8.6 Shibuya-Wulfman orbitals 120
8.7 Exercises 126

9. MOLECULAR INTEGRALS FROM HYPERSPHERICAL HARMONICS 127

9.1 Molecular integrals using exponential-type orbitals 127
9.2 Many center Sturmians 128
9.3 Overlap integrals involving Coulomb Sturmians 129
9.4 Shibuya-Wulfman integrals 132
9.5 Matrices representing kinetic energy and nuclear attraction 134
9.6 1-center densities in terms of 2k Sturmians 137
9.7 Interelectron repulsion integrals between two 1-center densities 140
9.8 2-center densities in terms of 2k Sturmians 143
9.9 Generalized scattering factors 150
9.10 Exercises 152

10. LAGRANGIANS FOR PARTICLES AND FIELDS 153

10.1 Introduction 153
10.2 Cyclic coordinates 156
10.3 Hamilton's unified formulation 158

10.4 Normal modes 161
10.5 Molecular vibrations and rotations 166
10.6 Lagrangian densities for fields 167
10.7 Electromagnetic potentials 169
10.8 Metric tensors 173
10.9 The Laplace-Beltrami operator 177
10.10 Geodesics 182
10.11 Huygens' principle 183
10.12 Exercises 184

11. COORDINATE TRANSFORMATIONS FOR N BODIES 187

11.1 Transforming the kinetic energy operator 187
11.2 The Laplace-Beltrami operator 191
11.3 Transformation of the kinetic energy operator 192
11.4 A simple example 193
11.5 Jacobi coordinates of a 3-body system with equal masses . 194
11.6 Normal mode transformations 195
11.7 A simple model for interatomic forces in molecules 196
11.8 A simple example 196
11.9 Separability in the harmonic approximation 200
11.10 The Morse potential 201
11.11 Exact solutions for the Morse oscillator 202
11.12 Morse oscillator eigenfunctions as a basis 203
11.13 Rotations and vibrations of diatomic molecules 205
11.14 The rotational and vibrational modes of molecules and clusters 206
11.15 Normal modes 208
11.16 A simple example 208
11.17 Exercises 210

12. SOME ILLUSTRATIVE EXAMPLES 211

12.1 Matrix elements of the kinetic energy operator 211
12.2 Matrix elements of the potential energy 211
12.3 Products of basis functions 213
12.4 Examples of basis sets 213
12.5 The problem of slow convergence 215
12.6 2-electron atoms: the results of Das et al. 216
12.7 Approximate relativistic corrections 219

12.8 Exercises 220

Appendix A THE D-DIMENSIONAL HARMONIC OSCILLATOR 221

A.1 Harmonic oscillators in one dimension 221
A.2 Creation and annihilation operators for harmonic oscillators 222
A.3 A collection of harmonic oscillators 225
A.4 d-dimensional isotropic harmonic oscillators 226
A.5 Fourier transforms of 3-dimensional harmonic oscillator wave functions 228
A.6 The hyperspherical Bessel transform of the radial function 229
A.7 Coupling coefficients for harmonic oscillator wave functions 230

Appendix B MOLECULAR INTEGRALS FOR SLATER-TYPE ORBITALS 233

B.1 Definition of STO's 233
B.2 Expansion of an arbitrary function of $s = kr$ in terms of Coulomb Sturmian radial functions 233
B.3 Evaluation of molecular integrals 236
B.4 STO overlap integrals 239
B.5 Interelectron repulsion integrals with STO's 240
B.6 Checks in the atomic case 242

Bibliography 247

Index 275

Preface

Everyone working in theoretical physics and chemistry is familiar with the beauty and utility of spherical harmonics. Solutions to the Schrödinger equation are very often obtained by separating the problem into an angular part and a radial part. The angular part of the problem is then solved using the elegant theorems involving spherical harmonics. For example, in electronic structure theory, the wavefunction is built up from radial functions and spherical harmonics, and in scattering problems, one uses expansions of plane waves in terms of spherical harmonics and spherical Bessel functions.

Remarkably, all of the theorems involving 3-dimensional harmonics have simple and beautiful d-dimensional generalizations. For example, a d-dimensional plane wave may be expanded in terms of hyperspherical harmonics and functions of the hyperradius which we call "hyperspherical Bessel functions".

Hyperspherical harmonics have shown themselves to be extremely useful, both in nuclear physics and in reactive scattering theory. However, their use has been confined to specialists with very strong backgrounds in mathematics. The aim of this book is to change the theory of hyperspherical harmonics from an esoteric field, mastered by specialists, into an easily-used tool with a place in the working kit of all theoretical physicists, theoretical chemists and mathematicians. The theory that we present in this book is accessible without knowledge of Lie-groups and representation theory, and can be understood with an ordinary knowledge of calculus. The book is accompanied by programs and exercises designed for teaching and practical use.

But why do we claim that the theory of hyperspherical harmonics ought to be part of the tool-kit of all theoretical workers in the physical sciences? Let us next try to answer this question.

In solving the Schrödinger equation for many-electron systems, the starting point is often the Hartree-Fock approximation, which is an independent-particle model. In such a model, each electron is thought of as moving in the attractive field of the nuclei and an averaged repulsive field due to all the other electrons. Electron correlation effects are then added using the formalism of configuration interaction.

In nuclear physics, however, the wave function of the nucleus is usually so highly correlated that an independent-particle model such as the Hartree-Fock approximation is not a good starting point. Instead, one tries to solve the many-nucleon Schrödinger equation directly, in a $(3N-6)$-dimensional space. In order to simplify such solutions, nuclear physicists have long relied on the theory of hyperspherical harmonics. But even in the study of electronic structure, one encounters systems that are so strongly correlated that the starting point must be a many-dimensional wave-equation. Furthermore, hyperspherical harmonics allow difficult molecular integrals involving exponential-type orbitals to be evaluated both rapidly and accurately.

In Chapter 1 of this book, we approach the theory of hyperspherical harmonics by starting from the theory of harmonic polynomials in a d-dimensional space. Harmonic polynomials are just homogeneous polynomials that satisfy the Laplace equation. The d-dimensional generalization is very easy: d-dimensional harmonic polynomials are just homogeneous polynomials of d Cartesian coordinates that also satisfy the generalized Laplace equation, and hyperspherical harmonics are harmonic polynomials in the Cartesian coordinates of the unit hypersphere.

Chapter 2 introduces generalized angular momentum in d-dimensional spaces. Hyperspherical harmonics are the eigenfunctions of the generalized angular momentum operator. They arise from irreducible representations of the rotation group $SO(d)$ and form complete basis sets, so that one can synthesize any smooth angular function. In this chapter, we also present a powerful hyperangular integration theorem that makes integration of polynomials on the unit sphere trivial.

In Chapter 3, we discuss the properties of Gegenbauer polynomials, which are sometimes called ultra-spherical polynomials. The Gegenbauer polynomials play the same role for hyperspherical harmonics as Legendre polynomials play in the theory of spherical harmonics. In fact, for $d = 3$, Gegenbauer polynomials are just Legendre polynomials. The reader may be familiar with the expansion of 3-dimensional plane waves in terms of Legendre polynomials multiplied by spherical Bessel functions. Analogously, a d-dimensional plane wave is a sum of hyperspherical Bessel functions times Gegenbauer polynomials. Gegenbauer polynomials also help us to generate

d-dimensional hyperspherical harmonics.

Chapter 4 introduces a number of tools for working with Fourier transforms in d-dimensional spaces. These tools will be used throughout the book. Of particular importance is the hyperspherical Bessel transform, which is shown to be no more difficult than the three dimensional spherical Bessel transform. It is also shown that the Green's function of the d-dimensional Laplace operator is the generating function for Gegenbauer polynomials, just as the Green's function of the 3-dimensional Laplacian generates the Legendre polynomials. Finally, we show properties of d-dimensional plane wave expansions both in terms of hyperspherical harmonic, and in terms of general complete basis sets.

In an extremely brilliant early paper, the Russian physicist V. Fock ([Fock, 1935], [Fock, 1958]) showed that the Schrödinger equation for hydrogenlike atoms in reciprocal space can be solved by means of 4-dimensional hyperspherical harmonics. To obtain these solutions, he projected 3-dimensional momentum-space onto the surface of a 4-dimensional unit sphere.

Fock's momentum-space solutions, when transformed back into direct space, are identical with the Coulomb Sturmian basis sets which were introduced into theoretical physics and theoretical chemistry by Per-Olov Löwdin and others ([Löwdin, 1955]). Coulomb Sturmian basis sets have very desirable completeness properties, and for this reason they are widely used. Fock's momentum-space calculation was soon generalized to d dimensions, and it was shown that solutions to the d-dimensional hydrogenlike Schrödinger equation can be expressed in terms of $(d+1)$-dimensional hyperspherical harmonics. The d-dimensional Coulomb Sturmians also have desirable completeness properties. Chapters 5 and 6 discuss the momentum space work of Fock and its generalizations to d dimensions.

The name "Sturmians" was due to M. Rotenberg ([Rotenberg, 1962],[Rotenberg, 1970]), who wished to emphasize the connection of these functions with Sturm-Liouville theory. Later, Osvaldo Goscinski ([Goscinski, 2003]) generalized the concept of Sturmian basis sets. He discussed many-particle isoenergetic basis sets which are solutions to the Schrödinger equation with weighted potentials. In such a scheme, the weighting factor attached to the potential takes on the role of the eigenvalue. In Chapter 7, we demonstrate the usefulness of Goscinskian configurations. A much more in-depth treatment of Goscinskian basis sets is found in our previous book ([Avery and Avery, 2006]).

There are many ways to construct hyperspherical harmonic basis sets.

In fact, they are determined only up to unitary transformations within the eigenspace of each principal eigenvalue λ. To achieve rapid convergence, it is important to choose hyperspherical basis sets that capture the geometry of the physical problem. Methods for doing this are discussed in Chapter 8.

In Chapter 9, we show how the use of 4-dimensional hyperspherical harmonics the otherwise difficult many-center interelectron repulsion integrals between exponential-type orbitals to be calculated rapidly and accurately. Currently computational quantum chemistry is dominated by a technology based on the use of Gaussian basis sets. This is due to the computational ease with which Gaussian many-center repulsion integrals can be calculated. However, Gaussian basis sets fail to represent accurately both the behavior of the electronic wave function at cusps, and its long-distance asymptotic behavior. Therefore many authors have addressed the difficult mathematical problems connected with the use of exponential-type orbitals (See, for example [Harris and Michels, 1967], [Weatherford, 1982] and [Weatherford and Jones, 1982]), but the problem is still open. We hope that our work will contribute to progress in this field. Molecular Sturmians and their integrals are discussed in depth in our book ([Avery, Rettrup and Avery, 2012]).

The Laplace-Beltrami operator is the generalization of the familiar Laplacian operator to curved spaces, and it plays a central role in the theory of hyperspherical harmonics. In Chapter 10 we derive an invariant form of the Laplace-Beltrami operator using variational methods. We show how the operator can be expressed in terms of the covariant and contravariant metric tensors of a space. In this chapter, variational methods are also applied to the equations governing the behavior of particles and fields.

The manifestly invariant form of the Laplace-Beltrami operator is used in Chapter 11 to transform the kinetic energy operator under arbitrary curvilinear coordinate changes. Normal mode transformations are discussed in connection with the rotational and vibrational spectra of molecules, and the use of exact Morse oscillator solutions as basis functions.

Finally, in Chapter 12, various types of basis sets are discussed, and a few calculational results are compared with experiment.

Computer programs and exercises

The programs that accompany this book allow the user to generate hyperspherical harmonics with ease, and to rapidly and automatically perform angular and hyperangular integration. The program for doing so is based on a powerful and general hyperangular integration theorem discussed in Chapter 2 of this book. To illustrate some of the main themes of the book, exercises are provided at the end of each chapter. These exercises are most easily solved if Mathematica ([Wolfram Research, 1991–2017]) is used as a computational aid.

The programs and instructions for how to use them, as well as electronic versions of the exercises and solutions, can be found the web-site:

`http://harmonics.kvante.org`

Contributions from other authors

Scientists with an interest in nuclear physics will find many nuclear applications of hyperspherical harmonics in the excellent recent textbook by Tapan Kumar Das entitled *Hyperspherical Expansion Techniques: Application to Problems in Physics* ([Das, 2016]). Among the topics treated by Professor Das are general many-body systems, the trinucleon system, potential harmonics, and the application of hyperspherical techniques to nuclear Bose-Einstein condensation.

Methods based on hyperspherical harmonics have been successfully applied to the analysis of reactive scattering and inelastic scattering in nuclear physics, and to nuclear decay. The following articles are representative of such studies: [Kezerashvili, 1983] and [Fang, 1977].

Vincenzo Aquilanti and his group at the University of Perugia in Italy have made important contributions to the fundamental mathematics of hyperspherical harmonics and hyperangular theory. They have applied the theory to solve problems in reactive scattering, electronic structure, and the chemistry and dynamics of clusters ([Aquilanti and Avery, 2001]–[Aquilanti and Calligiana, 2003]). In Argentina, Gustavo Gasaneo and his co-workers have also used hyperspherical harmonics and Coulomb Sturmians to treat scattering problems ([Gasaneo et al., 2009]–[Gasaneo and Macek, 2002]).

Applications of hyperspherical harmonics to molecular reactive scattering calculations have been developed by Prof. Aaron Kupperman's group at the California Institute of Technology [Kupperman and Hypes, 1986; Kuppermann, 2003], and as well by Prof. Vincenzo Aquilanti and his group at the University of Perugia in Italy ([Aquilanti et al., 1986], [Aquilanti et al., 1996]). The great importance of this work on the theory of molecular reactive scattering derives from the analysis of experimental studies of the detailed mechanism of chemical reactions. This field was pioneered by Dudley R. Herschbach, Yuan T. Lee and John C. Polanyi, who won the 1986 Nobel Prize in Chemistry for this work ([Lee, Gordon and Herschbach, 1971]).

Other work of note in the application to quantum chemistry calculations can be found for example in [Alijah et al., 1995], [Lepetit et al., 2006], [Barreto et al., 2007], [Mason and Schuh, 2008], and [Marcucci et al., 2009].

Much of the early work with hyperspherical harmonics has been done by Soviet nuclear physicists. The reason for this is that nuclear ground state wave functions are so highly correlated that independent particle approximations fail completely, and to obtain accurate results one must solve directly in a $(3N-6)$-dimensional space. The Russian authors have developed the Method of Trees, [Smorodinskii and Efros, 1973], [Kunetsov and Smorodinskii, 1976], [Klepikov, 1974] for generating symmetry-adapted hyperspherical harmonics, which they then use to calculate nuclear wave functions.

In the Method of Trees, one considers two subspaces, $(x_1, \ldots, x_{d_1})$ and $(x_{d_1+1}, \ldots, x_d)$, in each of which harmonic polynomials have been constructed. One then combines these harmonic polynomials in such a way as to produce a polynomial that is harmonic in the total space, making use of the properties of Jacobi polynomials. By successive application of this method, starting with 2-dimensional spaces, one can build up symmetry-adapted, orthonormal sets of hyperspherical harmonics appropriate for the nuclear N-body problem. For a review of this work, (see [Coletti, 1998] and [Aquilanti et al., 1996]).

Acknowledgments

We are extremely grateful to Professor Vincenzo Aquilanti and his coworkers at the University of Perugia, Italy for their decades-long hospitality, inspiration and collaboration in the field of hyperspherical harmonics. The pioneering work of the Perugia group is one of the essential foundations of this book.

One of us, (J.S.A.) gratefully acknowledges inspiration and knowledge gained from numerous visits to the laboratory of Professor Dudley R. Herschbach at Harvard University. These visits were made possible by grants from the Carlsberg Foundation of Denmark.

We would also like to express our sincere thanks to Ph.D. Stud. Spencer Kelly of the Department of Theoretical Physics, Imperial College of Science and Technology, London. We are grateful to him for his careful reading of several chapters of this book and for his numerous and valuable corrections and suggestions.

Chapter 1

HARMONIC FUNCTIONS

1.1 Definitions

Let us consider a d-dimensional Euclidean space with Cartesian coordinates $x_1, x_2, x_3, \ldots, x_d$. In this space, we can define a hyperradius r by the Euclidean radius:

$$r^2 \equiv \sum_{j=1}^{d} x_j^2 \tag{1.1}$$

We can also define the generalized Laplacian operator Δ by

$$\Delta \equiv \sum_{j=1}^{d} \frac{\partial^2}{\partial x_j^2} \tag{1.2}$$

A homogeneous polynomial of order n in the coordinates $x_1, x_2, x_3, \ldots, x_d$ is defined to be a polynomial of the form

$$f_n = a x_1^{n_1} x_2^{n_2} \cdots x_d^{n_d} + b x_1^{n'_1} x_2^{n'_2} \cdots x_d^{n'_d} + \cdots \tag{1.3}$$

where $a, b, c, \ldots$, are the coefficients of the polynomial, and

$$\begin{aligned} n_1 + n_2 + \cdots + n_d &= n \\ n'_1 + n'_2 + \cdots + n'_d &= n \\ \vdots \qquad\qquad & \;\vdots\; \vdots \end{aligned} \tag{1.4}$$

and where the numbers $n_1, n_2, \ldots, n_d$, $n'_1, n'_2, \ldots, n'_d$, are positive integers or zero. (For example, $f_2 = x^2 + 2xy$ is a homogeneous polynomial of order 2.)

A polynomial f_n is homogeneous and of order n if and only if the following equation holds

$$\sum_{j=1}^{d} x_j \frac{\partial f_n}{\partial x_j} = n f_n \tag{1.5}$$

(See exercise 10). We can see that (1.5) is true for a homogeneous polynomial, because for each monomial,

$$\sum_{j=1}^{d} x_j \frac{\partial}{\partial x_j} x_1^{n_1} x_2^{n_2} \cdots x_d^{n_d} = x_1^{n_1} x_2^{n_2} \cdots x_d^{n_d} \sum_{j=1}^{d} n_j \tag{1.6}$$

But from (1.4) we know that $n_1 + n_2 + \cdots + n_d = n$, and if each term of f_n has this property, then it holds for the whole polynomial.

It is interesting to notice that equation (1.5) holds for a wider class of functions than homogeneous polynomials. Let f_n be a homogeneous polynomial of order n, and let a be a number, not necessarily integral or positive or real. Then, making use of the chain rule and equation (1.5),

$$\begin{aligned}
\sum_{j=1}^{d} x_j \frac{\partial}{\partial x_j} (f_n^a) &= \sum_{j=1}^{d} x_j \left(\frac{\partial f_n^a}{\partial x_j} \right) \\
&= \sum_{j=1}^{d} x_j a f_n^{a-1} \frac{\partial}{\partial x_j} f_n \\
&= a f_n^{a-1} \sum_{j=1}^{d} x_j \frac{\partial}{\partial x_j} f_n \\
&= a f_n^{a-1} n f_n \\
&= an\, f_n^a
\end{aligned} \tag{1.7}$$

Now let f_n be a homogeneous polynomial of order n and $g_{n'}$ be a homogeneous polynomial of order n', and let a and b be any two numbers, not necessarily integers or positive or real. Then

$$\begin{aligned}
\sum_{j=1}^{d} x_j \frac{\partial}{\partial x_j} (f_n^a\, g_{n'}^b) &= \sum_{j=1}^{d} x_j \left(\frac{\partial f_n^a}{\partial x_j}\, g_{n'}^b + f_n^a\, \frac{\partial g_{n'}^b}{\partial x_j} \right) \\
&= (an + bn')(f_n^a\, g_{n'}^b)
\end{aligned} \tag{1.8}$$

In general, we will call a function f_s a *homogeneous function of order* s for some complex number s, if

$$\sum_{j=1}^{d} x_j \frac{\partial}{\partial x_j} f_s = s f_s \tag{1.9}$$

Homogeneous polynomials are special cases of homogeneous functions. Several properties of homogeneous functions flow from the definition in equation (1.9): The product of any two homogeneous functions is another homogeneous function whose order is the sum of the orders of the two terms in the product, since

$$\sum_{j=1}^{d} x_j \frac{\partial}{\partial x_j} (f_s\, f_{s'}) = f_{s'} \sum_{j=1}^{d} x_j \frac{\partial}{\partial x_j} f_s + f_s \sum_{j=1}^{d} x_j \frac{\partial}{\partial x_j} f_{s'}$$
$$= (s + s')\,(f_s\, f_{s'}) \tag{1.10}$$

The quotient of two homogeneous functions is a homogeneous function whose order is the difference between the orders of the two terms in the quotient, because

$$\sum_{j=1}^{d} x_j \frac{\partial}{\partial x_j} \left(\frac{f_s}{f_{s'}}\right) = \sum_{j=1}^{d} x_j \frac{\partial}{\partial x_j} (f_s\, f_{-s'})$$
$$= (s - s') \left(\frac{f_s}{f_{s'}}\right) \tag{1.11}$$

Finally, the sum of (or difference between) two homogeneous functions of the same order is a homogeneous function of that order:

$$\sum_{j=1}^{d} x_j \frac{\partial}{\partial x_j} (f_s \pm g_s) = s\,(f_s \pm g_s) \tag{1.12}$$

Besides homogeneous functions, we can also define harmonic functions, which are homogeneous functions that also satisfy the generalized Laplace equation.

$$\Delta h_s = 0 \tag{1.13}$$

Harmonic polynomials are special cases of harmonic functions. For harmonic polynomials, the generalized Laplace equation is defined everywhere, but for general harmonic functions, we must be aware of the fact that the generalized Laplace equation may have singularities. As we shall see later, harmonic polynomials and harmonic functions are closely related to both hyperspherical harmonics and to irreducible representations of rotation groups.

1.2 The canonical decomposition of a homogeneous polynomial

We would like to show that any homogeneous polynomial of order n can be expressed in the form:

$$f_n = h_n + r^2 h_{n-2} + r^4 h_{n-4} + \cdots \tag{1.14}$$

where r is the hyperradius (equation (1.1)) and where $h_n, h_{n-2}, ...$ are harmonic polynomials of order $n, n-2, \ldots$ respectively. This is called the canonical decomposition of f_n.

From (1.9), it follows that for any homogeneous function f_s of order s,

$$\Delta\left(r^\beta f_s\right) = \beta(\beta + d + 2s - 2)r^{\beta-2} f_s + r^\beta \Delta f_s \tag{1.15}$$

This relationship can be proved directly by carrying out the differentiation:

$$\begin{aligned}\Delta\left(r^\beta f_s\right) &= \sum_{j=1}^{d} \frac{\partial}{\partial x_j}\left(\beta r^{\beta-1}\frac{\partial r}{\partial x_j} f_s + r^\beta \frac{\partial f_s}{\partial x_j}\right)\\ &= \sum_{j=1}^{d}\left[\beta(\beta-2)r^{\beta-4}x_j^2 f_s + \beta r^{\beta-2} f_s + 2\beta r^{\beta-2} x_j \frac{\partial f_s}{\partial x_j} + r^\beta \frac{\partial^2 f_s}{\partial x_j^2}\right]\\ &= \beta(\beta + d + 2s - 2)r^{\beta-2} f_s + r^\beta \Delta f_s\end{aligned} \tag{1.16}$$

If the homogeneous function in (1.15) is also harmonic, then (since $\Delta h_s = 0$)

$$\Delta\left(r^\beta h_s\right) = \beta(\beta + d + 2s - 2)r^{\beta-2} h_s \tag{1.17}$$

In the special case where f_n is a homogeneous polynomial of order n, we can apply Δ repeatedly to both sides of equation (1.14) to obtain

$$\begin{aligned}\Delta f_n &= 2(d + 2n - 4)h_{n-2} + 4(d + 2n - 6)r^2 h_{n-4} + \cdots\\ \Delta^2 f_n &= 8(d + 2n - 6)(d + 2n - 8)h_{n-4} + \cdots\end{aligned} \tag{1.18}$$

and in general

$$\Delta^\nu f_n = \sum_{k=\nu}^{\lfloor n/2\rfloor} \frac{(2k)!!}{(2k-2\nu)!!}\frac{(d + 2n - 2k - 2)!!}{(d + 2n - 2k - 2\nu - 2)!!} r^{2k-2\nu} h_{n-2k} \tag{1.19}$$

where $\lfloor n/2\rfloor$ denotes the integral part of $n/2$, while $n!! \equiv n(n-2)(n-4)\cdots 2$ when n is even or $n!! \equiv n(n-2)(n-4)\cdots 1$ when n is odd, with $(-1)!! \equiv 1$. For example, when n is even and $\nu = n/2$, (1.19) becomes:

$$\Delta^{n/2} f_n = \frac{n!!(d + n - 2)!!}{(d-2)!!} h_0 \tag{1.20}$$

from which we obtain the important relationship

$$h_0 = \frac{(d-2)!!}{n!!(d + n - 2)!!} \Delta^{n/2} f_n \tag{1.21}$$

The reader is invited to prove equation (1.19) by induction. Show that if the relation is assumed for a positive integral value of ν, it follows for $\nu + 1$.

Then, since it holds for $\nu = 0$ and $\nu = 1$, it holds for all positive integral values of ν.

Equations (1.18) or (1.19) constitute a set of simultaneous equations which can be solved to yield h_n, h_{n-2}, h_{n-4},... in terms of f_n, Δf_n, $\Delta^2 f_n$,.... For example, when $n = 2$, (1.18) becomes $\Delta f_2 = 2dh_0$, so that

$$h_0 = \frac{1}{2d}\Delta f_2 \tag{1.22}$$

From (1.14) we have

$$f_2 = h_2 + r^2 h_0 \tag{1.23}$$

so that

$$h_2 = f_2 - \frac{r^2}{2d}\Delta f_2 \tag{1.24}$$

We are now in a position to write down the canonical decomposition of any homogeneous polynomial of order 2. For example, suppose that

$$f_2 = x^2 + 2xy \tag{1.25}$$

Then from (1.22) we have:

$$h_0 = \frac{1}{2d}\Delta(x^2 + 2xy) = \frac{1}{d} \tag{1.26}$$

and

$$h_2 = x^2 + 2xy - \frac{r^2}{d} \tag{1.27}$$

Doing the same for $n = 3$, we obtain

$$\begin{aligned} f_3 &= h_3 + r^2 h_1 \\ h_1 &= \frac{1}{2(d+2)}\Delta f_3 \\ h_3 &= f_3 - \frac{r^2}{2(d+2)}\Delta f_3 \end{aligned} \tag{1.28}$$

while for $n = 4$ we have

$$\begin{aligned} f_4 &= h_4 + r^2 h_2 + r^4 h_0 \\ h_0 &= \frac{1}{8d(d+2)}\Delta^2 f_4 \\ h_2 &= \frac{1}{2(d+4)}\Delta f_4 - \frac{r^2}{4d(d+4)}\Delta^2 f_4 \\ h_4 &= f_4 - \frac{r^2}{2(d+4)}\Delta f_4 + \frac{r^4}{8(d+2)(d+4)}\Delta^2 f_4 \end{aligned} \tag{1.29}$$

and for $n = 5$,

$$\begin{aligned}
f_5 &= h_5 + r^2 h_3 + r^4 h_1 \\
h_1 &= \frac{1}{8(d+2)(d+4)} \Delta^2 f_5 \\
h_3 &= \frac{1}{2(d+6)} \Delta f_5 - \frac{r^2}{4(d+2)(d+6)} \Delta^2 f_5 \\
h_5 &= f_5 - \frac{r^2}{2(d+6)} \Delta f_5 + \frac{r^4}{8(d+4)(d+6)} \Delta^2 f_5
\end{aligned} \tag{1.30}$$

and so on.

Alternatively we can write:

$$h_n = f_n + a_2 r^2 \Delta f_n + a_4 r^4 \Delta^2 f_n + \cdots \tag{1.31}$$

and then apply Δ to both sides of (1.31), making use of the fact that $\Delta h_n = 0$ to solve for a_2, a_4, a_6, etc. This gives:

$$\Delta h_n = 0 = \Delta f_n + a_2 \Delta(r^2 \Delta f_n) + a_4 \Delta(r^4 \Delta^2 f_n) + \cdots \tag{1.32}$$

Since $\Delta^\nu f_n$ is a homogeneous polynomial of order $n - 2\nu$, we have from (1.15),

$$\begin{aligned}
\Delta(r^2 \Delta f_n) &= 2(d+n-4)\Delta f_n + r^2 \Delta^2 f_n \\
\Delta(r^4 \Delta^2 f_n) &= 4(d+n-6) r^2 \Delta^2 f_n + r^4 \Delta^3 f_n \\
\vdots \quad & \quad \vdots \qquad\qquad \vdots
\end{aligned} \tag{1.33}$$

and so on. Inserting (1.33) into (1.32), we obtain a set of conditions which the coefficients a_j must satisfy in order for Δh_n to vanish:

$$\begin{aligned}
1 + 2(d+2n-4)a_2 &= 0 \\
a_2 + 4(d+2n-6)a_4 &= 0 \\
a_4 + 6(d+2n-8)a_6 &= 0 \\
\vdots \qquad\qquad & \vdots\ \vdots
\end{aligned} \tag{1.34}$$

etc. From (1.34) we obtain

$$\begin{aligned}
a_2 &= \frac{-1}{2(d+2n-4)} = \frac{(-1)^1 (d+2n-6)!!}{2!!(d+2n-4)!!} \\
a_4 &= \frac{1}{8(d+2n-4)(d+2n-6)} = \frac{(-1)^2 (d+2n-8)!!}{4!!(d+2n-4)!!}
\end{aligned} \tag{1.35}$$

and so on, so that (1.31) becomes

$$h_n = f_n - \frac{r^2}{2(d+2n-4)}\Delta f_n + \frac{r^4}{8(d+2n-4)(d+2n-6)}\Delta^2 f_n - \cdots \tag{1.36}$$

which can be compared with equations (1.24) and (1.28)–(1.30). In general,

$$a_{2k} = \frac{(-1)^k(d+2n-2k-4)!!}{(2k)!!(d+2n-4)!!} \tag{1.37}$$

so that finally we obtain

$$h_n = \sum_{k=0}^{\lfloor n/2 \rfloor} \frac{(-1)^k(d+2n-2k-4)!!}{(2k)!!(d+2n-4)!!} r^{2k}\Delta^k f_n \tag{1.38}$$

which is an explicit expression for the harmonic polynomial of highest order in the canonical decomposition of the homogeneous polynomial f_n. It is possible to derive a generalization of (1.38) by using (1.19) together with (1.38). Equation (1.19) can be rewritten in the form

$$\Delta^\nu f_n = \frac{(2\nu)!!(d+2n-2\nu-2)!!}{(d+2n-4\nu-4)!!} h_{n-2\nu} + \cdots \tag{1.39}$$

so that

$$h_{n-2\nu} = \frac{(d+2n-4\nu-4)!!}{(2\nu)!!(d+2n-2\nu-2)!!}\Delta^\nu f_n + \cdots \tag{1.40}$$

We now notice that $\Delta^\nu f_n$ is a homogeneous polynomial of order $n-2\nu$. We can project out the harmonic polynomial of highest order in $\Delta^\nu f_n$ by replacing n by $n-2\nu$ in Equation (1.38). If we do so, the complete expression in (1.40) becomes

$$\begin{aligned} h_{n-2\nu} &= \frac{(d+2n-4\nu-4)!!}{(2\nu)!!(d+2n-2\nu-2)!!} \\ &\times \sum_{k=0}^{\lfloor n/2-\nu \rfloor} \frac{(-1)^k(d+2n-4\nu-2k-4)!!}{(2k)!!(d+2n-4\nu-4)!!} r^{2k}\Delta^k\left(\Delta^\nu f_n\right) \end{aligned} \tag{1.41}$$

Apart from the constant in front of the series, this equation is identical to Equation (1.38) with n replaced by $n-2\nu$. The constant is determined by examining the $n = 2\nu$ term (where it must equal the expression in Equation (1.21)) and the $\nu = 0$ term in which case it must be unity. The coefficients in the series are required to take on the values shown in order that the polynomial should be harmonic. This generalization of (1.38) gives us explicit expressions for all of the harmonic polynomials in the canonical decomposition of f_n, not just the harmonic polynomial of highest order. Therefore the remaining terms, indicated by dots in (1.40), must be given by the terms corresponding to $k = 1, 2, 3, \ldots$ in (1.41).

1.3 More general canonical decompositions

It is interesting to notice that the canonical decomposition of any homogeneous function of the form

$$f_{n+m} = (x_1 + ix_2)^m f_n(x_3, x_4, x_5, \ldots, x_d) \tag{1.42}$$

is possible, regardless of the value of m, provided that f_n is a homogeneous polynomial in coordinates that do not include x_1 and x_2. The value of m may be negative, non-integral, or even complex. This is because

$$\begin{aligned} \frac{\partial^2}{\partial x_1^2}(x_1 + ix_2)^m &= \frac{\partial}{\partial x_1}\left[m(x_1 + ix_2)^{m-1}\right] \\ &= m(m-1)(x_1 + ix_2)^{m-2} \end{aligned} \tag{1.43}$$

while

$$\begin{aligned} \frac{\partial^2}{\partial x_2^2}(x_1 + ix_2)^m &= \frac{\partial}{\partial x_2}\left[i\ m(x_1 + ix_2)^{m-1}\right] \\ &= -m(m-1)(x_1 + ix_2)^{m-2} \end{aligned} \tag{1.44}$$

so that

$$\left(\frac{\partial^2}{\partial x_1^2} + \frac{\partial^2}{\partial x_2^2}\right)(x_1 + ix_2)^m = 0 \tag{1.45}$$

Thus, when the generalized Laplacian operator

$$\Delta \equiv \left(\frac{\partial^2}{\partial x_1^2} + \frac{\partial^2}{\partial x_2^2} + \cdots + \frac{\partial^2}{\partial x_d^2}\right) \tag{1.46}$$

acts on $f_{n+m} = (x_1 + ix_2)^m f_n(x_3, x_4, x_5, \ldots, x_d)$, the contribution of the factor $(x_1 + ix_2)^m$ vanishes, and the canonical decomposition proceeds as though this factor were not there. More generally, we can divide the d coordinates into two disjoint sets $\mathbf{x}_a = \{x_1, x_2, \ldots, x_j\}$ and $\mathbf{x}_b = \{x_{j+1}, x_{j+2}, \ldots, x_d\}$ and similarly divide the generalized d-dimensional generalized Laplacian operator into two parts, each acting exclusively on one of the subspaces defined by the division.

$$\begin{aligned} \Delta_a &\equiv \left(\frac{\partial^2}{\partial x_1^2} + \frac{\partial^2}{\partial x_2^2} + \cdots + \frac{\partial^2}{\partial x_j^2}\right) \\ \Delta_b &\equiv \left(\frac{\partial^2}{\partial x_{j+1}^2} + \frac{\partial^2}{\partial x_{j+2}^2} + \cdots + \frac{\partial^2}{\partial x_d^2}\right) \end{aligned} \tag{1.47}$$

Then if $h(x_1, x_2, \ldots, x_j)$ is an harmonic function, i.e. if

$$\left(\frac{\partial^2}{\partial x_1^2} + \frac{\partial^2}{\partial x_2^2} + \cdots + \frac{\partial^2}{\partial x_j^2}\right) h(x_1, x_2, \ldots, x_j) = 0 \tag{1.48}$$

and if $f_n(x_{j+1}, x_{j+2}, \ldots, x_d)$ is a homogeneous polynomial, a canonical decomposition can be carried through for the homogeneous function $h(x_1, x_2, \ldots, x_j)f_n(x_{j+1}, x_{j+2}, \ldots, x_d)$. We can see that this is the case because for any two functions A and B of the Cartesian coordinates $x_1, x_2, \ldots, x_d$

$$\begin{aligned}\Delta(AB) &= \sum_{i=1}^{d} \frac{\partial}{\partial x_i}\left(\frac{\partial A}{\partial x_i}B + A\frac{\partial B}{\partial x_i}\right) \\ &= B\,\Delta A + A\,\Delta B + 2\sum_{i=1}^{d} \frac{\partial A}{\partial x_i}\frac{\partial B}{\partial x_i} \qquad (1.49)\end{aligned}$$

From (1.48) and (1.49) it follows that

$$\begin{aligned}&\Delta\left[h(x_1, x_2, \ldots, x_j)f_n(x_{j+1}, x_{j+2}, \ldots, x_d)\right] \\ &= h(x_1, x_2, \ldots, x_j)\Delta\left[f_n(x_{j+1}, x_{j+2}, \ldots, x_d)\right] \qquad (1.50)\end{aligned}$$

In other words, under the action of generalized Laplacian operator, the harmonic function $h(x_1, x_2, \ldots, x_j)$ behaves as though it were a constant. The canonical decomposition of the homogeneous polynomial $f_n(x_{j+1}, x_{j+2}, \ldots, x_d)$ proceeds in the usual way, except that every term in the decomposition is multiplied by the harmonic function $h(x_1, x_2, \ldots, x_j) = h(\mathbf{x}_a)$.

$$h(\mathbf{x}_a)f_n(\mathbf{x}_b) = h(\mathbf{x}_a)h_n(\mathbf{x}_b) + r_b^2 h(\mathbf{x}_a)h_{n-2}(\mathbf{x}_b) + r_b^4 h(\mathbf{x}_a)h_{n-4}(\mathbf{x}_b) + \cdots \qquad (1.51)$$

with $r_b^2 \equiv x_{j+1}^2 + x_{j+2}^2 + \cdots + x_d^2$. From (1.49) it follows that terms such as $h(\mathbf{x}_a)h_n(\mathbf{x}_b)$ and $h(\mathbf{x}_a)h_{n-2}(\mathbf{x}_b)$ are harmonic. In fact, it follows from (1.49) that any product of harmonic functions of disjoint sets of coordinates is harmonic. For example

$$(x_1 + ix_2)^a(x_3 + ix_4)^b(x_5 + ix_6)^c(x_7 + ix_8)^d \cdots \qquad (1.52)$$

is harmonic, regardless of the values of $a, b, c, d, \ldots$.

1.4 Exercises

(1) In a 3-dimensional space, give an example of a homogeneous polynomial of order 3 which is not harmonic. Show that equation (1.15) holds for this polynomial for $\beta = 1$, $\beta = 2$ and $\beta = 3$.

(2) Show that $h_2 = x^2 + 2xy - \frac{1}{3}(x^2 + y^2 + z^2)$ is an harmonic polynomial of order 2. Give an example of an harmonic polynomial of order 4.

(3) Let x, y, z and t be the Cartesian coordinates of a 4-dimensional space. Give an example of a homogeneous polynomial of order 2 in this space which is not harmonic. Find the canonical decomposition of this polynomial. Do the same for a homogeneous polynomials of order 3 in the same space.
(4) Show that $z^7(x+iy)^{-4}$ is a homogeneous function of order 3.
(5) Show that $\sqrt{x}$ is a homogeneous function of order 1/2.
(6) Show that $(x+iy)^{(1+i/2)}y^{-3/2}$ is a homogeneous function of order $i/2 - 1/2$.
(7) Show that $(xyz + x^3)^a$ is a homogeneous function of order $3a$.
(8) Show that $(xz^2 + x^3)^a y^b$ is a homogeneous function of order $3a + b$.
(9) Type `xcoords` to obtain a list of coordinates $x_1, x_2, \ldots, x_{18}$. Use this to construct a generalized Laplacian operator that works up to $d = 18$.
(10) Show that a polynomial f_n is homogeneous and of order n if and only if the following equation holds

$$\sum_{j=1}^{d} x_j \frac{\partial f_n}{\partial x_j} = n f_n \tag{1.53}$$

Use a pen and paper rather than Mathematica.
(11) In a 3-dimensional space use equation (1.38) to find the harmonic polynomial of highest order in the canonical decomposition of the 4th order homogeneous polynomial x^4.
(12) Show that the function $(x_1+ix_2)^a(x_3+ix_4)^b(x_5+ix_6)^c(x_7+ix_8)^d$ is harmonic regardless of the values of a, b, c and d.

Chapter 2

GENERALIZED ANGULAR MOMENTUM

2.1 Eigenfunctions of generalized angular momentum

The generalized Laplacian operator Δ can be written in the form

$$\Delta \equiv \sum_{j=1}^{d} \frac{\partial^2}{\partial x_j^2} = \frac{1}{r^{d-1}} \frac{\partial}{\partial r} r^{d-1} \frac{\partial}{\partial r} - \frac{\Lambda^2}{r^2} \tag{2.1}$$

where r is the hyperradius and Λ^2 is the generalized angular momentum operator, defined by

$$\Lambda^2 \equiv \sum_{s>t}^{d} \Lambda_{s,t}^2 \tag{2.2}$$

where

$$\Lambda_{s,t} \equiv \frac{1}{i} \left(x_s \frac{\partial}{\partial x_t} - x_t \frac{\partial}{\partial x_s} \right) \tag{2.3}$$

A discussion of the derivation of equation (2.1) can be found in Chapter 10, and generalized angular momentum is treated in detail in [Avery, 1989; Avery, 2000; Avery and Avery, 2006; Avery, 2009; Avery, Rettrup and Avery, 2012; Vilinken, 1968; Wen and Avery, 1985]. Notice that $\Lambda_{st} f(r) = 0$, where $f(r)$ is any function of the hyperradius, and therefore $\Lambda^2 f(r) = 0$. We would like to show that if h_λ is an harmonic polynomial of order λ, then

$$\left[\Lambda^2 - \lambda(\lambda + d - 2)\right] r^{-\lambda} h_\lambda = 0 \tag{2.4}$$

In other words, we wish to show that $r^{-\lambda} h_\lambda$ is an eigenfunction of the generalized angular momentum operator Λ^2 belonging to the eigenvalue $\lambda(\lambda + d - 2)$. To show this, we first notice that

$$r^{-\lambda} h_\lambda \equiv Y_\lambda(\mathbf{u}) \tag{2.5}$$

is a pure function of the angular coordinates $\mathbf{u} \equiv \mathbf{x}/r$, independent of the hyperradius, so that

$$\frac{\partial}{\partial r} Y_\lambda(\mathbf{u}) = 0 \tag{2.6}$$

(See Exercise 9). Then, since $h_\lambda = r^\lambda Y_\lambda(\mathbf{u})$ and $\Delta h_\lambda = 0$, we have

$$\begin{aligned} \Delta h_\lambda &= \left[\frac{1}{r^{d-1}}\frac{\partial}{\partial r} r^{d-1} \frac{\partial}{\partial r} - \frac{\Lambda^2}{r^2}\right] r^\lambda Y_\lambda(\mathbf{u}) \\ &= \left[\lambda(\lambda + d - 2) - \Lambda^2\right] r^{\lambda-2} Y_\lambda(\mathbf{u}) = 0 \end{aligned} \tag{2.7}$$

Dividing through by $r^{\lambda-2}$, we obtain

$$\left[\Lambda^2 - \lambda(\lambda + d - 2)\right] Y_\lambda(\mathbf{u}) = 0 \tag{2.8}$$

and from (2.5) we can see that this relation is the same as (2.4).

2.2 Hyperspherical harmonics

A hyperspherical harmonic is defined to be an angular function which is an eigenfunction of the generalized angular momentum operator Λ^2. We have shown that hyperspherical harmonics can be generated from harmonic polynomials h_λ by dividing them by r^λ, so that they become pure angular functions. For the case where $d = 3$, equation (2.8) can be given a more familiar form by noticing that the operator Λ^2 then corresponds to L^2, while λ corresponds to l. Then, introducing the index m to label the different linearly independent eigenfunctions of L^2 belonging to the same value of l, we have

$$\left[L^2 - l(l+1)\right] Y_{l,m}(\mathbf{u}) = 0 \tag{2.9}$$

which is the familiar relationship satisfied by spherical harmonics in 3-dimensional space. In general, we shall use the index μ to label the different linearly independent eigenfunctions of Λ^2 belonging to the same value of λ, so that hyperspherical harmonics will be written as $Y_{\lambda,\mu}(\mathbf{u})$.

Since Λ^2 acts only on the hyperangles and commutes with r, equation (2.4) implies that

$$\left[\Lambda^2 - \lambda(\lambda + d - 2)\right] h_\lambda = 0 \tag{2.10}$$

The canonical decomposition of a homogeneous polynomial f_n thus corresponds to a resolution of the polynomial into eigenfunctions of Λ^2. If we let $\lambda = n - 2\nu$ in equation (1.41), it follows that

$$\begin{aligned} O_\lambda[f_n] &= \frac{(d + 2\lambda - 4)!!r^{n-\lambda}}{(n-\lambda)!!(d+n+\lambda-2)!!} \\ &\times \sum_{k=0}^{\lfloor \lambda/2 \rfloor} \frac{(-1)^k (d + 2\lambda - 2k - 4)!!}{(2k)!!(d + 2\lambda - 4)!!} r^{2k} \Delta^{k+n/2-\lambda/2} f_n \end{aligned} \tag{2.11}$$

is the projection of f_n onto the part of Hilbert space spanned by the set of eigenfunctions of Λ^2 belonging to eigenvalue $\lambda(\lambda + d - 2)$. We can thus write the canonical decomposition of f_n in the form

$$f_n = \sum_{\lambda=n,n-2,\ldots} O_\lambda[f_n] \tag{2.12}$$

where O_λ is the projection operator appearing in (2.11). When n is even, f_n contains only even eigenfunctions of Λ^2, while when n is odd, it contains only odd eigenfunctions.

2.3 Generalized solid angle

We now introduce a generalized solid angle element $d\Omega$ defined by

$$dx_1 dx_2..dx_d = r^{d-1} dr \; d\Omega_d \tag{2.13}$$

The total solid angle can be evaluated by noticing that

$$\int_0^\infty dr \; r^{d-1} e^{-r^2} \int d\Omega_d = \prod_{j=1}^d \int_{-\infty}^\infty dx_j e^{-x_j^2} \tag{2.14}$$

But

$$\int_0^\infty dr \; r^{d-1} e^{-r^2} = \frac{1}{2}\Gamma\left(d/2\right) \tag{2.15}$$

and

$$\int_{-\infty}^\infty dx_j e^{-x_j^2} = \sqrt{\pi} \tag{2.16}$$

Combining equations (2.14)–(2.16), we find that the total solid angle is given by

$$I(0) \equiv \int d\Omega_d = \frac{2\pi^{d/2}}{\Gamma(d/2)} = \begin{cases} \dfrac{2\pi^{d/2}}{(d/2-1)!} & d \text{ even} \\[2ex] \dfrac{2^{(d+1)/2}\pi^{(d-1)/2}}{(d-2)!!} & d \text{ odd} \end{cases} \tag{2.17}$$

For the first few values of d, equation (2.17) gives

$$\begin{aligned} d&=1 & I(0)&=2\\ d&=2 & I(0)&=2\pi\\ d&=3 & I(0)&=4\pi\\ d&=4 & I(0)&=2\pi^2\\ d&=5 & I(0)&=\frac{8\pi^2}{3}\\ d&=6 & I(0)&=\pi^3\\ d&=7 & I(0)&=\frac{16\pi^3}{15}\\ d&=8 & I(0)&=\frac{\pi^4}{3}\\ d&=9 & I(0)&=\frac{32\pi^4}{105} \end{aligned} \tag{2.18}$$

2.4 Hyperangular integration

We shall now show that if f_n is a homogeneous polynomial as defined by equation (1.3), then

$$\int d\Omega_d\ f_n = \begin{cases} I(0)r^n h_0 & \text{if n is even} \\ 0 & \text{if n is odd} \end{cases} \tag{2.19}$$

where $I(0)$ and h_0 are given by equations (2.17) and (1.21). Equation (2.19) follows directly from (1.14) and (2.10) and from the assumption that Λ^2 is Hermitian with respect to integration over the generalized solid angle element $d\Omega$. Since two eigenfunctions of an Hermitian operator are orthogonal if they correspond to different eigenvalues, it follows that

$$\int d\Omega_d\ h_\lambda h_{\lambda'} = 0 \qquad \text{if } \lambda \neq \lambda' \tag{2.20}$$

But h_0 is a constant, and therefore (2.20) implies that

$$\int d\Omega_d\ h_\lambda = 0 \qquad \text{if } \lambda \neq 0 \tag{2.21}$$

Then from (2.21) and (1.14), we have that for even n,

$$\begin{aligned} \int d\Omega_d\ f_n &= \int d\Omega_d\ (h_n + r^2 h_{n-2} + \cdots + r^n h_0) \\ &= r^n h_0 \int d\Omega_d = I(0) r^n h_0 \end{aligned} \tag{2.22}$$

while for odd n, the integral is always zero.

The Hermiticity of Λ^2 with respect to integration over $d\Omega_d$ can be established from a generalized form of Green's first identity:

$$\int_0^\infty dr\ r^{d-1} \int d\Omega_d \left[\Phi_1 \Delta \Phi_2 - \Phi_2 \Delta \Phi_1\right] = 0 \tag{2.23}$$

Equation (2.23) holds if Φ_1 and Φ_2 fall off sufficiently rapidly for large values of r. We now let

$$\Phi_j = R(r) W_j(\mathbf{u}) \qquad j = 1, 2 \tag{2.24}$$

where the functions $W_j(\mathbf{u})$ are real. Then (2.23) can be rewritten in the form

$$\int_0^\infty dr\ r^{d-1} R^2 \int d\Omega_d \left[W_1 \Lambda^2 W_2 - W_2 \Lambda^2 W_1\right] = 0 \tag{2.25}$$

where we have made use of (2.1) and noticed that the terms involving derivatives with respect to r cancel out. From (2.25) it follows that

$$\int d\Omega_d \left[W_1 \Lambda^2 W_2 - W_2 \Lambda^2 W_1\right] = 0 \tag{2.26}$$

The Hermiticity of Λ^2 follows from (2.26) and from the fact that Λ^2 is real.

2.5 A general theorem for hyperangular integration

Theorem

Let

$$I(\mathbf{n}) \equiv \int d\Omega_d \left(\frac{x_1}{r}\right)^{n_1} \left(\frac{x_2}{r}\right)^{n_2} \cdots \left(\frac{x_d}{r}\right)^{n_d} \tag{2.27}$$

where $x_1, x_2, \ldots, x_d$ are the Cartesian coordinates of a d-dimensional space, $d\Omega_d$ is the generalized solid angle, r is the hyperradius, and $Re[n_j] > -1,\ \ j = 1, 2, 3, \ldots, d$. Then

$$I(\mathbf{n}) = \frac{2}{\Gamma\left(\frac{d+n}{2}\right)} \prod_{j=1}^{d} \frac{1}{2}\left(1 + e^{i\pi n_j}\right) \Gamma\left(\frac{n_j + 1}{2}\right) \tag{2.28}$$

where

$$n \equiv \sum_{j=1}^{d} n_j \tag{2.29}$$

Proof

Consider the integral

$$\int_0^\infty dr\ r^{d-1} e^{-r^2} \int d\Omega_d\ x_1^{n_1} x_2^{n_2} \cdots x_d^{n_d} = \prod_{j=1}^{d} \int_{-\infty}^{\infty} dx_j\ x_j^{n_j} e^{-x_j^2} \tag{2.30}$$

The right-hand side of (2.30) can be evaluated and has the form

$$\prod_{j=1}^{d}\int_{-\infty}^{\infty} dx_j\ x_j^{n_j} e^{-x_j^2} = \prod_{j=1}^{d}\frac{1}{2}\left(1+e^{i\pi n_j}\right)\Gamma\left(\frac{n_j+1}{2}\right) \tag{2.31}$$

provided that $Re[n_j] > -1, \quad j = 1, 2, 3, \ldots, d$. The left-hand side of (2.30) can be written as

$$\int_0^{\infty} dr\ r^{d+n-1}e^{-r^2}\int d\Omega_d\left(\frac{x_1}{r}\right)^{n_1}\left(\frac{x_2}{r}\right)^{n_2}\cdots\left(\frac{x_d}{r}\right)^{n_d} = \frac{I(\mathbf{n})}{2}\Gamma\left(\frac{d+n}{2}\right) \tag{2.32}$$

Combining equations (2.30)–(2.32), we obtain (2.28), Q.E.D.

Comments

If all the n_j's are zero or positive integers, then (2.28) reduces to

$$I(\mathbf{n}) = \begin{cases} \dfrac{\pi^{d/2}}{2^{(n/2-1)}\Gamma\left(\frac{d+n}{2}\right)}\displaystyle\prod_{j=1}^{d}(n_j-1)!! & \text{if all the } n_j\text{'s are even} \\ 0 & \text{otherwise} \end{cases} \tag{2.33}$$

When $d = 3$, (2.33) becomes

$$\int d\Omega_3\left(\frac{x_1}{r}\right)^{n_1}\left(\frac{x_2}{r}\right)^{n_2}\left(\frac{x_3}{r}\right)^{n_3} = \begin{cases} \dfrac{4\pi}{(n+1)!!}\displaystyle\prod_{j=1}^{3}(n_j-1)!! & \text{all } n_j\text{'s even} \\ 0 & \text{otherwise} \end{cases} \tag{2.34}$$

while when $d = 4$, (2.33) tells us that

$$\int d\Omega_4\left(\frac{x_1}{r}\right)^{n_1}\left(\frac{x_2}{r}\right)^{n_2}\left(\frac{x_3}{r}\right)^{n_3}\left(\frac{x_4}{r}\right)^{n_4} = \begin{cases} \dfrac{4\pi^2}{(n+2)!!}\displaystyle\prod_{j=1}^{4}(n_j-1)!! & \text{all } n_j\text{'s even} \\ 0 & \text{otherwise} \end{cases} \tag{2.35}$$

Let us now consider a general polynomial (not necessarily homogeneous) of the form:

$$P(\mathbf{x}) = \sum_{\mathbf{n}} c_{\mathbf{n}}\ x_1^{n_1}x_2^{n_2}\cdots x_d^{n_d} \tag{2.36}$$

Then we have

$$\int d\Omega_d \; P(\mathbf{x}) = \sum_{\mathbf{n}} c_{\mathbf{n}} \int d\Omega \; x_1^{n_1} x_2^{n_2} \cdots x_d^{n_d} = \sum_{\mathbf{n}} c_{\mathbf{n}} \; r^n I(\mathbf{n}) \qquad (2.37)$$

It can be seen that equation (2.28) can be used to evaluate the generalized angular integral of any polynomial whatever, regardless of whether or not it is homogeneous.

The utility of these relationships is very great indeed. They provide a method for developing the theory of angular momentum and hyperangular momentum which supplements the usual group-theoretical methods. For example, Clebsch-Gordan coefficients and generalized Clebsch-Gordan coefficients may be generated by means of (2.33), rather than through equations derived from group theory, which become quite complicated when they are generalized to d dimensions. Furthermore d-dimensional partial differential equations may be solved by this means. Matrix elements between basis functions can be readily evaluated by means of the hyperangular integration theorem (2.33), leaving only a single hyperradial integral to be evaluated, an operation which can, if necessary, be performed numerically. We will give examples of this procedure later in the book.

2.6 Chains of subgroups

Table 2.1 shows the first few 4-dimensional hyperspherical harmonics with

$$(u_1, u_2, u_3, u_4) \equiv \frac{1}{r}(x_1, x_2, x_3, x_4) \qquad (2.38)$$

We will explain in Chapter 3 how such a table can be constructed. If we look at the table, we can see that there are $(\lambda+1)^2$ different eigenfunctions of Λ^2 corresponding to each value of λ. From these, we have constructed an orthonormal set of hyperspherical harmonics, i.e. a set whose members fulfill the orthonormality conditions

$$\int d\Omega_d \; Y_{\lambda',\mu'}(\mathbf{u}) Y_{\lambda,\mu}(\mathbf{u}) = \delta_{\lambda',\lambda}\delta_{\mu',\mu} \qquad (2.39)$$

In equation (2.39), μ stands for the set of indices which label the different hyperspherical harmonics corresponding to the same value of λ. The hyperspherical harmonics shown in Table 2.1 are simultaneous eigenfunctions of $\Lambda^2_{(4)}$, $\Lambda^2_{(3)}$, and $\Lambda^2_{(2)}$, where

$$\Lambda^2_{(d)} \equiv \sum_{s>t}^{d} \Lambda_{s,t} \qquad (2.40)$$

Table 2.1: 4-dimensional hyperspherical harmonics

λ	l	m	$\sqrt{2}\pi\ Y_{\lambda,l,m}(\mathbf{u})$
0	0	0	1
1	1	1	$-\sqrt{2}(u_1+iu_2)$
1	1	0	$-2u_3$
1	1	-1	$\sqrt{2}(u_1-iu_2)$
1	0	0	$2u_4$
2	2	2	$\sqrt{3}(u_1+iu_2)^2$
2	2	1	$-2\sqrt{3}u_3(u_1+iu_2)$
2	2	0	$-\sqrt{2}(u_1^2+u_2^2-2u_3^2)$
2	2	-1	$2\sqrt{3}u_3(u_1-iu_2)$
2	2	-2	$\sqrt{3}(u_1-iu_2)^2$
2	1	1	$-2\sqrt{3}\ u_4(u_1+iu_2)$
2	1	0	$2\sqrt{6}\ u_4u_3$
2	1	-1	$2\sqrt{3}\ u_4(u_1-iu_2)$
2	0	0	$3u_4^2-u_1^2-u_2^2-u_3^2$

In other words, the indices are organized according to the chain of subgroups

$$SO(4) \supset SO(3) \supset SO(2) \tag{2.41}$$

In Fock's treatment of hydrogenlike atoms, which we will discuss later, these 4-dimensional hyperspherical harmonics are shown to be closely related to the Fourier transforms of hydrogenlike atomic orbitals.

2.7 Degeneracy of hyperspherical harmonics

The degeneracy of the familiar 3-dimensional harmonics corresponding to a particular value of l is

$$\sum_{\mu} 1 = \sum_{m=-l}^{l} = 2l+1 \tag{2.42}$$

For the 4-dimensional hyperspherical harmonics $Y_{\lambda,l,m}(\mathbf{u})$ the degeneracy is

$$\sum_{\mu} 1 = \sum_{l=0}^{\lambda}(2l+1) = (\lambda+1)^2 \tag{2.43}$$

For the 5-dimensional hyperspherical harmonics $Y_{\lambda,l,\mu}(\mathbf{u})$ the degeneracy is

$$\sum_{\mu} 1 = \sum_{l=0}^{\lambda}(l+1)^2 = \frac{(3+2\lambda)(2+\lambda)!}{6\lambda!} \tag{2.44}$$

For the 6-dimensional hyperspherical harmonics $Y_{\lambda,l,\mu}(\mathbf{u})$ the degeneracy is

$$\sum_{\mu} 1 = \sum_{l=0}^{\lambda} \frac{(3+2l)(2+l)!}{6l!} = \frac{(2+\lambda)(3+\lambda)!}{12\lambda!} \tag{2.45}$$

and, more generally, for d-dimensional hyperspherical harmonics $Y_{\lambda,l,\mu}(\mathbf{u})$ one finds the degeneracy to be

$$\sum_{\mu} 1 = \frac{(d+2\lambda-2)(d+\lambda-3)!}{\lambda!(d-2)!} \tag{2.46}$$

Notice that for $\lambda = 0$ the degeneracy is always 1, while for $\lambda = 1$ it is always d.

2.8 Representations of $\Lambda_{s,t}$

For $d \geq 4$, the commutation rules for the rotation operators $\Lambda_{r,s}$, are as follows: If r, s, t and u are four indices, none of which is equal to any of the others,

$$\begin{aligned} [\Lambda_{r,s}, \Lambda_{s,t}] &= i\Lambda_{t,r} = -i\Lambda_{r,t} \\ [\Lambda_{r,s}, \Lambda_{t,u}] &= 0 \end{aligned} \tag{2.47}$$

In the case where $d = 4$ and $\lambda = 1$, we know from equation (2.46) that the degeneracy is 4. Thus any set of 4 orthonormal hyperspherical harmonics corresponding to $\lambda = 1$ can be used to construct the representation. The set of $\lambda = 1$ hyperspherical harmonics, constructed according to the standard method, is

$$\left\{ \frac{-1}{\pi}(u_1 + iu_2), \frac{\sqrt{2}}{\pi}u_3, \frac{1}{\pi}(u_1 - iu_2), \frac{\sqrt{2}}{\pi}u_4 \right\} \tag{2.48}$$

However, in finding representations of the rotation operators $\Lambda_{s,t}$, it is more convenient to use the basis

$$\left\{ \frac{\sqrt{2}}{\pi} u_1, \frac{\sqrt{2}}{\pi} u_2, \frac{\sqrt{2}}{\pi} u_3, \frac{\sqrt{2}}{\pi} u_4 \right\} \tag{2.49}$$

which is related to the basis shown in (2.48) by a unitary transformation. In terms of the basis shown in equation (2.49), the matrices representing

$$\Lambda_{s,t} \equiv \frac{1}{i}\left(x_s \frac{\partial}{\partial x_t} - x_t \frac{\partial}{\partial x_s} \right) \qquad s \neq t \tag{2.50}$$

can be found by first noting that

$$\Lambda_{s,t} u_k = \frac{1}{i}\left(u_s \delta_{kt} - u_t \delta_{ks} \right) \tag{2.51}$$

which has the form of a rotation operator, since it changes the direction of the vector u_k without changing its length. Using equation (2.33), we then have

$$\begin{aligned} \int d\Omega_d \; u_j \Lambda_{s,t} u_k &= \frac{1}{i} \int d\Omega_d \; u_j \left(u_s \delta_{kt} - u_t \delta_{ks} \right) \\ &= \frac{\pi^{d/2}}{i\Gamma(d/2+1)} \left(\delta_{js} \delta_{kt} - \delta_{jt} \delta_{ks} \right) \end{aligned} \tag{2.52}$$

The representations thus obtained for the rotation operators, based on the functions shown in equation (2.49), are

$$\Lambda_{1,2} = \begin{pmatrix} 0 & -i & 0 & 0 \\ i & 0 & 0 & 0 \\ 0 & 0 & 0 & 0 \\ 0 & 0 & 0 & 0 \end{pmatrix} \tag{2.53}$$

$$\Lambda_{1,3} = \begin{pmatrix} 0 & 0 & -i & 0 \\ 0 & 0 & 0 & 0 \\ i & 0 & 0 & 0 \\ 0 & 0 & 0 & 0 \end{pmatrix} \tag{2.54}$$

$$\Lambda_{1,4} = \begin{pmatrix} 0 & 0 & 0 & -i \\ 0 & 0 & 0 & 0 \\ 0 & 0 & 0 & 0 \\ i & 0 & 0 & 0 \end{pmatrix} \tag{2.55}$$

$$\Lambda_{2,3} = \begin{pmatrix} 0 & 0 & 0 & 0 \\ 0 & 0 & -i & 0 \\ 0 & i & 0 & 0 \\ 0 & 0 & 0 & 0 \end{pmatrix} \tag{2.56}$$

$$\Lambda_{2,4} = \begin{pmatrix} 0 & 0 & 0 & 0 \\ 0 & 0 & 0 & -i \\ 0 & 0 & 0 & 0 \\ 0 & i & 0 & 0 \end{pmatrix} \tag{2.57}$$

$$\Lambda_{3,4} = \begin{pmatrix} 0 & 0 & 0 & 0 \\ 0 & 0 & 0 & 0 \\ 0 & 0 & 0 & -i \\ 0 & 0 & i & 0 \end{pmatrix} \tag{2.58}$$

The matrix representations of the rotation operators are Hermitian, and they obey the same multiplication table as the operators themselves. This, for example, we know that

$$[\Lambda_{1,2}, \Lambda_{3,4}] = 0 \tag{2.59}$$

while

$$[\Lambda_{1,2}, \Lambda_{2,3}] = -i\Lambda_{1,3} \tag{2.60}$$

and we can verify by matrix multiplication that the same relationships hold for their representations. Furthermore, we can verify by matrix multiplication, that

$$\Lambda_{1,2}^2 + \Lambda_{1,3}^2 + \Lambda_{1,4}^2 + \Lambda_{2,3}^2 + \Lambda_{2,4}^2 + \Lambda_{3,4}^2 = \Lambda^2 = \begin{pmatrix} 3 & 0 & 0 & 0 \\ 0 & 3 & 0 & 0 \\ 0 & 0 & 3 & 0 \\ 0 & 0 & 0 & 3 \end{pmatrix} \tag{2.61}$$

which agrees with the fact that when $\lambda = 1$ and $d = 4$, the eigenvalue of Λ^2 is $\lambda(\lambda + d - 2) = 3$.

When $d = 5$ and $\lambda = 1$, representations of $\Lambda_{i,j}$ can be found using the basis

$$\left\{ \sqrt{\frac{15}{8\pi^2}}\, u_1, \sqrt{\frac{15}{8\pi^2}}\, u_2, \sqrt{\frac{15}{8\pi^2}}\, u_3, \sqrt{\frac{15}{8\pi^2}}\, u_4, \sqrt{\frac{15}{8\pi^2}}\, u_5 \right\} \tag{2.62}$$

which is the 5-dimensional generalization of equation (2.49). In the case where $d = 5$, there are $d(d-1)/2 = 10$ different rotation operators to be represented: $\Lambda_{1,2}$, $\Lambda_{1,3}$, $\Lambda_{1,4}$, $\Lambda_{1,5}$, $\Lambda_{2,3}$, $\Lambda_{2,4}$, $\Lambda_{2,5}$, $\Lambda_{3,4}$, $\Lambda_{3,5}$ and $\Lambda_{4,5}$. We find, for example, that

$$\Lambda_{1,2} = \begin{pmatrix} 0 & -i & 0 & 0 & 0 \\ i & 0 & 0 & 0 & 0 \\ 0 & 0 & 0 & 0 & 0 \\ 0 & 0 & 0 & 0 & 0 \\ 0 & 0 & 0 & 0 & 0 \end{pmatrix} \tag{2.63}$$

$$\Lambda_{1,3} = \begin{pmatrix} 0 & 0 & -i & 0 & 0 \\ 0 & 0 & 0 & 0 & 0 \\ i & 0 & 0 & 0 & 0 \\ 0 & 0 & 0 & 0 & 0 \\ 0 & 0 & 0 & 0 & 0 \end{pmatrix} \tag{2.64}$$

and so on. No surprises here. Thus, no matter how high d might be, there is no difficulty in obtaining $\lambda = 1$ representations of $\Lambda_{i,j}$. The case where $\lambda = 0$ is trivial. For $\lambda = 2$ and for higher integral values of λ, representations based on hyperspherical harmonics can readily be obtained, but it requires more space to write down the results, because the size, n, of each matrix is determined by equation (2.46) and the number of matrices is equal to $n(n-1)/2$.

2.9 Exercises

(1) Type `?solidangle` to obtain the definition of the function, and use it to find the total solid angle when $d = 2, 3, \ldots, 12$.

(2) Type `?y4`, `?y4c`, `?conjugate`, and `?hz4` to obtain the definitions of the functions. Use these functions to generate a table of 4-dimensional hyperspherical harmonics up to $\lambda = 2$ and show that the functions are orthonormal.

(3) Type `?y5` and `?hz5` to obtain the definitions of the functions. Use these functions to generate a table of 5-dimensional hyperspherical harmonics up to $\lambda = 2$ and show that the functions are orthonormal.

(4) Type `?y6` and `?hz6` to obtain the definitions of the functions. Use these functions to generate a table of 6-dimensional hyperspherical harmonics up to $\lambda = 2$ and show that the functions are orthonormal.

(5) Show that each of the hyperspherical harmonics generated in the previous three exercises is an harmonic polynomial in the variables $u_1, u_2, ..., u_6$.

(6) Using the 4-dimensional hyperspherical harmonics with $\lambda = 1$ shown in equation (2.49), generate the representations of the rotation operators shown in equations (2.53)–(2.58).

(7) Show that the matrix representations generated in the previous exercise obey equations (2.59)–(2.61).

(8) Using the 4-dimensional hyperspherical harmonics generated in Exercise 2, with $\lambda = 2$, generate a representation of $\Lambda_{1,2}$.

(9) Show the if

$$\begin{aligned} x_1 &= r\sin\theta\cos\phi \\ x_2 &= r\sin\theta\sin\phi \\ x_3 &= r\cos\theta \end{aligned}$$

and if $\mathbf{u} \equiv r^{-1}\mathbf{x}$, then

$$\frac{\partial f(\mathbf{u})}{\partial r} = 0$$

Could a similar argument be made for spaces of higher dimension? Use a pencil and paper, rather than Mathematica.

Chapter 3

GEGENBAUER POLYNOMIALS

3.1 The generating function for Gegenbauer polynomials

In the theory of hyperspherical harmonics, Gegenbauer polynomials play a role which is analogous to the role played by Legendre polynomials in the theory of the familiar 3-dimensional spherical harmonics. In fact, the Legendre polynomials are a special case of Gegenbauer polynomials. [Avery, 1989], [Avery, 2000], [Avery and Avery, 2006], [Avery, 2009], [Avery, Rettrup and Avery, 2012], [Vilinken, 1968], [Wen and Avery, 1985]

We can define Gegenbauer polynomials in terms of a generating function. To do this, we first introduce the unit vectors:

$$\mathbf{u} \equiv \frac{1}{r}(x_1, x_2, \ldots, x_d) \equiv \frac{1}{r}\mathbf{x} \tag{3.1}$$

and

$$\mathbf{u}' \equiv \frac{1}{r'}(x'_1, x'_2, \ldots, x'_d) \equiv \frac{1}{r'}\mathbf{x}' \tag{3.2}$$

Then we can write:

$$\frac{1}{|\mathbf{x} - \mathbf{x}'|^{d-2}} = \frac{1}{r_>^{d-2}(1 + \epsilon^2 - 2\epsilon \mathbf{u} \cdot \mathbf{u}')^\alpha} \tag{3.3}$$

where $\alpha \equiv d/2 - 1$ and

$$\epsilon \equiv \frac{r_<}{r_>} \equiv \begin{cases} \dfrac{r}{r'} & \text{if } r' > r \\[2ex] \dfrac{r'}{r} & \text{if } r > r' \end{cases} \tag{3.4}$$

If we expand the function shown in (3.3) as a Taylor series in ϵ, then the Gegenbauer polynomial $C_\lambda^\alpha(\mathbf{u} \cdot \mathbf{u}')$ is defined as the coefficient of ϵ^λ in the series. In other words, the Gegenbauer polynomials are defined in such a way that

$$\frac{1}{|\mathbf{x}-\mathbf{x}'|^{d-2}} = \frac{1}{r_>^{d-2}} \sum_{\lambda=0}^{\infty} \left(\frac{r_<}{r_>}\right)^{\lambda} C_\lambda^\alpha(\mathbf{u}\cdot\mathbf{u}') \tag{3.5}$$

For the case where $d = 3$, this series reduces to the familiar series defining Legendre polynomials:

$$\frac{1}{|\mathbf{x}-\mathbf{x}'|} = \frac{1}{r_>} \sum_{l=0}^{\infty} \left(\frac{r_<}{r_>}\right)^{l} P_l(\mathbf{u}\cdot\mathbf{u}') \tag{3.6}$$

from which one can see that

$$P_l(\mathbf{u}\cdot\mathbf{u}') = C_l^{1/2}(\mathbf{u}\cdot\mathbf{u}') \tag{3.7}$$

The first few Gegenbauer polynomials are

$$\begin{aligned}
C_0^\alpha(\mathbf{u}\cdot\mathbf{u}') &= 1\\
C_1^\alpha(\mathbf{u}\cdot\mathbf{u}') &= 2\alpha\mathbf{u}\cdot\mathbf{u}'\\
C_2^\alpha(\mathbf{u}\cdot\mathbf{u}') &= 2\alpha(\alpha+1)(\mathbf{u}\cdot\mathbf{u}')^2 - \alpha\\
C_3^\alpha(\mathbf{u}\cdot\mathbf{u}') &= \frac{4}{3}\alpha(\alpha+1)(\alpha+2)(\mathbf{u}\cdot\mathbf{u}')^3 - 2\alpha(\alpha+1)\mathbf{u}\cdot\mathbf{u}'
\end{aligned} \tag{3.8}$$

and in general,

$$C_\lambda^\alpha(\mathbf{u}\cdot\mathbf{u}') = \sum_{t=0}^{\lfloor\lambda/2\rfloor} \frac{(-1)^t(\alpha)_{\lambda-t}(2\mathbf{u}\cdot\mathbf{u}')^{\lambda-2t}}{t!(\lambda-2t)!} \tag{3.9}$$

where $\alpha \equiv d/2 - 1$ and

$$\begin{aligned}
(\alpha)_j &\equiv \alpha(\alpha+1)(\alpha+2)\cdots(\alpha+j-1)\\
&= \frac{\Gamma(\alpha+j)}{\Gamma(\alpha)}\\
&= \frac{(d+2j-4)!!}{2^j(d-4)!!}
\end{aligned} \tag{3.10}$$

In (3.10) we have used the fact that $\Gamma(n+1) = (2n)!!/2^n$. Solving the set of equations (3.8) for $(\mathbf{u}\cdot\mathbf{u}')^n$, one can also obtain the inverse relations:

$$\begin{aligned}
(\mathbf{u}\cdot\mathbf{u}')^0 &= C_0^\alpha(\mathbf{u}\cdot\mathbf{u}')\\
(\mathbf{u}\cdot\mathbf{u}')^1 &= \frac{1}{2\alpha}C_1^\alpha(\mathbf{u}\cdot\mathbf{u}')\\
(\mathbf{u}\cdot\mathbf{u}')^2 &= \frac{1}{2(\alpha+1)}C_2^\alpha(\mathbf{u}\cdot\mathbf{u}') + \frac{\alpha}{2\alpha(\alpha+1)}C_0^\alpha(\mathbf{u}\cdot\mathbf{u}')
\end{aligned} \tag{3.11}$$

and in general [Hua, 1963]

$$(\mathbf{u}\cdot\mathbf{u}')^n = \frac{n!}{2^n} \sum_{s=0}^{\lfloor n/2\rfloor} \frac{\alpha+n-2s}{s!(\alpha)_{n-s+1}} C_{n-2s}^\alpha(\mathbf{u}\cdot\mathbf{u}') \tag{3.12}$$

3.2 Properties of Gegenbauer polynomials

If we choose the origin of our coordinate system in such a way that $\mathbf{x}' = 0$, then

$$\Delta \frac{1}{|\mathbf{x}-\mathbf{x}'|^{d-2}} = \frac{1}{r^{d-1}}\frac{\partial}{\partial r} r^{d-1}\frac{\partial}{\partial r} r^{2-d} = 0 \qquad \text{if } r \neq 0 \tag{3.13}$$

From (3.5) and (3.13) it follows that for $r < r'$,

$$\Delta \frac{1}{|\mathbf{x}-\mathbf{x}'|^{d-2}} = \sum_{\lambda=0}^{\infty} \frac{1}{r'^{\lambda+d-2}} \Delta \left[r^\lambda C_\lambda^\alpha(\mathbf{u}\cdot\mathbf{u}')\right] = 0 \tag{3.14}$$

Since this relationship must hold for all values of r', each term in the series must vanish separately, and thus

$$\Delta \left[r^\lambda C_\lambda^\alpha(\mathbf{u}\cdot\mathbf{u}')\right] = \left[\frac{1}{r^{d-1}}\frac{\partial}{\partial r} r^{d-1}\frac{\partial}{\partial r} - \frac{\Lambda^2}{r^2}\right] r^\lambda C_\lambda^\alpha(\mathbf{u}\cdot\mathbf{u}') = 0 \tag{3.15}$$

But

$$\frac{1}{r^{d-1}}\frac{\partial}{\partial r} r^{d-1}\frac{\partial}{\partial r} r^\lambda = \lambda(\lambda+d-2) r^{\lambda-2} \tag{3.16}$$

so that

$$\left[\Lambda^2 - \lambda(\lambda+d-2)\right] C_\lambda^\alpha(\mathbf{u}\cdot\mathbf{u}') = 0 \tag{3.17}$$

In other words, the Gegenbauer polynomial $C_\lambda^\alpha(\mathbf{u}\cdot\mathbf{u}')$ is an eigenfunction of Λ^2 belonging to the eigenvalue $\lambda(\lambda+d-2)$. It must therefore be possible to express $C_\lambda^\alpha(\mathbf{u}\cdot\mathbf{u}')$ as a linear combination of hyperspherical harmonics belonging to the same eigenvalue. Thus we can write:

$$C_\lambda^\alpha(\mathbf{u}\cdot\mathbf{u}') = \sum_\mu a_{\lambda,\mu}(\mathbf{u}) Y_{\lambda,\mu}(\mathbf{u}') \tag{3.18}$$

Since $C_\lambda^\alpha(\mathbf{u}\cdot\mathbf{u}')$ is a function of the scalar, $\mathbf{u}\cdot\mathbf{u}'$, it must be invariant under rotations of the coordinate system. In other words, if R is an element of the rotation group, $SO(d)$, then

$$R\, C_\lambda^\alpha(\mathbf{u}\cdot\mathbf{u}') = C_\lambda^\alpha(\mathbf{u}\cdot\mathbf{u}') \tag{3.19}$$

Thus, under the same rotation,

$$\sum_\mu a_{\lambda,\mu}(R^{-1}\mathbf{u}) Y_{\lambda,\mu}(R^{-1}\mathbf{u}') = \sum_\mu a_{\lambda,\mu}(\mathbf{u}) Y_{\lambda,\mu}(\mathbf{u}') \tag{3.20}$$

Under the rotation R, $Y_{\lambda,\mu}(\mathbf{u}')$ is transformed into a linear combination of hyperspherical harmonics belonging to the same value of λ:

$$Y_{\lambda,\mu}(R^{-1}\mathbf{u}') = \sum_{\mu'} Y_{\lambda,\mu'}(\mathbf{u}') D^\lambda_{\mu',\mu}(R) \tag{3.21}$$

The transformation matrix is required to be unitary, since we wish the transformed hyperspherical harmonics to fulfill the orthonormality relation (2.39). Therefore (from the unitarity requirement)

$$\sum_{\mu} D^{\lambda *}_{\mu'',\mu}(R) D^{\lambda}_{\mu',\mu}(R) = \delta_{\mu'',\mu'} \tag{3.22}$$

In order for $C^{\alpha}_{\lambda}(\mathbf{u}\cdot\mathbf{u}')$ to be invariant, the coefficients $a_{\lambda,\mu}(\mathbf{u})$ must have a transformation law which is the inverse of (3.21), so that we must have

$$a_{\lambda,\mu}(R^{-1}\mathbf{u}) = \sum_{\mu''} a_{\lambda,\mu''}(\mathbf{u}) D^{\lambda *}_{\mu'',\mu}(R) \tag{3.23}$$

Taking the complex conjugate of (3.23) and comparing it to (3.21), we can see that $a^*_{\lambda,\mu}(\Omega)$ must have the same transformation law under rotations as $Y_{\lambda,\mu}(\Omega)$. This can only hold for all values of R in $SO(d)$ if

$$a_{\lambda,\mu}(\mathbf{u}) = K_\lambda Y^*_{\lambda,\mu}(\mathbf{u}) \tag{3.24}$$

where K_λ is a constant. Substituting (3.24) into (3.18), we have the sum rule

$$C^{\alpha}_{\lambda}(\mathbf{u}\cdot\mathbf{u}') = K_\lambda \sum_{\mu} Y^*_{\lambda,\mu}(\mathbf{u}) Y_{\lambda,\mu}(\mathbf{u}') \tag{3.25}$$

which is the d-dimensional generalization of the familiar sum rule for 3-dimensional spherical harmonics:

$$P_l(\mathbf{u}\cdot\mathbf{u}') = \frac{4\pi}{2l+1} \sum_{m} Y^*_{l,m}(\mathbf{u}) Y_{l,m}(\mathbf{u}') \tag{3.26}$$

The constant K_λ remains to be determined, and we shall do this below. Combining (3.25) with the orthonormality relation (2.39), we can write

$$\begin{aligned}
&\int d\Omega_s\; C^{\alpha}_{\lambda}(\mathbf{u}\cdot\mathbf{u}') C^{\alpha}_{\lambda'}(\mathbf{u}\cdot\mathbf{u}'') \\
&= \int d\Omega_d\; [K_\lambda \sum_{\mu} Y^*_{\lambda,\mu}(\mathbf{u}) Y_{\lambda,\mu}(\mathbf{u}')][K_{\lambda'} \sum_{\mu'} Y_{\lambda',\mu'}(\mathbf{u}) Y^*_{\lambda',\mu'}(\mathbf{u}'')] \\
&= \delta_{\lambda',\lambda} K^2_\lambda \sum_{\mu} Y^*_{\lambda,\mu}(\mathbf{u}') Y_{\lambda,\mu}(\mathbf{u}'') \\
&= \delta_{\lambda',\lambda} K_\lambda C^{\alpha}_{\lambda}(\mathbf{u}'\cdot\mathbf{u}'')
\end{aligned} \tag{3.27}$$

Equation (3.25) implies that if we multiply any function of the hyperangles $F(\Omega)$ by $C^{\alpha}_{\lambda}(\mathbf{u}\cdot\mathbf{u}')/K_\lambda$ and integrate over $d\Omega$ we will obtain a projection of the function onto the part of Hilbert space spanned by the eigenfunctions of Λ^2 belonging to the eigenvalue $\lambda(\lambda + d - 2)$:

$$\begin{aligned} O_\lambda[F(\mathbf{u})] &= \frac{1}{K_\lambda} \sum_\mu Y_{\lambda,\mu}(\mathbf{u}) \int d\Omega'_d \, Y^*_{\lambda,\mu}(\mathbf{u}') F(\mathbf{u}') \\ &= \frac{1}{K_\lambda} \int d\Omega'_d \, C^\alpha_\lambda(\mathbf{u} \cdot \mathbf{u}') F(\mathbf{u}') \end{aligned} \tag{3.28}$$

If the function $F(\Omega)$ already happens to lie entirely within this part of Hilbert space, then of course it is unaffected by this projection.

3.3 Determination of the constant K_λ

In Chapter 2, we saw that the degeneracy of a set of d-dimensional hyperspherical harmonics corresponding to the principal quantum number λ is given by

$$\sum_\mu 1 = \frac{(d + 2\lambda - 2)(\lambda + d - 3)!}{\lambda!(d-2)!} \tag{3.29}$$

If we set $\mathbf{u} = \mathbf{u}'$ in (3.25) we have

$$C^\alpha_\lambda(1) = K_\lambda \sum_\mu Y^*_{\lambda,\mu}(\mathbf{u}) Y_{\lambda,\mu}(\mathbf{u}) \tag{3.30}$$

Integrating both sides of (3.30) over solid angle then yields

$$\int d\Omega_d \, C^\alpha_\lambda(1) = K_\lambda \sum_\mu \int d\Omega_d \, Y^*_{\lambda,\mu}(\mathbf{u}) Y_{\lambda,\mu}(\mathbf{u}) \tag{3.31}$$

so that

$$C^\alpha_\lambda(1) I(0) = K_\lambda \sum_\mu 1 = K_\lambda \frac{(d + 2\lambda - 2)(\lambda + d - 3)!}{\lambda!(d-2)!} \tag{3.32}$$

Our next task is to evaluate $C^\alpha_\lambda(1)$. In order to do so, we return to the definition of their generating function, equations (3.3) and (3.5). Setting $\mathbf{u} \cdot \mathbf{u}' = 1$ in these equations, expanding the left-hand side of (3.3) as a Taylor series in ϵ, we obtain

$$C^\alpha_\lambda(1) = \frac{(\lambda + d - 3)!}{\lambda!(d-3)!} \tag{3.33}$$

Finally, substituting (3.33) into (3.32) and solving for K_λ, we have

$$K_\lambda = \frac{(\lambda + d - 3)! I(0)}{\lambda!(d-3)!} \frac{\lambda!(d-2)!}{(d + 2\lambda - 2)(\lambda + d - 3)!} \tag{3.34}$$

which simplifies to

$$K_\lambda = \frac{(d-2)I(0)}{d+2\lambda-2} \tag{3.35}$$

where $I(0)$ is the total solid angle in the d-dimensional space. For the first few values of d, we have

$$\begin{aligned}
d&=3 \qquad K_\lambda = \frac{4\pi}{2\lambda+1} \equiv \frac{4\pi}{2l+1}\\
d&=4 \qquad K_\lambda = \frac{4\pi^2}{(2\lambda+2)}\\
d&=5 \qquad K_\lambda = \frac{8\pi^2}{(2\lambda+3)}\\
d&=6 \qquad K_\lambda = \frac{4\pi^3}{(2\lambda+4)}\\
d&=7 \qquad K_\lambda = \frac{16\pi^3}{3(2\lambda+5)}\\
d&=8 \qquad K_\lambda = \frac{2\pi^4}{(2\lambda+6)}\\
d&=9 \qquad K_\lambda = \frac{32\pi^4}{15(2\lambda+7)}
\end{aligned} \tag{3.36}$$

Thus, in the familiar case where $d=3$, equation (3.27) becomes

$$\int d\Omega\; P_l(\mathbf{u}\cdot\mathbf{u}')P_{l'}(\mathbf{u}\cdot\mathbf{u}'') = \delta_{l',l}\,\frac{4\pi}{2l+1}P_l(\mathbf{u}'\cdot\mathbf{u}'') \tag{3.37}$$

while (3.28) can be written as

$$\begin{aligned}
O_l[F(\Omega)] &= \frac{(2l+1)}{4\pi}\sum_{m=-l}^{l} Y_{l,m}(\Omega)\int d\Omega'\; Y^*_{l,m}(\Omega')F(\Omega')\\
&= \frac{(2l+1)}{4\pi}\int d\Omega'\; P_l(\mathbf{u}\cdot\mathbf{u}')F(\Omega')
\end{aligned} \tag{3.38}$$

3.4 The generalization of $P_l^m(\mathbf{u}\cdot\mathbf{u}')$

We have seen that Gegenbauer polynomials are the d-dimensional generalizations of the Legendre polynomials. It is therefore logical to ask whether the associated Legendre polynomials also have a d-dimensional generalization. Associated Legendre polynomials occur in the theory of the familiar spherical harmonics in a 3-dimensional space. As is well known, these ordinary spherical harmonics are given by

$$Y_{l,m}(\theta,\phi) = Ne^{im\phi}P_l^m(\cos\theta) \tag{3.39}$$

where N is a normalizing constant. We can rewrite this as

$$Y_{l,m}(\hat{\mathbf{x}}) = \frac{N}{r^l}\, h_m(x,y) P_l^m(z) \tag{3.40}$$

where

$$h_m(x,y) \equiv (x+iy)^m \tag{3.41}$$

is an harmonic polynomial of order m in the 2-dimensional space spanned by x and y. In other words, by means of the associated Legendre polynomials, we can start with an harmonic polynomial in a 2-dimensional space and then construct an harmonic polynomial in a 3-dimensional space. Can this procedure be generalized to higher dimensions? We can see that this is the case by means of the following argument: Suppose that $h_s(x_1, x_2, \ldots, x_{d-1})$ is an harmonic polynomial which is independent of x_d and of order s. Then

$$\Delta h_s = \Delta' h_s = 0 \tag{3.42}$$

where Δ' is the generalized Laplacian operator in the $(d-1)$-dimensional space and where

$$\Delta = \Delta' + \frac{\partial^2}{\partial x_d^2} \tag{3.43}$$

Multiplying by x^{n-s} for $n \geq s$, we obtain a homogeneous polynomial of order n. From (3.42) and (3.43) it follows that for $n \geq s$

$$\begin{aligned}
\Delta^k(x_d^{n-s} h_s(x_1, \ldots, x_{d-1})) &= \frac{\partial^{2k}}{\partial x_d^{2k}} x_d^{n-s} h_s(x_1, \ldots, x_{d-1}) \\
&= \frac{(n-s)!}{(n-s-2k)!} x_d^{n-s-2k} h_s(x_1, \ldots, x_{d-1})
\end{aligned} \tag{3.44}$$

We now make use of equation (3.44) together with (1.38) to project out the harmonic polynomial of highest order contained in the homogeneous polynomial

$$f_n = x_d^{n-s} h_s(x_1, \ldots, x_{d-1}) \tag{3.45}$$

This gives us

$$\begin{aligned}
h_n &= O_n[x_d^{n-s} h_s(x_1, \ldots, x_{d-1})] \\
&= \sum_{k=0}^{\lfloor (n-s)/2 \rfloor} \frac{(-1)^k (d+2n-2k-4)!!}{(2k)!!(d+2n-4)!!} r^{2k} \Delta^k f_n \\
&= \sum_{k=0}^{\lfloor (n-s)/2 \rfloor} \frac{(-1)^k (d+2n-2k-4)!!}{(2k)!!(d+2n-4)!!} \frac{(n-s)!}{(n-s-2k)!} r^{2k} x_d^{n-s-2k} \\
&\quad \times h_s(x_1, \ldots, x_{d-1})
\end{aligned} \tag{3.46}$$

We next take outside the k-sum the terms that are independent of k. This gives us:

$$h_n = \frac{(n-s)!}{(d+2n-4)!!} \times \sum_{k=0}^{\lfloor (n-s)/2 \rfloor} \frac{(-1)^k (d+2n-2k-4)!!}{(2k)!!(n-s-2k)!} r^{2k} x_d^{n-s-2k} h_s(x_1, \ldots, x_{d-1}) \tag{3.47}$$

Thus we finally obtain:

$$h_n = r^{n-s} h_s(x_1, \ldots, x_{d-1}) \frac{(n-s)!}{(d+2n-4)!!} \times \sum_{k=0}^{\lfloor (n-s)/2 \rfloor} \frac{(-1)^k (d+2n-2k-4)!!}{(2k)!!(n-s-2k)!} \left(\frac{x_d}{r}\right)^{n-s-2k} \tag{3.48}$$

We shall now try to show that the series in (3.48) defines a function that can be expressed in terms of Gegenbauer polynomials, and that in fact this function is the d-dimensional generalization of an associated Legendre polynomial. From (3.9) we have

$$C_\lambda^\alpha(\mathbf{u} \cdot \mathbf{u}') = \sum_{k=0}^{\lfloor \lambda/2 \rfloor} \frac{(-1)^k \ (d+2\lambda-2k-4)!!(\mathbf{u} \cdot \mathbf{u}')^{\lambda-2k}}{(2k)!!(\lambda-2k)!(d-4)!!} \tag{3.49}$$

Since

$$\alpha \equiv \frac{d}{2} - 1 \tag{3.50}$$

it follows that $d = 2(\alpha+1)$. Thus if we make the replacements $\alpha \to \alpha + s$ and $\lambda \to n - s$, it is equivalent to making the replacements $d \to d + 2s$, with $2\lambda \to 2n - 2s$ and $(d+2\lambda-2k-4)!! \to (d+2n-2k-4)!!$. Therefore

$$C_{n-s}^{\alpha+s}\left(\frac{x_d}{r}\right) = \sum_{k=0}^{\lfloor (n-s)/2 \rfloor} \frac{(-1)^k (d+2n-2k-4)!!}{(2k)!!(n-s-2k)!(d+2s-4)!!} \left(\frac{x_d}{r}\right)^{n-s-2k} \tag{3.51}$$

Finally, combining (3.48) and (3.51), we can write:

$$h_n = r^{n-s} h_s(x_1, \ldots, x_{d-1}) \frac{(n-s)!(d+2s-4)!!}{(d+2n-4)!!} C_{n-s}^{\alpha+s}\left(\frac{x_d}{r}\right) \tag{3.52}$$

where r is the hyperradius appropriate to a d-dimensional space. Equation (3.52) is comparable to equation (3.40).

3.5 The standard tree

Equation (3.52) allows us to generate an orthonormal set of hyperspherical harmonics in a d-dimensional space, starting with such a set in a $(d-1)$-dimensional space. For example, let us consider the case where $d=4$ and $d-1=3$. We already know the familiar 3-dimensional spherical harmonics $Y_{l,m}(\theta,\phi)$. From these, we can generate a set of harmonic polynomials by multiplying by $r_{(3)}^l$:

$$h_l(x_1,x_2,x_3) = r_{(3)}^l Y_{l,m}(\theta,\phi) \tag{3.53}$$

The subscript (3) in equation (3.53) serves to remind us that the radius here corresponds to $d=3$ and is defined by

$$r_{(3)}^2 \equiv x_1^2 + x_2^2 + x_3^2 \tag{3.54}$$

From equations (3.53) and (3.52), we can see that

$$O_\lambda[x_4^{\lambda-l} h_l(x_1,x_2,x_3)] \sim r_{(4)}^{\lambda-l} r_{(3)}^l h_l(x_1,x_2,x_3) C_{\lambda-l}^{1+l}\left(\frac{x_4}{r_{(4)}}\right) \tag{3.55}$$

In equation (3.55) the subscript (4) indicates that $r_{(4)}$ is the hyperradius appropriate to a 4-dimensional space:

$$r_{(4)}^2 \equiv x_1^2 + x_2^2 + x_3^2 + x_4^2 \tag{3.56}$$

If we define the angle χ by

$$\frac{x_4}{r_{(4)}} \equiv \cos\chi \tag{3.57}$$

Then, since

$$\begin{aligned} r_{(4)}^2 &= r_{(3)}^2 + x_4^2 \\ 1 &= \frac{r_{(3)}^2}{r_{(4)}^2} + \frac{x_4^2}{r_{(4)}^2} = \frac{r_{(3)}^2}{r_{(4)}^2} + \cos^2\chi \end{aligned} \tag{3.58}$$

and

$$\cos^2\chi + \sin^2\chi = \left(\frac{x_4}{r_{(4)}^2}\right)^2 + \sin^2\chi = 1 \tag{3.59}$$

it follows that

$$\frac{r_{(3)}}{r_{(4)}} = \sin\chi \tag{3.60}$$

Dividing equation (3.55) by $r_{(4)}^\lambda$, we obtain a set of functions

$$Y_{\lambda,l,m}(\chi,\theta,\phi) \sim \sin^l\chi C_{\lambda-l}^{1+l}(\cos\chi) Y_{l,m}(\theta,\phi) \tag{3.61}$$

In general, if we have a set of functions which satisfy

$$\left[\Lambda_{d-1}^2 - l(l+d-3)\right] Y_{l,\mu}(\Omega_{d-1}) = 0 \tag{3.62}$$

then the set of functions

$$Y_{\lambda,l,\mu}(\Omega_d) \sim \left(\frac{r_{(d-1)}}{r_{(d)}}\right)^l C_{\lambda-l}^{\alpha+l}\left(\frac{x_d}{r_{(d)}}\right) Y_{l,\mu}(\Omega_{d-1}) \tag{3.63}$$

will satisfy

$$\left[\Lambda_d^2 - \lambda(\lambda+d-2)\right] Y_{\lambda,l,\mu}(\Omega_d) = 0 \tag{3.64}$$

Equations (3.63) and (3.64) imply that a set of mutually orthogonal hyperspherical harmonics can be written down explicitly in the form:

$$Y_{\mu_1,\mu_2,\cdots,}(u_1,u_2,\cdots) \sim e^{im\phi} \prod_{j=1}^{d-2} C_{\mu_j-\mu_{j+1}}^{\alpha_j+\mu_{j+1}}(\cos\theta_j)\,(\sin\theta_j)^{\mu_{j+1}} \tag{3.65}$$

where

$$2\alpha_j \equiv d-j-1 \tag{3.66}$$

and where the angles $\theta_1, \theta_2, \ldots, \theta_{d-2}$ are defined by the relationship:

$$\begin{aligned}
x_1 &= r\ \sin\theta_1 \sin\theta_2 \cdots \sin\theta_{d-2}\cos\phi \\
x_2 &= r\ \sin\theta_1 \sin\theta_2 \cdots \sin\theta_{d-2}\sin\phi \\
x_3 &= r\ \sin\theta_1 \sin\theta_2 \cdots \cos\theta_{d-2} \\
\vdots\ &\ \vdots\ \ \vdots \\
x_{d-1} &= r\ \sin\theta_1 \cos\theta_2 \\
x_d &= r\ \cos\theta_1
\end{aligned} \tag{3.67}$$

The indices μ_j are integers which satisfy

$$\mu_1 \geq \mu_2 \geq \cdots \geq |\mu_{d-1}| \tag{3.68}$$

In the case of hyperspherical harmonics of this type, the chain of indices μ_j is organized according to the following chain of subgroups of $SO(d)$:

$$SO(d) \supset SO(d-1) \supset SO(d-2) \supset \cdots \supset SO(2) \tag{3.69}$$

The 4-dimensional hyperspherical harmonics shown in Table 2.1 are of this type, where the unit vectors u_j are related to the angles by

$$\begin{aligned}
u_1 &= \sin\chi \sin\theta \cos\phi \\
u_2 &= \sin\chi \sin\theta \sin\phi \\
u_3 &= \sin\chi \cos\theta \\
u_4 &= \cos\chi
\end{aligned} \tag{3.70}$$

It should be emphasized, however, that this is not the only way to organize the indices μ_j. In physical applications, it is desirable to adapt the symmetry of the hyperspherical harmonics to the symmetry of the problem being treated. In the Method of Trees, which we will discuss later, the chain of subgroups of $SO(d)$ shown in (3.69) is called the *standard tree*. Other possible chains of subgroups correspond to other trees.

3.6 Normalization

Equation (3.27) can be used to normalize the hyperspherical harmonics. If we let $\mathbf{u}' = \mathbf{u}''$ and $\lambda = \lambda'$ in (2.28), then the integral becomes

$$\int d\Omega_d \, |C_\lambda^\alpha(\mathbf{u}\cdot\mathbf{u}')|^2 = K_\lambda C_\lambda^\alpha(1) \tag{3.71}$$

Combining equations (3.33), (3.34) and (3.71), we have:

$$\int d\Omega_d \, |C_\lambda^\alpha(\mathbf{u}\cdot\mathbf{u}')|^2 = \frac{I(0)(d-2)(\lambda+d-3)!}{\lambda!(d+2\lambda-2)(d-3)!} \tag{3.72}$$

For example, when $d = 4$, (3.71) becomes:

$$\int d\Omega_4 \, |C_\lambda^1(\mathbf{u}\cdot\mathbf{u}')|^2 = 2\pi^2 \tag{3.73}$$

For $l = 0$, the hyperspherical harmonics in Table 2.1 can be written in the form:

$$Y_{\lambda,0,0}(\mathbf{u}) = N_\lambda C_\lambda^1(u_4) \tag{3.74}$$

where N_λ is a normalizing constant. If we let $\mathbf{u}' = (0,0,0,1)$ in (3.73), so that $\mathbf{u}\cdot\mathbf{u}' = u_4$, then (3.73) and (3.74) yield the normalization condition:

$$\int d\Omega_4 \, |Y_{\lambda,0,0}(\mathbf{u})|^2 = N_\lambda^2 \int d\Omega_4 \, |C_\lambda^1(u_4)|^2 = 2\pi^2 N_\lambda^2 = 1 \tag{3.75}$$

so that

$$Y_{\lambda,0,0}(\mathbf{u}) = \frac{1}{\sqrt{2}\pi} C_\lambda^1(u_4) \tag{3.76}$$

In general, if we let $u_d = \cos\theta$, then the element of solid angle in a d-dimensional space is given by

$$d\Omega_d = \sin^{d-2}\theta \; d\theta \; d\Omega_{d-1} \tag{3.77}$$

where $d\Omega_{d-1}$ is the element of solid angle in a $(d-1)$-dimensional space. The reason for this will be discussed in Chapter 10 in connection with the volume element and the Jacobian for the transformation from Cartesian coordinates to hyperspherical polar coordinates. Then from (2.17), (3.72) and (3.77), we have:

$$\begin{aligned}\int d\Omega_d\, |C_\lambda^\alpha(u_d)|^2 &= \frac{2\pi^{(d-1)/2}}{\Gamma(d/2-1/2)}\int_0^\pi d\theta\ \sin^{d-2}\theta\, |C_\lambda^\alpha(\cos\theta)|^2\\ &= \frac{2\pi^{d/2}(d-2)(\lambda+d-3)!}{\Gamma(d/2)\lambda!(d+2\lambda-2)(d-3)!}\end{aligned} \tag{3.78}$$

so that

$$\int_0^\pi d\theta\ \sin^{d-2}\theta\, |C_\lambda^\alpha(\cos\theta)|^2 = \frac{\pi^{1/2}\Gamma(d/2-1/2)(d-2)(\lambda+d-3)!}{\Gamma(d/2)\lambda!(d+2\lambda-2)(d-3)!} \tag{3.79}$$

where $\alpha = d/2-1$. From (3.77) and (3.79), it follows that the normalization constant $N_{\lambda,\mu}$ of equation (3.65) is given by

$$N_{\lambda,\mu}^{-2} = 2\pi\prod_{j=1}^{d-2}\frac{\pi^{1/2}\Gamma(\alpha_j+\mu_{j+1}+1/2)(\alpha_j+\mu_{j+1})(2\alpha_j+\mu_j+\mu_{j+1}-1)!}{\Gamma(\alpha_j+\mu_{j+1}+1)(\mu_j-\mu_{j+1})!(\alpha_j+\mu_j)(2\alpha_j+2\mu_{j+1}-1)!} \tag{3.80}$$

Equations (3.65) and (3.80) imply that if $Y_{l,\mu}(u_1,u_2,\ldots,u_{d-1})$ is a correctly normalized hyperspherical harmonic in a $(d-1)$-dimensional space, and if

$$\begin{aligned}u_d &= \frac{x_d}{r} = \cos\theta\\ \alpha &= \frac{d}{2}-1\\ \lambda &\geq l\end{aligned} \tag{3.81}$$

then

$$\begin{aligned}Y_{\lambda,l,\mu}(u_1,u_2,\ldots,u_d) &= N\sin^l\theta\ C_{\lambda-l}^{\alpha+l}(\cos\theta)Y_{l,\mu}(\Omega_{d-1})\\ &= N\sin^l\theta\ C_{\lambda-l}^{\alpha+l}(u_d)Y_{l,\mu}(u_1,u_2,\ldots,u_{d-1})\end{aligned} \tag{3.82}$$

will be a properly normalized hyperspherical harmonic in d dimensions, provided that

$$N = \left(\frac{\Gamma(\alpha+l+1)(\lambda-l)!(\alpha+\lambda)(2\alpha+2l-1)!}{\sqrt{\pi}\Gamma(\alpha+l+1/2)(\alpha+l)(2\alpha+\lambda+l-1)!}\right)^{1/2} \tag{3.83}$$

This can be found by dividing the expression for $Y_{\lambda,l,\mu}(u_1,u_2,\ldots,u_d)$ by $Y_{l,\mu}(u_1,u_2,\ldots,u_{d-1})$. Note that $Y_{l,\mu}(u_1,u_2,\ldots,_{d-1})$ can be expressed using (3.65) and (3.80) with $\lambda\to l$ and the products in each equation, starting with $j=2$. Below we list the first few 6-dimensional hyperspherical harmonics, with the chain of subgroups ordered according to the standard tree. The reader can verify, using our general angular integration theorem (2.27)–(2.29), that the harmonics listed in the table satisfy the orthonormality relation

$$\int d\Omega_6 \, Y^*_{\lambda',l',\mu'}(\hat{\mathbf{u}}) Y_{\lambda,l,\mu}(\hat{\mathbf{u}}) = \delta_{\lambda',\lambda}\delta_{l',l}\delta_{\mu',\mu} \tag{3.84}$$

Notice also that the number of hyperspherical harmonics corresponding to particular values of λ are given by equation (2.45), or alternatively (2.46). The hyperspherical harmonics constructed by means of the standard tree are related through a unitary transformation to hyperspherical harmonics corresponding to non-standard trees. One can also verify that

$$\Delta[r^\lambda Y_{\lambda,l,\mu}(\hat{\mathbf{u}})] = 0 \tag{3.85}$$

In the important special case where $d = 4$, equation 3.82 becomes

$$Y_{\lambda,l,m}(\Omega_4) = N_{\lambda,l} \, (1 - u_4^2)^{l/2} C^{1+l}_{\lambda-l}(u_4) Y_{l,m}(u_1, u_2, u_3) \tag{3.86}$$

where

$$N_{\lambda,l} = (-1)^\lambda i^l (2l)!! \sqrt{\frac{2(\lambda+1)(\lambda-l)!}{\pi(\lambda+l+1)!}} \tag{3.87}$$

Table 3.1: The first few 6-dimensional hyperspherical harmonics, with the subgroups ordered according to the standard tree

$(\lambda, l, \mu_3, \mu_2, \mu_1)$	$\pi^{3/2} Y_{\lambda,l,\mu}$
$(0,0,0,0,0)$	1
$(1,0,0,0,0)$	$\sqrt{6}\, u_6$
$(1,1,0,0,0)$	$\sqrt{6}\, u_5$
$(1,1,1,0,0)$	$\sqrt{6}\, u_4$
$(1,1,1,1,0)$	$\sqrt{6}\, u_3$
$(1,1,1,1,1)$	$-\sqrt{3}\,(u_1 + iu_2)$
$(1,1,1,1,-1)$	$\sqrt{3}\,(u_1 - iu_2)$
$(2,0,0,0,0)$	$-\frac{2}{\sqrt{5}}(u_1^2 + u_2^2 + u_3^2 + u_4^2 + u_5^2 - 5u_6^2)$
$(2,1,0,0,0)$	$4\sqrt{3}\, u_5 u_6$
$(2,1,1,0,0)$	$4\sqrt{3}\, u_4 u_6$
$(2,1,1,1,0)$	$4\sqrt{3}\, u_3 u_6$
$(2,1,1,1,1)$	$-2\sqrt{6}\,(u_1 + iu_2) u_6$
$(2,1,1,1,-1)$	$2\sqrt{6}\,(u_1 - iu_2) u_6$

Table 3.2: 6-dimensional hyperspherical harmonics (continued)

$(\lambda, l, \mu_1, \mu_2, \mu_3)$	$\pi^{3/2} Y_{\lambda,l,\mu}$
$(2, 2, 0, 0, 0)$	$-\sqrt{\frac{6}{5}}\,(u_1^2 + u_2^2 + u_3^2 + u_4^2 - 4u_5^2)$
$(2, 2, 1, 0, 0)$	$4\sqrt{3}\, u_4 u_5$
$(2, 2, 1, 1, 0)$	$4\sqrt{3}\, u_3 u_5$
$(2, 2, 1, 1, 1)$	$-2\sqrt{6}\,(u_1 + iu_2)u_5$
$(2, 2, 1, 1, -1)$	$2\sqrt{6}\,(u_1 - iu_2)u_5$
$(2, 2, 2, 0, 0)$	$-\sqrt{2}\,(u_1^2 + u_2^2 + u_3^2 - 3u_4^2)$
$(2, 2, 2, 1, 0)$	$-4\sqrt{3}\, u_3 u_4$
$(2, 2, 2, 1, 1)$	$-2\sqrt{6}\,(u_1 + iu_2)u_4$
$(2, 2, 2, 1, -1)$	$2\sqrt{6}\,(u_1 - iu_2)u_4$
$(2, 2, 2, 2, 0)$	$-2(u_1^2 + u_2^2 - 2u_3^2)$
$(2, 2, 2, 2, 1)$	$-2\sqrt{6}\,(u_1 + iu_2)u_3$
$(2, 2, 2, 2, -1)$	$2\sqrt{6}\,(u_1 - iu_2)u_3$
$(2, 2, 2, 2, 2)$	$\sqrt{6}\,(u_1 + iu_2)^2$
$(2, 2, 2, 2, -2)$	$\sqrt{6}\,(u_1 - iu_2)^2$

3.7 Nonstandard trees

In physical applications, the standard hyperspherical harmonics are not necessarily the most convenient ones to use. It may be more appropriate to use harmonics that are adapted to the symmetry of the problem being studied. Therefore, we shall now turn our attention to nonstandard trees. To make the argument as general as possible, let us consider a general fork, joining two subspaces, whose dimensions are respectively d_1 and d_2, where $d_1 + d_2 = d$.

Suppose that we have constructed the harmonic polynomials $h_{l_1}(\mathbf{x}_1)$ and $h_{l_2}(\mathbf{x}_2)$ in the two subspaces. We would like to use them as building blocks for constructing an harmonic polynomial of order λ in the d-dimensional space whose position vector is

$$\mathbf{x} = (\mathbf{x}_1, \mathbf{x}_2) \tag{3.88}$$

If we let

$$\begin{aligned} r_1^2 &\equiv \mathbf{x}_1 \cdot \mathbf{x}_1 \\ r_2^2 &\equiv \mathbf{x}_2 \cdot \mathbf{x}_2 \\ r^2 &\equiv \mathbf{x} \cdot \mathbf{x} \end{aligned} \tag{3.89}$$

Then

$$f_{\lambda,l_1,l_2} = r_1^{\beta} r_2^{\beta'} h_{l_1}(\mathbf{x}_1) h_{l_2}(\mathbf{x}_2) \tag{3.90}$$

will be a homogeneous polynomial of order λ in the coordinates $x_1, x_2, \ldots, x_d$ provided that β and β' are even and

$$\beta + \beta' + l_1 + l_2 = \lambda \tag{3.91}$$

If we act on f_{λ,l_1,l_2} with the projection operator shown in equation (1.38), we will obtain an harmonic polynomial of order λ. The simplest case is that for which $\beta = \beta' = 0$ and $l_1 + l_2 = \lambda$. In this simple case, the homogeneous polynomial shown in equation (3.90) is already harmonic, because

$$\Delta[h_{l_1}(\mathbf{x}_1) h_{l_2}(\mathbf{x}_2)] = h_{l_2}(\mathbf{x}_2)\Delta_1 h_{l_1}(\mathbf{x}_1) + h_{l_1}(\mathbf{x}_1)\Delta_2 h_{l_2}(\mathbf{x}_2) = 0 \tag{3.92}$$

where

$$\begin{aligned} \Delta &= \Delta_1 + \Delta_2 \\ \Delta_1 &= \sum_{j=1}^{d_1} \frac{\partial^2}{\partial x_j^2} \\ \Delta_2 &= \sum_{j=d_1+1}^{d} \frac{\partial^2}{\partial x_j^2} \end{aligned} \tag{3.93}$$

Table 3.3: The first few 6-dimensional hyperspherical harmonics, with the subgroups ordered according to a non-standard tree

$(\lambda, l_1, m_1, l_2, m_2)$	$\pi^{3/2} Y_{\lambda,l,\mu}$
$(0,0,0,0,0)$	1
$(1,1,0,0,0)$	$\sqrt{6}\, u_3$
$(1,1,1,0,0)$	$-\sqrt{3}\,(u_1 + iu_2)$
$(1,1,-1,0,0)$	$\sqrt{3}\,(u_1 - iu_2)$
$(1,0,0,1,0)$	$\sqrt{6}\, u_6$
$(1,0,0,1,1)$	$-\sqrt{3}\,(u_4 + iu_5)$
$(1,0,0,1,-1)$	$\sqrt{3}\,(u_4 - iu_5)$
$(2,1,0,1,0)$	$4\sqrt{3}\, u_3 u_6$
$(2,1,0,1,1)$	$-2\sqrt{6}\, u_3(u_4 + iu_5)$
$(2,1,0,1,-1)$	$2\sqrt{6}\, u_3(u_4 - iu_5)$
$(2,1,1,1,0)$	$-2\sqrt{6}\,(u_1 + iu_2)u_6$
$(2,1,1,1,1)$	$2\sqrt{3}\,(u_1 + iu_2)(u_4 + iu_5)$
$(2,1,1,1,-1)$	$-2\sqrt{3}\,(u_1 + iu_2)(u_4 - iu_5)$

Table 3.4: 6-dimensional hyperspherical harmonics corresponding to a non-standard tree (continued)

$(\lambda, l_1, m_1, l_2, m_2)$	$\pi^{3/2} Y_{\lambda,l,\mu}$
$(2,1,-1,1,0)$	$2\sqrt{6}\ (u_1 - iu_2)u_6$
$(2,1,-1,1,1)$	$-2\sqrt{3}\ (u_1 - iu_2)(u_4 + iu_5)$
$(2,1,-1,1,-1)$	$2\sqrt{3}(u_1 - iu_2)(u_4 - iu_5)$
$(2,2,2,0,0)$	$\sqrt{6}\ (u_1 + iu_2)^2$
$(2,2,1,0,0)$	$-2\sqrt{6}\ u_3(u_1 + iu_2)$
$(2,2,0,0,0)$	$2(u_1^2 + u_2^2 - 2u_3^2)$
$(2,2,-1,0,0)$	$2\sqrt{6}\ u_3(u_1 - iu_2)$
$(2,2,-2,0,0)$	$\sqrt{6}\ (u_1 - iu_2)^2$
$(2,0,0,2,2)$	$\sqrt{6}\ (u_4 + iu_5)^2$
$(2,0,0,2,1)$	$-2\sqrt{6}\ u_6(u_4 + iu_5)$
$(2,0,0,2,0)$	$(u_4^2 + u_5^2 - 2u_6^2)$
$(2,0,0,2,-1)$	$2\sqrt{6}\ u_6(u_4 - iu_5)$
$(2,0,0,2,-2)$	$\sqrt{6}\ (u_4 - iu_5)^2$
$\beta = 2$	$2(u_1^2 + u_2^2 + u_3^2 - u_4^2 - u_5^2 - u_6^2)$

The next-simplest case is that for which $\beta = 2$, $\beta' = 0$ and $l_1+l_2+2 = \lambda$. so that

$$f_{\lambda,l_1,l_2} = r_1^2 h_{l_1}(\mathbf{x}_1)h_{l_2}(\mathbf{x}_2) \tag{3.94}$$

Making use of equation (1.17), we have

$$\Delta[r_1^2 h_{l_1}(\mathbf{x}_1)h_{l_2}(\mathbf{x}_2)] = 2(d_1 + 2l_1)h_{l_1}(\mathbf{x}_1)h_{l_2}(\mathbf{x}_2) \tag{3.95}$$

and

$$\Delta^2[r_1^2 h_{l_1}(\mathbf{x}_1)h_{l_2}(\mathbf{x}_2)] = 0 \tag{3.96}$$

so that, using (1.38), we have

$$h_{\lambda,l_1,l_2}(\mathbf{x}) = \left[r_1^2 - \frac{(d_1 + 2l_1)}{(d + 2\lambda - 4)} r^2\right] h_{l_1}(\mathbf{x}_1)h_{l_2}(\mathbf{x}_2) \tag{3.97}$$

This can be rewritten in a more symmetrical form:

$$h_{\lambda,l_1,l_2}(\mathbf{x}) = \left[\frac{(d_2 + 2l_2)r_1^2 - (d_1 + 2l_1)r_2^2}{(d + 2\lambda - 4)}\right] h_{l_1}(\mathbf{x}_1)h_{l_2}(\mathbf{x}_2) \tag{3.98}$$

The reader might now be wondering why we do not also try a homogeneous polynomial of the form

$$f_{\lambda,l_1,l_2}(\mathbf{x}) = x_j^{\lambda-l_1-l_2} h_{l_1}(\mathbf{x}_1)h_{l_2}(\mathbf{x}_2) \tag{3.99}$$

since we could certainly project out from this an harmonic polynomial. Our reason for not doing this is that the desired symmetry properties of the non-standard hyperspherical harmonics would then be lost, but these are retained when we multiply by $r_1^{\beta} r_2^{\beta'}$.

To illustrate our discussion of the general fork, let us consider the set of 6-dimensional hyperspherical harmonics corresponding to the non-standard tree shown in Figure 3.2.

Almost all of the non-standard 6-dimensional hyperspherical harmonics in the table show below correspond to the case where $\beta = \beta' = 0$. The only exception is the final entry in the table, which corresponds to $\beta = 2$, $\beta' = 0$, and $l_1 = l_2 = 0$. The non-standard hyperspherical harmonics in the table form an orthonormal set, as can be verified using our general angular and hyperangular integration theorem, equations (2.27)–(2.29). A much more complete discussion of the *Method of Trees*, used to construct nonstandard hyperspherical harmonics, can be found in the Russian literature, and in the papers of Prof. Vincenzo Aquilanti and his co-workers.

3.8 Exercises

(1) Equations (3.3)–(3.5) define the Gegenbauer polynomials in terms of their generating function. Make a series expansion of $(1+\epsilon^2-2\epsilon x)^{-\alpha}$ in powers of ϵ about the point $\epsilon=0$. Show that the various terms in the series agree with the Gegenbauer polynomials generated by Mathematica for $\alpha=1/2,1,3/2$.

(2) Remembering that $\alpha=d/2-1$, show that for $d=3$, the Gegenbauer polynomials reduce to the Legendre polynomials.

(3) Make a table showing the first few Gegenbauer polynomials for arbitrary α, and compare it with equation (3.8).

(4) Check that equation (3.33) holds for $d=3,4,5,6,\ldots,12$.

(5) Type `?solidangle` to obtain a definition of this function. Use equation (3.35) to calculate K_λ for $d=3,4,5,6$ and $\lambda=0,1,\ldots,10$.

(6) Check that equation (3.51) holds.

(7) For a 3-dimensional space, use equation (3.52) to construct harmonic polynomials, starting with the 2-dimensional harmonic polynomials $(x\pm iy)^s$ for $s=0,1,2$. Use these harmonic polynomials to construct normalized 3-dimensional spherical harmonics up to l=2.

(8) Use equation (3.52) to construct a set of 4-dimensional hyperspherical harmonics.

(9) The two tables at the end of the chapter show 6-dimensional hyperspherical harmonics constructed according to the standard tree and according to a non-standard tree. For the case where $\lambda=1$, show that the standard and nonstandard harmonics in the tables are related by a unitary transformation.

(10) Show that when the normalization constant is defined by equation (3.83) a set of properly normalized 4-dimensional hyperspherical harmonics can be constructed starting with a set of 3-dimensional spherical harmonics.

(11) The operator `projh[d,lambda,f,xs]` in a d-dimensional space projects out the harmonic polynomial of highest order from an λth order homogeneous polynomial in the variables defined by the list `xs`. For the case where $d=5$, make a table of the highest-order harmonic polynomials `projh[5,`λ`,`x_1^λ`]` of x_1^λ for $\lambda=0,1,2,...,7$. Use these harmonic polynomials to construct hyperspherical harmonics $Y_{\lambda,\mu}(u_1,\ldots,u_5)=r^{-n}h_{\lambda,\mu}(x_1,\ldots,x_5)$, and normalize them using the integration over solid angle, `hz5`. Show that the resulting 5-dimensional hyperspherical harmonics form an orthonormal set.

Chapter 4

FOURIER TRANSFORMS IN d DIMENSIONS

4.1 Notation and basic properties

Let us introduce the abbreviated notation:

$$\int_{-\infty}^{\infty} dx_1 \int_{-\infty}^{\infty} dx_2 \cdots \int_{-\infty}^{\infty} dx_d \; f(x_1, x_2, \ldots, x_d) \equiv \int d\mathbf{x} \; f(\mathbf{x}) \tag{4.1}$$

and

$$e^{i(p_1 x_1 + p_2 x_2 + \cdots + p_d x_d)} \equiv e^{i\mathbf{p}\cdot\mathbf{x}} \tag{4.2}$$

Then the d-dimensional Fourier transform of the function $f(\mathbf{x})$ is given by

$$f^t(\mathbf{p}) = \frac{1}{(2\pi)^{d/2}} \int d\mathbf{x} \; e^{-i\mathbf{p}\cdot\mathbf{x}} \; f(\mathbf{x}) \tag{4.3}$$

while the inverse transform is

$$f(\mathbf{x}) = \frac{1}{(2\pi)^{d/2}} \int d\mathbf{p} \; e^{i\mathbf{p}\cdot\mathbf{x}} \; f^t(\mathbf{p}) \tag{4.4}$$

We would like to show that the scalar product of two functions in direct space is equal to the scalar product of their Fourier transforms in reciprocal space. From (4.4) we have

$$g(\mathbf{x})^* = \frac{1}{(2\pi)^{d/2}} \int d\mathbf{p}' \; e^{-i\mathbf{p}'\cdot\mathbf{x}} \; g^t(\mathbf{p}')^* \tag{4.5}$$

so that

$$\int d\mathbf{x} \; g(\mathbf{x})^* f(\mathbf{x}) = \frac{1}{(2\pi)^d} \int d\mathbf{p} \int d\mathbf{p}' \; g^t(\mathbf{p}')^* f^t(\mathbf{p}) \int d\mathbf{x} \; e^{i(\mathbf{p}-\mathbf{p}')\cdot\mathbf{x}} \tag{4.6}$$

However,

$$\frac{1}{(2\pi)^d} \int d\mathbf{x} \; e^{i(\mathbf{p}-\mathbf{p}')\cdot\mathbf{x}} = \delta(\mathbf{p}-\mathbf{p}') \tag{4.7}$$

so that

$$\int d\mathbf{x}\; g(\mathbf{x})^* f(\mathbf{x}) = \int d\mathbf{p}\; g^t(\mathbf{p})^* f^t(\mathbf{p}) \tag{4.8}$$

Equation (4.8) implies that if we have an orthonormal set of functions $\{\phi_j(\mathbf{x})\}$ in direct space, so that

$$\int d\mathbf{x}\; \phi_{j'}^*(\mathbf{x})\phi_j(\mathbf{x}) = \delta_{j',j} \tag{4.9}$$

then their Fourier transforms form an orthonormal set in reciprocal space:

$$\int d\mathbf{p}\; \phi_{j'}^{t*}(\mathbf{p})\phi_j^t(\mathbf{p}) = \delta_{j',j} \tag{4.10}$$

From (4.8) it also follows that

$$\int d\mathbf{x}\; \phi_{j'}^*(\mathbf{x})V(\mathbf{x})\phi_j(\mathbf{x}) = \int d\mathbf{p}\; \phi_{j'}^{t*}(\mathbf{p})(V\phi_j)^t(\mathbf{p}) \tag{4.11}$$

where

$$(V\phi_j)^t(\mathbf{p}) \equiv \frac{1}{(2\pi)^{d/2}} \int d\mathbf{x}\; e^{-i\mathbf{p}\cdot\mathbf{x}}\; V(\mathbf{x})\phi_j(\mathbf{x}) \tag{4.12}$$

Also, from (4.4) and (4.8) we have

$$\int d\mathbf{p}\; \phi_{j'}^{t*}(\mathbf{p})\; \Delta\phi_j^t(\mathbf{p}) = -\int d\mathbf{p}\; \phi_{j'}^{t*}(\mathbf{p})\; p^2\phi_j^t(\mathbf{p}) \tag{4.13}$$

where

$$p^2 \equiv \mathbf{p}\cdot\mathbf{p} \tag{4.14}$$

4.2 Expansions of a *d*-dimensional plane wave

In a 3-dimensional space, it is often useful to expand a plane wave in terms of spherical harmonics and spherical Bessel functions. One might therefore ask whether a similar expansion could be made for a d-dimensional plane wave. To make such an expansion, we first write [Avery, 1989; Avery, 2000; Avery and Avery, 2006; Avery, 2009; Avery, Rettrup and Avery, 2012; Vilinken, 1968; Wen and Avery, 1985]

$$e^{i\mathbf{p}\cdot\mathbf{x}} = \sum_{n=0}^{\infty} \frac{(ipr)^n}{n!}(\mathbf{u}_p\cdot\mathbf{u})^n \tag{4.15}$$

where $\mathbf{u}$ is defined by (3.1), and where

$$\mathbf{u}_p \equiv \frac{1}{p}(p_1, p_2, \ldots, p_d) \tag{4.16}$$

This can be derived if we write $e^{i\mathbf{p}\cdot\mathbf{x}} = e^{ipr\mathbf{u_p}\cdot\mathbf{u}}$ and perform a Taylor series expansion about $\mathbf{u_p}\cdot\mathbf{u} = 0$. From (3.12) we then have

$$e^{i\mathbf{p}\cdot\mathbf{x}} = \sum_{n=0}^{\infty} \left(\frac{ipr}{2}\right)^n \sum_{s=0}^{\lfloor n/2 \rfloor} \frac{\alpha+n-2s}{s!(\alpha)_{n-s+1}} C^{\alpha}_{n-2s}(\mathbf{u}_p\cdot\mathbf{u}) \tag{4.17}$$

so that

$$\begin{aligned} O_\lambda\left[e^{i\mathbf{p}\cdot\mathbf{x}}\right] &= C^{\alpha}_{\lambda}(\mathbf{u}_p\cdot\mathbf{u}) \sum_{n=0}^{\infty} \left(\frac{ipr}{2}\right)^n \sum_{s=0}^{\lfloor n/2 \rfloor} \frac{\alpha+n-2s}{s!(\alpha)_{n-s+1}} \delta_{s,(n-\lambda)/2} \\ &= C^{\alpha}_{\lambda}(\mathbf{u}_p\cdot\mathbf{u}) \sum_{n=\lambda,\lambda+2,\ldots}^{\infty} \left(\frac{ipr}{2}\right)^n \frac{\alpha+\lambda}{(n/2-\lambda/2)!(\alpha)_{n/2+\lambda/2+1}} \end{aligned} \tag{4.18}$$

Remembering the definition of $(\alpha)_j$, equation (3.10), we can write:

$$(\alpha)_{n/2+\lambda/2+1} = \frac{(d+n+\lambda-2)!!}{2^{n/2+\lambda/2+1}(d-4)!!} \tag{4.19}$$

so that

$$O_\lambda\left[e^{i\mathbf{p}\cdot\mathbf{x}}\right] = i^\lambda(d+2\lambda-2)(d-4)!!\,j^d_\lambda(pr)C^{\alpha}_{\lambda}(\mathbf{u}_p\cdot\mathbf{u}) \tag{4.20}$$

where (with $2t = n-\lambda$)

$$j^d_\lambda(pr) \equiv \sum_{t=0}^{\infty} \frac{(-1)^t(pr)^{2t+\lambda}}{(2t)!!(d+2t+2\lambda-2)!!} \tag{4.21}$$

From (4.20) it follows that we can expand a d-dimensional plane wave in terms of Gegenbauer polynomials and in terms of the functions $j^d_\lambda(pr)$:

$$e^{i\mathbf{p}\cdot\mathbf{x}} = \sum_{\lambda=0}^{\infty} i^\lambda(d+2\lambda-2)(d-4)!!\,j^d_\lambda(pr)C^{\alpha}_{\lambda}(\mathbf{u}_p\cdot\mathbf{u}) \tag{4.22}$$

When $d = 3$ and $\lambda = l$, equation (4.22) reduces to the familiar expansion:

$$e^{i\mathbf{p}\cdot\mathbf{x}} = \sum_{l=0}^{\infty} i^l(2l+1)j_l(pr)P_l(\mathbf{u}_p\cdot\mathbf{u}) \tag{4.23}$$

Thus we can see that the functions $j^d_\lambda(pr)$ defined by (4.21) are the d-dimensional generalization of spherical Bessel functions, and it might be appropriate to call them *hyperspherical Bessel functions.* Like spherical Bessel functions, they can be expressed in terms of ordinary Bessel functions, the relationship being

$$j^d_\lambda(pr) = \frac{\Gamma(\alpha)2^{\alpha-1}J_{\alpha+\lambda}(pr)}{(d-4)!!(pr)^\alpha} \tag{4.24}$$

From equations (4.22) and (3.35), and from the sum rule (3.25), it follows that we can express a d-dimensional plane wave hyperspherical Bessel functions and hyperspherical harmonics:

$$e^{i\mathbf{p}\cdot\mathbf{x}} = (d-2)!!I(0)\sum_{\lambda=0}^{\infty} i^{\lambda} j_{\lambda}^{d}(pr)\sum_{\mu} Y_{\lambda,\mu}^{*}(\mathbf{u}_p)Y_{\lambda,\mu}(\mathbf{u}) \tag{4.25}$$

where $I(0)$ is the total solid angle, equation (2.17). When $d = 3$ and $\lambda = l$, equation (4.25) reduces to the familiar expression:

$$e^{i\mathbf{p}\cdot\mathbf{x}} = 4\pi\sum_{l=0}^{\infty} i^{l} j_{l}(pr)\sum_{m} Y_{l,m}^{*}(\mathbf{u}_p)Y_{l,m}(\mathbf{u}) \tag{4.26}$$

4.3 The Green's function of Δ

In this section, we will show that the generating function for Gegenbauer polynomials is proportional to the Green's function of the d-dimensional Laplacian operator. This relationship is an extremely important one, since it allows us to write down integral solutions to inhomogeneous partial differential equations involving the d-dimensional Laplacian.

The expansion of a plane wave in terms of Gegenbauer polynomials and hyperspherical Bessel functions, equation (4.22), allows us to take the Fourier transform of the function

$$f(|\mathbf{x}|) = \frac{1}{|\mathbf{x}|^{d-2}} = \frac{1}{r^{d-2}} \tag{4.27}$$

Making use of the fact that (from (3.28))

$$\frac{1}{K_\lambda}\int d\Omega_d\; C_{\lambda}^{\alpha}(\mathbf{u}_p\cdot\mathbf{u}) = O_{\lambda}[1] = \delta_{\lambda,0} \tag{4.28}$$

we find that the d-dimensional Fourier transform of this function is given by

$$\begin{aligned}
f^t(p) &= \frac{1}{(2\pi)^{d/2}}\int d\mathbf{x}\; e^{-i\mathbf{p}\cdot\mathbf{x}}\;\frac{1}{r^{d-2}} \\
&= \frac{1}{(2\pi)^{d/2}}\int d\mathbf{x}\sum_{\lambda=0}^{\infty}(-i)^{\lambda}(d+2\lambda-2)(d-4)!!j_{\lambda}^{d}(pr)C_{\lambda}^{\alpha}(\mathbf{u}_p\cdot\mathbf{u})\frac{1}{r^{d-2}} \\
&= \frac{1}{(2\pi)^{d/2}}\sum_{\lambda=0}^{\infty}(-i)^{\lambda}(d+2\lambda-2)(d-4)!! \\
&\times\int_0^{\infty} dr\; r\; j_{\lambda}^{d}(pr)\int d\Omega\; C_{\lambda}^{\alpha}(\mathbf{u}_p\cdot\mathbf{u}) \\
&= \frac{I(0)}{(2\pi)^{d/2}}(d-2)!!\int_0^{\infty} dr\; r\; j_0^{d}(pr)
\end{aligned} \tag{4.29}$$

The hyperradial integral in equation (4.29) is oscillatory, but its principal value is given by

$$\lim_{\zeta \to 0} \int_0^\infty dr\; r\; j_0^d(pr) e^{-\zeta r} = \frac{1}{(d-4)!!\; p^2} \tag{4.30}$$

Using the principal value, we obtain:

$$f^t(p) = \frac{I(0)(d-2)}{(2\pi)^{d/2}} \frac{1}{p^2} \tag{4.31}$$

and thus

$$f(r^{d-2}) = \frac{1}{|\mathbf{x}|^{d-2}} = \frac{I(0)(d-2)}{(2\pi)^d} \int d\mathbf{p}\; \frac{1}{p^2}\; e^{i\mathbf{p}\cdot\mathbf{x}} \tag{4.32}$$

Then, shifting the origin of the coordinate system, we have

$$\frac{1}{|\mathbf{x}-\mathbf{x}'|^{d-2}} = \frac{I(0)(d-2)}{(2\pi)^d} \int d\mathbf{p} \frac{1}{p^2}\; e^{i\mathbf{p}\cdot(\mathbf{x}-\mathbf{x}')} \tag{4.33}$$

and

$$\Delta \frac{1}{|\mathbf{x}-\mathbf{x}'|^{d-2}} = -\frac{I(0)(d-2)}{(2\pi)^d} \int d\mathbf{p}\; e^{i\mathbf{p}\cdot(\mathbf{x}-\mathbf{x}')} \tag{4.34}$$

Finally, remembering that

$$\frac{1}{(2\pi)^d} \int d\mathbf{p}\; e^{i\mathbf{p}\cdot(\mathbf{x}-\mathbf{x}')} = \delta(\mathbf{x}-\mathbf{x}') \tag{4.35}$$

we have

$$\Delta \frac{1}{|\mathbf{x}-\mathbf{x}'|^{d-2}} = -I(0)(d-2)\; \delta(\mathbf{x}-\mathbf{x}') \tag{4.36}$$

Thus, the generating function for the Gegenbauer polynomials is seen to be proportional to the Green's function of the d-dimensional Laplacian operator Δ.

4.4 Hyperspherical Bessel transforms

From (4.20) it follows that if we have a function of the form

$$f(\mathbf{x}) = R_\lambda(r) Y_{\lambda,\mu}(\mathbf{u}) \tag{4.37}$$

then its Fourier transform is given by

$$f^t(\mathbf{p}) = S_\lambda(p) Y_{\lambda,\mu}(\mathbf{u}_p) \tag{4.38}$$

where

$$Y_{\lambda,\mu}(\mathbf{u}_p) = \frac{1}{K_\lambda} \int d\Omega_d\; C_\lambda^\alpha(\mathbf{u}_p \cdot \mathbf{u})\; Y_{\lambda,\mu}(\mathbf{u}) \tag{4.39}$$

and

$$\begin{aligned}
S_\lambda(p) &= \frac{(-i)^\lambda (d-2)!!I(0)}{(2\pi)^{d/2}} \int_0^\infty dr\ r^{d-1} j_\lambda^d(pr) R_\lambda(r) \\
R_\lambda(r) &= \frac{i^\lambda (d-2)!!I(0)}{(2\pi)^{d/2}} \int_0^\infty dp\ p^{d-1} j_\lambda^d(pr) S_\lambda(p) \\
I(0) &= \frac{2\pi^{d/2}}{\Gamma[d/2]}
\end{aligned} \tag{4.40}$$

In (4.40), $S_\lambda(p)$ is the *hyperspherical Bessel transform* of $R_\lambda(r)$. In general, if $f_\lambda(\mathbf{x})$ is an eigenfunction of Λ^2, then $f_\lambda^t(\mathbf{p})$ has the same angular dependence in reciprocal space as $f_\lambda(\mathbf{x})$ has in direct space. It follows that if G is any subgroup of $SO(d)$, and if $f_\lambda(\mathbf{x})$ is a basis function of an irreducible representation of G in direct space, then $f_\lambda^t(\mathbf{p})$ will be a basis function for the same irreducible representation of G in reciprocal space. In other words, symmetry properties with respect to rotation are preserved under Fourier transformation. In the special case where $d = 3$, equations (4.37)–(4.40) become

$$f(\mathbf{x}) = R_l(r) Y_{l,m}(\mathbf{u}) \tag{4.41}$$

$$f^t(\mathbf{p}) = S_l(p) Y_{l,m}(\mathbf{u}_p) \tag{4.42}$$

where

$$Y_{l,m}(\mathbf{u}_p) = \frac{(2l+1)}{4\pi} \int d\Omega_3\ P_l(\mathbf{u}_p \cdot \mathbf{u})\ Y_{l,m}(\mathbf{u}) \tag{4.43}$$

and

$$\begin{aligned}
S_l(p) &= (-i)^l \sqrt{\frac{2}{\pi}} \int_0^\infty dr\ r^2 j_l(pr) R_l(r) \\
R_l(r) &= i^l \sqrt{\frac{2}{\pi}} \int_0^\infty dp\ p^2 j_l^d(pr) S_l(p) \\
I(0) &= 4\pi
\end{aligned} \tag{4.44}$$

4.5 An alternative expansion of a plane wave

Suppose that we have a complete set of orthonormal functions $\{\phi_j(\mathbf{x})\}$ in a d-dimensional space. The completeness condition (in the sense of distributions) can be written in the form

$$\sum_j \phi_j^*(\mathbf{x}) \phi_j(\mathbf{x}') = \delta(\mathbf{x} - \mathbf{x}') \tag{4.45}$$

Multiplying both sides of (4.45) by $e^{-i\mathbf{p}\cdot\mathbf{x}'}$ and integrating over $d\mathbf{x}'$, we obtain:

$$\sum_j \phi_j^*(\mathbf{x}) \int d\mathbf{x}'\ e^{-i\mathbf{p}\cdot\mathbf{x}'}\ \phi_j(\mathbf{x}') = \int d\mathbf{x}'\ e^{-i\mathbf{p}\cdot\mathbf{x}'}\ \delta(\mathbf{x}-\mathbf{x}') = e^{-i\mathbf{p}\cdot\mathbf{x}} \tag{4.46}$$

so that

$$\begin{aligned} e^{-i\mathbf{p}\cdot\mathbf{x}} &= \sum_j \phi_j^*(\mathbf{x}) \int d\mathbf{x}'\ e^{-i\mathbf{p}\cdot\mathbf{x}'}\ \phi_j(\mathbf{x}') \\ &= (2\pi)^{d/2} \sum_j \phi_j^*(\mathbf{x})\phi_j^t(\mathbf{p}) \end{aligned} \tag{4.47}$$

Then

$$\begin{aligned} f^t(\mathbf{p}) &= \frac{1}{(2\pi)^{d/2}} \int d\mathbf{x}\ e^{-i\mathbf{p}\cdot\mathbf{x}} f(\mathbf{x}) \\ &= \sum_j \phi_j^t(\mathbf{p}) \int d\mathbf{x}\ \phi_j^*(\mathbf{x}) f(\mathbf{x}) \end{aligned} \tag{4.48}$$

and

$$\begin{aligned} f(\mathbf{x}) &= \frac{1}{(2\pi)^{d/2}} \int d\mathbf{p}\ e^{i\mathbf{p}\cdot\mathbf{x}} f^t(\mathbf{p}) \\ &= \sum_j \phi_j(\mathbf{x}) \int d\mathbf{p}\ \phi_j^*(\mathbf{p}) f^t(\mathbf{p}) \end{aligned} \tag{4.49}$$

It follows from (4.47) that if the set of functions $\{\phi_j(\mathbf{x})\}$ is chosen in such a way that they are basis functions of irreducible representations of a group G, and if I_γ^ν is the set of indices j such that $\phi_j(\mathbf{x})$ transforms like the νth basis function of the γth irreducible representation of G, then

$$P_\gamma^\nu \left[e^{-i\mathbf{p}\cdot\mathbf{x}}\right] = (2\pi)^{d/2} \sum_{j\in I_\gamma^\nu} \phi_j^*(\mathbf{x})\phi_j^t(\mathbf{p}) \tag{4.50}$$

If we multiply (4.47) on the left by $e^{i\mathbf{p}'\cdot\mathbf{x}}$ and integrate over dx, we obtain

$$\begin{aligned} \int d\mathbf{x}\ e^{i(\mathbf{p}'-\mathbf{p})\cdot\mathbf{x}} &= (2\pi)^{d/2} \sum_j \int d\mathbf{x}\ e^{i\mathbf{p}'\cdot\mathbf{x}} \phi_j^*(\mathbf{x})\phi_j^t(\mathbf{p}) \\ &= (2\pi)^d \sum_j \phi_j^{*t}(\mathbf{p}')\phi_j^t(\mathbf{p}) \end{aligned} \tag{4.51}$$

so that (in the sense of distributions)

$$\sum_j \phi_j^{*t}(\mathbf{p}')\phi_j^t(\mathbf{p}) = \delta(\mathbf{p}-\mathbf{p}') \tag{4.52}$$

Let us now try to make the meaning of completeness relations like (4.45) and (4.52) a little more precise: Suppose that there exists a Hilbert space H with an orthonormal basis $\{\phi_j(\mathbf{x})\}$. Then, for any $f \in H$, we can write

$$f(\mathbf{x}) = \sum_j \phi_j(\mathbf{x}) \int d\mathbf{x}' \; \phi_j^*(\mathbf{x}') f(\mathbf{x}') \tag{4.53}$$

But we could equally well have written

$$f(\mathbf{x}) = \int d\mathbf{x}' \; f(\mathbf{x}') \; \delta(\mathbf{x} - \mathbf{x}') \tag{4.54}$$

Thus we can see that the sum on the left-hand side of (4.45) is acting like a Dirac delta function; but the relationship is only known to hold within H. Similar considerations hold for (4.52). In the discussion above, we imagined the set of functions $\{\phi_j(\mathbf{x})\}$ to be symmetry-adapted, and we let I_γ^ν stand for a domain within which all the functions transform like the νth basis function of the γth irreducible representation of the symmetry group G. Then if

$$P_\gamma^\nu \left[f(\mathbf{x}) \right] = \sum_{j \in I_\gamma^\nu} \phi_j(\mathbf{x}) \int d\mathbf{x}' \; \phi_j^*(\mathbf{x}') f(\mathbf{x}') = f(\mathbf{x}) \tag{4.55}$$

we can conclude that $f(\mathbf{x})$ lies entirely within the domain I_γ^ν and that it transforms like the νth basis function of the γth irreducible representation of G. What about its Fourier transform, $f_n^t(\mathbf{x})$? From (4.48),

$$\begin{aligned} f_n^t(\mathbf{p}) &= \sum_j \phi_j^t(\mathbf{p}) \int d\mathbf{x}' \; \phi_j^*(\mathbf{x}') f_n(\mathbf{x}') \\ &= \sum_{j \in I_\gamma^\nu} \phi_j^t(\mathbf{p}) \int d\mathbf{x}' \; \phi_j^*(\mathbf{x}') f_n(\mathbf{x}') \end{aligned} \tag{4.56}$$

The second line of (4.56) follows from the fact the the $d\mathbf{x}'$ integral vanishes unless $\phi_j^*(\mathbf{x}')$ lies within the domain I_γ^ν. Thus, if $f_n(\mathbf{x})$ lies within the domain I_γ^ν, then $f_n^t(\mathbf{p})$ will lie within the corresponding domain in reciprocal space. One can express this by saying that symmetry properties are preserved under Fourier transformation.

4.6 The Fourier convolution theorem

Let

$$f(\mathbf{x}) = \frac{1}{(2\pi)^{d/2}} \int d\mathbf{p}' \; e^{i\mathbf{p}' \cdot \mathbf{x}} \; f^t(\mathbf{p}') \tag{4.57}$$

and

$$g(\mathbf{x}) = \frac{1}{(2\pi)^{d/2}} \int d\mathbf{p}'' \; e^{i\mathbf{p}''\cdot\mathbf{x}} \; g^t(\mathbf{p}'') \tag{4.58}$$

Then we can write

$$\begin{aligned} &\int d\mathbf{x} \; e^{-i\mathbf{p}\cdot\mathbf{x}} f(\mathbf{x}) g(\mathbf{x}) \\ &= \frac{1}{(2\pi)^d} \int d\mathbf{p}' \int d\mathbf{p}'' f^t(\mathbf{p}') g^t(\mathbf{p}'') \int d\mathbf{x} \; e^{i(\mathbf{p}'+\mathbf{p}''-\mathbf{p})\cdot\mathbf{x}} \\ &= \int d\mathbf{p}' \int d\mathbf{p}'' f^t(\mathbf{p}') g^t(\mathbf{p}'') \; \delta(\mathbf{p}' + \mathbf{p}'' - \mathbf{p}) \end{aligned} \tag{4.59}$$

so that

$$\int d\mathbf{x} \; e^{-i\mathbf{p}\cdot\mathbf{x}} f(\mathbf{x}) g(\mathbf{x}) = \int d\mathbf{p}' \; f^t(\mathbf{p}') g^t(\mathbf{p} - \mathbf{p}') \tag{4.60}$$

Thus we see that in a d-dimensional space, the Fourier convolution theorem has exactly the same form as in 3 dimensions. In a similar way, it is easy to show that

$$\int d\mathbf{p} \; e^{i\mathbf{p}\cdot\mathbf{x}} f^t(\mathbf{p}) g^t(\mathbf{p}) = \int d\mathbf{x}' \; f(\mathbf{x}') g(\mathbf{x} - \mathbf{x}') \tag{4.61}$$

4.7 Harmonic analysis for non-Euclidean spaces

It is interesting to ask whether something analogous to Fourier transform theory can be developed for spaces whose metric is non-Euclidean. For example, we might think of the surface of a very large hypersphere of hyperradius r, embedded in a d-dimensional space. Since the hyperradius is very large, the surface is locally almost flat, but nevertheless it has a slight curvature. On this surface, the unit vector $\mathbf{u} \equiv \mathbf{x}/r$ plays the role which $\mathbf{x}$ would play in a Euclidean space. Just as we can write

$$f(\mathbf{x}) = \frac{1}{(2\pi)^{d/2}} \int d\mathbf{p} \; e^{i\mathbf{p}\cdot\mathbf{x}} \; f^t(\mathbf{p}) \tag{4.62}$$

where

$$f^t(\mathbf{p}) = \frac{1}{(2\pi)^{d/2}} \int d\mathbf{x} \; e^{-i\mathbf{p}\cdot\mathbf{x}} \; f(\mathbf{x}) \tag{4.63}$$

in a Euclidean space, so, on our very large hypersphere, we can write

$$f(\mathbf{u}) = \sum_{\lambda,\mu} Y_{\lambda,\mu}(\mathbf{u}) \; a_{\lambda,\mu} \tag{4.64}$$

where. from the orthonormality of the hyperspherical harmonics, we have

$$a_{\lambda,\mu} = \int d\Omega_d \; Y^*_{\lambda,\mu}(\mathbf{u}) f(\mathbf{u}) \tag{4.65}$$

Provided that $f(\mathbf{u})$ can be expanded as a polynomial, our general hyperangular integration theorem can be used to carry out the integration in (4.65). More generally, we can try to find the set of hyperspherical harmonics appropriate for any non-Euclidean space, and these can be used as a plane-wave-like basis for an analogue to Fourier transform theory.

4.8 Exercises

(1) In a 6-dimensional space, the hyperradius has the form: $r^2 = x_1^2 + x_2^2 + x_3^2 + x_4^2 + x_5^2 + x_6^2$. Use equation (4.3) to calculate the Fourier transform of e^{-r^2}.

(2) Use equation (4.4) to calculate the inverse Fourier transform, and verify that the starting point is reached.

(3) According to equation (4.8), the scalar product of two functions is invariant under Fourier transformation. Show that this relationship holds in a 6-dimensional space for $g(\mathbf{x}) = e^{-ar^2}$ while $f(\mathbf{x}) = e^{-br^2}$ and where r is the 6-dimensional hyperradius.

(4) Make a series expansion of the function e^{ia} about the point $a = 0$. Compare this expansion with equation (4.15) with the substitution $a = \mathbf{p} \cdot \mathbf{r}$.

(5) The hyperspherical Bessel function is defined by the series shown in equation (4.21) and also by equation (4.24). Make the expansion shown in (4.21) up to 25 terms. Write `?jd` to obtain the definition of the function, and compare with equation (4.24). Make plots comparing the series with `jd[d,λ,pr]` for various values of λ and d.

(6) Using equation (4.25), calculate the radial part of the Fourier transform of $e^{-ar}\, Y_{\lambda,\mu}(\mathbf{u})$ (i.e., the hyperspherical Bessel transform) for $d = 3, 4, ..., 7$ and $\lambda = 0, 1$, where r is the appropriate hyperradius. Show that the inverse hyperspherical Bessel transform brings one back to the starting point.

(7) For $\lambda = 0$, calculate the hyperspherical Bessel transforms of re^{-ar} for $d = 3, 4, \ldots, 7$ and show that the results agree with those obtained by differentiating the transforms of e^{-ar} with respect to a.

(8) Use equation (4.36) with $d = 3$ to show that the Green's function solution $\phi(\mathbf{x}) = \int d^3x' \rho(\mathbf{x}') \frac{1}{|\mathbf{x}-\mathbf{x}'|}$ satisfies Poisson's equation, $\Delta\phi(\mathbf{x}) = -4\pi\rho(\mathbf{x})$. Use a pen and paper rather than Mathematica.

Chapter 5

FOCK'S TREATMENT OF HYDROGENLIKE ATOMS AND ITS GENERALIZATION

5.1 Fock's original treatment

From equation (2.43), we can see that for $d = 4$, the number of hyperspherical harmonics corresponding to a given value of λ is $(\lambda + 1)^2$. If we let $\lambda + 1 = n$, where n is the value of the principal quantum number of the hydrogenlike atomic orbitals, we can see that this degree of degeneracy is the same as that of hydrogenlike atoms. (Notice that the meaning of n here differs from its meaning in previous chapters.) It is therefore interesting to ask whether there is a relationship between 4-dimensional hyperspherical harmonics and hydrogenlike orbitals. V. Fock (1935) was able to show that such a relationship does indeed exist. His argument is as follows: [Fock, 1935], [Fock, 1958], [Bandar and Itzaksen, 1966]

In 3-dimensional direct space, the Schrödinger equation (in atomic units) for an electron moving in the potential $V(\mathbf{x})$ is

$$\left[-\frac{1}{2}\nabla^2 + V(\mathbf{x})\right]\psi(\mathbf{x}) = E\psi(\mathbf{x}) \tag{5.1}$$

We can let

$$\psi(\mathbf{x}) = \frac{1}{(2\pi)^{3/2}}\int d\mathbf{p}\; e^{i\mathbf{p}\cdot\mathbf{x}}\psi^t(\mathbf{p}) \tag{5.2}$$

where

$$\psi^t(\mathbf{p}) = \frac{1}{(2\pi)^{3/2}}\int d\mathbf{x}\; e^{-i\mathbf{p}\cdot\mathbf{x}}\psi(\mathbf{x}) \tag{5.3}$$

Substituting (5.2) into (5.1), we have

$$\frac{1}{(2\pi)^{3/2}}\int d\mathbf{p}\;\left[\frac{p^2}{2} + V(\mathbf{x}) - E\right] e^{i\mathbf{p}\cdot\mathbf{x}}\psi^t(\mathbf{p}) = 0 \tag{5.4}$$

We now multiply on the left by $e^{-i\mathbf{p}'\cdot\mathbf{x}}/(2\pi)^{3/2}$ and integrate over $d\mathbf{x}$. This gives:

$$\left[\frac{p'^2}{2} - E\right] \psi^t(\mathbf{p}') = \frac{-1}{(2\pi)^{3/2}} \int d\mathbf{p}\; V^t(\mathbf{p}' - \mathbf{p})\; \psi^t(\mathbf{p}) \tag{5.5}$$

(see Exercise 5), which is the 1-particle Schrödinger equation in reciprocal space. For hydrogenlike atoms,

$$V(\mathbf{x}) = -\frac{Z}{r} \tag{5.6}$$

so that from (4.31),

$$V^t(\mathbf{p}) = -\sqrt{\frac{2}{\pi}}\frac{Z}{p^2} \tag{5.7}$$

Letting

$$-2E = k^2 \tag{5.8}$$

and combining (5.5), (5.6) and (5.7), we obtain

$$\left[p'^2 + k^2\right] \psi^t(\mathbf{p}') = \frac{Z}{\pi^2} \int d\mathbf{p}\; \frac{1}{|\mathbf{p}' - \mathbf{p}|^2}\; \psi^t(\mathbf{p}) \tag{5.9}$$

Fock then made the transformation:

$$\begin{aligned}
u_1 &= \frac{2kp_1}{k^2 + p^2} \equiv \sin\chi\ \sin\theta\ \cos\phi \\
u_2 &= \frac{2kp_2}{k^2 + p^2} \equiv \sin\chi\ \sin\theta\ \sin\phi \\
u_3 &= \frac{2kp_3}{k^2 + p^2} \equiv \sin\chi\ \cos\theta \\
u_4 &= \frac{k^2 - p^2}{k^2 + p^2} \equiv \cos\chi
\end{aligned} \tag{5.10}$$

Here θ and ϕ are the polar angles of the vector $\mathbf{p}$:

$$\begin{aligned}
p_1 &= p\ \sin\theta\ \cos\phi \\
p_2 &= p\ \sin\theta\ \sin\phi \\
p_3 &= p\ \cos\theta
\end{aligned} \tag{5.11}$$

while

$$\chi \equiv \cos^{-1}\left(\frac{k^2 - p^2}{k^2 + p^2}\right) = \sin^{-1}\left(\frac{2kp}{k^2 + p^2}\right) \tag{5.12}$$

is an angle introduced by Fock in order to transform the integral d^3p into an integral over solid angle in a 4-dimensional space. Fock's transformation maps the 3-dimensional p-space onto the surface of a unit sphere in a 4-dimensional space. It is easy to verify from (5.10) that

$$u_1^2 + u_2^2 + u_3^2 + u_4^2 = 1 \tag{5.13}$$

From equation (3.77) or from the Jacobian of the transformation from Cartesian coordinates to 4-dimensional hyperspherical coordinates, one finds that the element of solid angle in the 4-dimensional space is given by

$$\begin{aligned} d\Omega_4 &= \sin^2\chi \ \sin\theta \ d\chi d\theta d\phi \\ &= \left(\frac{2kp}{k^2+p^2}\right)^2 \sin\theta \ d\chi d\theta d\phi \end{aligned} \tag{5.14}$$

Comparing this with

$$d\mathbf{p} = p^2 dp \ \sin\theta \ d\theta d\phi \tag{5.15}$$

and making use of the fact that

$$\frac{d\chi}{dp} = \frac{2k}{k^2+p^2} \tag{5.16}$$

we have

$$\begin{aligned} d\Omega_4 &= \left(\frac{2k}{k^2+p^2}\right)^3 d\mathbf{p} \\ d\mathbf{p} &= \left(\frac{k^2+p^2}{2k}\right)^3 d\Omega_4 \end{aligned} \tag{5.17}$$

Also, from (5.10), we have:

$$\begin{aligned} \mathbf{p}\cdot\mathbf{p}' &= \frac{4k^2}{(k^2+p^2)(k^2+p'^2)(\mathbf{u}\cdot\mathbf{u}' - u_4u_4')} \\ \frac{1}{|\mathbf{p}-\mathbf{p}'|^2} &= \frac{4k^2}{(k^2+p^2)(k^2+p'^2)}\frac{1}{|\mathbf{u}-\mathbf{u}'|^2} \end{aligned} \tag{5.18}$$

Inserting (5.17) and (5.18) into (5.9), we obtain:

$$\left[p'^2 + k^2\right]^2 \psi^t(\mathbf{p}') = \frac{Z}{2k\pi^2}\int d\Omega_4 \ \frac{(k^2+p^2)^2}{|\mathbf{u}'-\mathbf{u}|^2} \ \psi^t(\mathbf{p}) \tag{5.19}$$

We now let

$$\psi^t(\mathbf{p}) = \frac{4k^{5/2}}{(k^2+p^2)^2} \ \Phi(\mathbf{u}) \tag{5.20}$$

(As shown in Section 5.3 below, the factor $4k^{5/2}$ in the numerator is needed to normalize $\psi^t(\mathbf{p})$). Equation (5.19) then takes on the simple form

$$\Phi(\mathbf{u}') = \frac{Z}{2k\pi^2}\int d\Omega_4 \ \frac{1}{|\mathbf{u}'-\mathbf{u}|^2} \ \Phi(\mathbf{u}) \tag{5.21}$$

From equation (3.5), with $d = 4$ and $\alpha = d/2 - 1 = 1$, we have

$$\frac{1}{|\mathbf{u}' - \mathbf{u}|^2} = \sum_{\lambda=0}^{\infty} C_\lambda^1(\mathbf{u} \cdot \mathbf{u}') \tag{5.22}$$

so that (5.21) becomes

$$\Phi(\mathbf{u}') = \frac{Z}{2k\pi^2} \sum_{\lambda=0}^{\infty} \int d\Omega_4 \; C_\lambda^1(\mathbf{u} \cdot \mathbf{u}') \; \Phi(\mathbf{u}) \tag{5.23}$$

Remembering equation (3.28) we can rewrite this in the form

$$\Phi(\mathbf{u}') = \frac{Z}{2k\pi^2} \sum_{\lambda=0}^{\infty} K_\lambda \; O_\lambda[\Phi(\mathbf{u})] \tag{5.24}$$

For $d = 4$,

$$K_\lambda = \frac{I(0)}{\lambda + 1} = \frac{2\pi^2}{\lambda + 1} \tag{5.25}$$

so that equation (5.24) becomes:

$$\Phi(\mathbf{u}') = \frac{Z}{k} \sum_{\lambda=0}^{\infty} \frac{1}{\lambda + 1} \; O_\lambda[\Phi(\mathbf{u}')] \tag{5.26}$$

If $\Phi(\mathbf{u})$ is an eigenfunction of Λ^2. so that

$$O_{\lambda'}[\Phi(\mathbf{u}')] = \delta_{\lambda'\lambda} \Phi(\mathbf{u}) \tag{5.27}$$

then (5.26) will be satisfied provided that

$$\frac{Z}{k(\lambda + 1)} = 1 \tag{5.28}$$

or, from (5.8),

$$E = -\frac{k^2}{2} = -\frac{Z^2}{2(\lambda + 1)^2} = -\frac{Z^2}{2n^2} \qquad \lambda = 0, 1, 2, \ldots \qquad n = 1, 2, 3, \ldots \tag{5.29}$$

where we have made the identification $\lambda + 1 = n$. We can see that Fock's treatment gives the usual energy levels for hydrogenlike atoms. For the transformed wave function $\Phi(\mathbf{u})$, any 4-dimensional hyperspherical harmonic will do, but for most applications, it is convenient to use hyperspherical harmonics of the type shown in Table 2.1. Thus we obtain the Fourier transformed hydrogenlike orbitals:

$$\psi_{n,l,m}^t(\mathbf{p}) = \frac{4k^{5/2}}{(k^2 + p^2)^2} Y_{n-1,l,m}(\mathbf{u}) \equiv M(p) Y_{n-1,l,m}(\mathbf{u})$$

$$M(p) \equiv \frac{4k^{5/2}}{(k^2 + p^2)^2} \tag{5.30}$$

For the first few values of n, l and m, (5.30) yields:

$$
\begin{aligned}
\psi^t_{1,0,0}(\mathbf{p}) &= \frac{2\sqrt{2}\, k^{5/2}}{(k^2+p^2)^2\pi} \\
\psi^t_{2,0,0}(\mathbf{p}) &= \frac{4\sqrt{2}\, k^{5/2}(k^2-p^2)}{(k^2+p^2)^3\pi} \\
\psi^t_{2,1,-1}(\mathbf{p}) &= -\frac{8ik^{7/2}(p_1-ip_2)}{(k^2+p^2)^3\pi} \\
\psi^t_{2,1,0}(\mathbf{p}) &= -\frac{8i\sqrt{2}\, k^{7/2}}{(k^2+p^2)\pi} \\
\psi^t_{2,1,1}(\mathbf{p}) &= \frac{8ik^{7/2}(p_1+ip_2)}{(k^2+p^2)^3\pi} \\
\vdots \quad & \vdots \quad \vdots
\end{aligned}
\tag{5.31}
$$

To see how Fock's reciprocal space solutions to the hydrogenlike wave equation are related to the familiar hydrogenlike orbitals, we can make a table of hydrogenlike orbitals with Z/n replaced by the constant k. The radial functions become

$$
\begin{aligned}
R'_{1,0}(r) &= 2k^{3/2}e^{-kr} \\
R'_{2,0}(r) &= 2k^{3/2}(1-kr)e^{-kr} \\
R'_{2,1}(r) &= \frac{2k^{3/2}}{\sqrt{3}}\; kr\; e^{-kr} \\
R'_{3,0}(r) &= 2k^{3/2}\left(1-2kr+\frac{2k^2r^2}{3}\right)e^{-kr} \\
\vdots \quad & \vdots \quad \vdots
\end{aligned}
\tag{5.32}
$$

and so on, and the corresponding wave functions will be

$$
\chi_{n,l,m}(\mathbf{x}) = R'_{n,l}(r)Y_{l,m}(\mathbf{u}) \tag{5.33}
$$

As you can verify, taking the Fourier transforms of the wave functions defined by equations (5.32) and (5.33), and making the substitutions shown in equation (5.10), we obtain the Fourier transformed solutions of V. Fock, equation (5.30). But this set of solutions is not quite the same as a set of familiar hydrogenlike orbitals because Z/n is everywhere replaced by the constant k. A set of Fock's solutions corresponding to a particular value of k is called a set of *Coulomb Sturmians*. Such a set obeys a potential-weighted orthonormality relation, as we will discuss in detail in Chapters 6 and 7.

5.2 Generalization of Fock's treatment

It is interesting to ask whether the Fock transformation can be generalized in such a way as to solve the d-dimensional hydrogenlike Schrödinger equation

$$\left[-\frac{1}{2}\Delta - \frac{Z}{r}\right]\psi(\mathbf{x}) = E\psi(\mathbf{x}) \tag{5.34}$$

Here Δ is the generalized Laplacian operator, Z is a constant, and r is the hyperradius, as defined by equation (1.1). Interest in solutions to (5.34) derives from the fact that the solutions can be used as basis functions in d-dimensional problems, such as reactive scattering. Equation (5.34) can be solved in direct space, but it is interesting to use instead a generalization of Fock's method, since this method leads directly to the Fourier transform of the wave function.

The generalization of Fock's method is quite straightforward. We begin by expressing the wave function in terms of its Fourier transform:

$$\psi(\mathbf{x}) = \frac{1}{(2\pi)^{d/2}}\int d\mathbf{p}\; e^{i\mathbf{p}\cdot\mathbf{x}}\psi^t(\mathbf{p}) \tag{5.35}$$

where

$$\psi^t(\mathbf{p}) = \frac{1}{(2\pi)^{d/2}}\int d\mathbf{x}\; e^{-i\mathbf{p}\cdot\mathbf{x}}\psi(\mathbf{x}) \tag{5.36}$$

and where we have used the abbreviated notation of equations (4.1) and (4.2). Inserting (5.35) into (5.34) yields:

$$\frac{1}{(2\pi)^{d/2}}\int d\mathbf{p}\;\left[\frac{p^2}{2} + V(\mathbf{x}) - E\right] e^{i\mathbf{p}\cdot\mathbf{x}}\psi^t(\mathbf{p}) = 0 \tag{5.37}$$

where $V(\mathbf{x}) = -Z/r$. Multiplying on the left by $e^{-i\mathbf{p}'\cdot\mathbf{x}}$ and integrating over dx, we obtain the d-dimensional reciprocal-space Schrödinger equation:

$$\left[\frac{p'^2}{2} - E\right]\psi^t(\mathbf{p}') = -\frac{1}{(2\pi)^{d/2}}\int d\mathbf{p}\; V^t(\mathbf{p}'-\mathbf{p})\;\psi^t(\mathbf{p}) \tag{5.38}$$

Then, as before, letting

$$k^2 \equiv -2E \tag{5.39}$$

we have

$$\left[p'^2 + k^2\right]\psi^t(\mathbf{p}') = -\frac{2}{(2\pi)^{d/2}}\int d\mathbf{p}\; V^t(\mathbf{p}'-\mathbf{p})\psi^t(\mathbf{p}) \tag{5.40}$$

We must now calculate the d-dimensional Fourier transform of the potential $-Z/r$. From equation (4.40) we have

$$V^t(p) = -\frac{(d-2)!!I(0)Z}{(2\pi)^{d/2}} \int_0^\infty dr\; r^{d-2} j_0^d(pr) \tag{5.41}$$

The principal value of the oscillatory integral in (5.41) is found by adding a convergence factor $e^{-\zeta r}$ and then finding the limit as ζ approaches zero. This yields

$$\lim_{\zeta\to 0}\int_0^\infty dr\; r^{d-2} j_0^d(pr)\; e^{-\zeta r} = \frac{(d-3)!!}{p^{d-1}} \tag{5.42}$$

Noticing that

$$(d-2)!!(d-3)!! = (d-2)! \tag{5.43}$$

we can combine (5.42) and (5.43) to write the d-dimensional Fourier transform of $V(r) = -Z/r$ as

$$V^t(p) = -\frac{(d-2)!I(0)Z}{(2\pi)^{d/2}p^{d-1}} \tag{5.44}$$

Notice that for $d = 3$, this reduces to (5.7). Substitution of (5.44) into (5.40) yields:

$$\left[p'^2 + k^2\right]\psi^t(\mathbf{p}') = \frac{2Z(d-2)!I(0)}{(2\pi)^d}\int d\mathbf{p}\;\frac{1}{|\mathbf{p}'-\mathbf{p}|^{d-1}}\psi^t(\mathbf{p}) \tag{5.45}$$

We now introduce the $(d+1)$-dimensional unit vector $\mathbf{u}$ and the angles $\theta_1, \theta_2, \ldots, \theta_d$ defined by

$$\begin{aligned}
u_1 &= \frac{2kp_1}{k^2+p^2} \equiv \sin\theta_1\;\sin\theta_2\cdots\;\sin\theta_{d-1}\;\cos\theta_d \\
u_2 &= \frac{2kp_2}{k^2+p^2} \equiv \sin\theta_1\;\sin\theta_2\cdots\;\sin\theta_{d-1}\;\sin\theta_d \\
u_3 &= \frac{2kp_3}{k^2+p^2} \equiv \sin\theta_1\;\sin\theta_2\cdots\;\cos\theta_{d-1} \\
&\vdots \\
u_d &= \frac{2kp_d}{k^2+p^2} \equiv \sin\theta_1\;\cos\theta_2 \\
u_{d+1} &= \frac{k^2-p^2}{k^2+p^2} \equiv \cos\theta_1
\end{aligned} \tag{5.46}$$

The angles $\theta_2, \theta_3, \ldots, \theta_d$ are the polar angles of the d-dimensional vector $\mathbf{p}$:

$$\begin{aligned}
p_1 &= p\ \sin\theta_2 \cdots\ \sin\theta_{d-1}\ \cos\theta_d \\
p_2 &= p\ \sin\theta_2 \cdots\ \sin\theta_{d-1}\ \sin\theta_d \\
p_3 &= p\ \sin\theta_2 \cdots\ \cos\theta_{d-1} \\
&\ \vdots \qquad \vdots \qquad \vdots \\
p_{d-1} &= p\ \sin\theta_2\ \cos\theta_3 \\
p_d &= p\ \cos\theta_2
\end{aligned} \tag{5.47}$$

From (5.47) it follows that the volume element in the d-dimensional p-space can be written in the form:

$$d\mathbf{p} \equiv dp_1 dp_2 \cdots dp_d = p^{d-1} dp\ \sin^{d-2}\theta_2 \sin^{d-3}\theta_3 \cdots \sin\theta_{d-2}\ d\theta_2 \cdots d\theta_d \tag{5.48}$$

while from (5.47) it follows that the element of solid angle in the $(d+1)$-dimensional space is

$$d\Omega_{d+1} = \sin^{d-1}\theta_1 \sin^{d-2}\theta_2 \cdots \sin\theta_{d-2}\ d\theta_1 d\theta_2 \cdots d\theta_d \tag{5.49}$$

(A discussion of the way in which the volume element transforms under curvilinear coordinate transformations can be found in Chapter 10.) From (5.46) we also have

$$\theta_1 = \cos^{-1}\left(\frac{k^2 - p^2}{k^2 + p^2}\right) = \sin^{-1}\left(\frac{2kp}{k^2 + p^2}\right) \tag{5.50}$$

so that

$$\frac{d\theta_1}{dp} = \frac{2k}{k^2 + p^2} \tag{5.51}$$

Combining equations (5.48)–(5.51), we obtain the relation:

$$d\mathbf{p} = dp_1 dp_2 \cdots dp_d = \left(\frac{k^2 + p^2}{2k}\right)^d d\Omega_{d+1} \tag{5.52}$$

Since

$$p_j = \frac{k^2 + p^2}{2k} u_j \qquad\qquad j = 1, 2, \ldots, d \tag{5.53}$$

it follows that

$$\frac{1}{|\mathbf{p} - \mathbf{p}'|^2} = \frac{4k^2}{(k^2 + p^2)(k^2 + p'^2)} \frac{1}{|\mathbf{u} - \mathbf{u}'|^2} \tag{5.54}$$

From (5.52) and (5.54), it follows that (5.45) can be rewritten in the form:

$$\begin{aligned}&\left[p'^2 + k^2\right] \psi^t(\mathbf{p}')\\&= \frac{Z(d-2)!I(0)}{(2\pi)^d} \int d\Omega_{d+1} \left(\frac{k^2+p^2}{2k}\right)^d \left[\frac{4k^2}{(k^2+p^2)(k^2+p'^2)}\right]^{d/2-1/2}\\&\times \frac{1}{|\mathbf{u}'-\mathbf{u}|^{d-1}} \psi^t(\mathbf{p})\end{aligned} \tag{5.55}$$

or

$$\left[p'^2 + k^2\right]^{d/2+1/2} \psi^t(\mathbf{p}') = \frac{Z(d-2)!I(0)}{k(2\pi)^d} \int d\Omega_{d+1} \frac{(k^2+p^2)^{d/2+1/2}}{|\mathbf{u}'-\mathbf{u}|^{d-1}} \psi^t(\mathbf{p}) \tag{5.56}$$

We now let

$$(k^2+p^2)^{d/2+1/2}\psi^t(\mathbf{p}) = N\Phi(\mathbf{u}) \tag{5.57}$$

where N is a normalizing constant. With this substitution, (5.56) becomes:

$$\Phi(\mathbf{u}') = \frac{Z(d-2)!I(0)}{k(2\pi)^d} \int d\Omega_{d+1} \frac{1}{|\mathbf{u}'-\mathbf{u}|^{d-1}} \Phi(\mathbf{u}) \tag{5.58}$$

For a space of dimension $(d+1)$,

$$\alpha = (d+1)/2 - 1 = d/2 - 1/2 \tag{5.59}$$

With this value of α, equation (3.5) yields:

$$\frac{1}{|\mathbf{u}'-\mathbf{u}|^{d-1}} = \sum_{\lambda=0}^{\infty} C_\lambda^\alpha(\mathbf{u}\cdot\mathbf{u}') \tag{5.60}$$

and thus (5.58) becomes

$$\Phi(\mathbf{u}') = \frac{Z(d-2)!I(0)}{k(2\pi)^d} \sum_{\lambda=0}^{\infty} \int d\Omega_{d+1}\, C_\lambda^\alpha(\mathbf{u}\cdot\mathbf{u}')\, \Phi(\mathbf{u}) \tag{5.61}$$

Comparing (5.61) with (2.17) and (3.28) and remembering that in (5.61), the integration is an integration over solid angle in a $(d+1)$-dimensional space, we can rewrite (5.61) in the form:

$$\Phi(\mathbf{u}) = \frac{Z(d-1)!\pi^{1/2}}{k2^{d-2}\Gamma(d/2)\Gamma(d/2+1/2)} \sum_{\lambda=0}^{\infty} \frac{1}{(d+2\lambda-1)} O_\lambda^{(d+1)}\Phi(\mathbf{u}) \tag{5.62}$$

Since

$$\frac{2^{d-1}}{\pi^{1/2}}\Gamma(d/2)\Gamma(d/2+1/2) = (d-1)! \tag{5.63}$$

equation (5.62) takes on the simple form:

$$\Phi(\mathbf{u}) = \frac{2Z}{k} \sum_{\lambda=0}^{\infty} \frac{1}{(d+2\lambda-1)} O_{\lambda}^{(d+1)} \Phi(\mathbf{u}) \tag{5.64}$$

If $\Phi(\Omega_{d+1})$ is a hyperspherical harmonic in the $(d+1)$-dimensional space, corresponding to the principal quantum number λ, then (5.64) will be satisfied provided that

$$k = \frac{2Z}{(d+2\lambda-1)} \tag{5.65}$$

If we remember that $k^2 = -2E$, and if we let $n = \lambda+1$, then (5.65) implies that

$$E = -\frac{2Z^2}{(d+2n-3)^2} \qquad n = 1, 2, 3, \ldots \tag{5.66}$$

The reciprocal-space wave function corresponding to this is given by (5.57), with

$$\Phi(\mathbf{u}) = Y_{n-1,\mu}(\mathbf{u}) \tag{5.67}$$

The degree of degeneracy of the energy levels is given by equation (2.46) with d replaced by $d+1$ and λ replaced by $n-1$:

$$\sum_{\mu} 1 = \frac{(d+2n-3)(d+n-3)!}{(n-1)!(d-1)!} \tag{5.68}$$

5.3 Normalization

The normalization constant N in (5.57) can be found in the following way: We would like the solutions $\psi^t(\mathbf{p})$ to obey

$$\int d\mathbf{p}\ |\psi^t(\mathbf{p})|^2 = 1 \tag{5.69}$$

and the hyperspherical harmonics are normalized in such a way that

$$\int d\Omega_{d+1}\ |\Phi(\mathbf{u})|^2 = 1 \tag{5.70}$$

Since $\psi^t(\mathbf{p})$ is a solution of (5.38), its Fourier transform is therefore a solution of (5.34). Thus $\psi^t(\mathbf{p})$ must obey the virial theorem:

$$\int d\mathbf{p}\ p^2\ |\psi^t(\mathbf{p})|^2 = -2E \int d\mathbf{p}\ |\psi^t(\mathbf{p})|^2 \tag{5.71}$$

From (5.69) and from the definition $k^2 = -2E$, we obtain

$$\frac{1}{k^2}\int d\mathbf{p}\; p^2\; |\psi^t(\mathbf{p})|^2 = \int d\mathbf{p}\; |\psi^t(\mathbf{p})|^2 = 1 \tag{5.72}$$

Finally, combining (5.52), (5.57), (5.70) and (5.72), we have

$$\frac{1}{k^2}\int d\mathbf{p}\; p^2\; |\psi^t(\mathbf{p})|^2 = \frac{2N^2}{(2k)^{d+2}}\int d\Omega_{d+1}|\Phi(\mathbf{u})|^2 = \frac{2N^2}{(2k)^{d+2}} = 1 \tag{5.73}$$

Thus, the solutions will be properly normalized if we let

$$N = \frac{1}{2^{1/2}}(2k)^{d/2+1} \tag{5.74}$$

Thus, remembering (5.57) and (5.67), we obtain

$$\psi^t(\mathbf{p}) = \frac{(2k)^{d/2+1}}{2^{1/2}(k^2+p^2)^{d/2+1/2}} Y_{n-1,\mu}(\mathbf{u}) \tag{5.75}$$

When $d = 3$, (5.75) reduces to (5.30).

5.4 Exercises

(1) Show that equation (5.13) follows from (5.10).

(2) Equation (5.19) is an integral equation for the Fourier transformed solution to the wave equation for hydrogenlike atoms. Show that with the substitution indicated by equation (5.20), this integral equation takes on the simple form shown in (5.21). Use a pen and paper rather than Mathematica.

(3) Equations (5.26) and (5.27) imply that any set of eigenfunctions of the generalized angular momentum operator in a 4-dimensional space will yield Fourier transformed solutions to the hydrogenlike wave equation. Write down a set of solutions using equation (5.30) and the standard hyperspherical harmonics. Write down a set of standard 4-dimensional hyperspherical harmonics up to $\lambda = 1$. Propose an alternative set, and show that these are related to the original solutions by a unitary transformation. Will the corresponding Coulomb Sturmians also be related by a unitary transformation?

(4) By taking the Fourier transform of the Coulomb Sturmians shown in equations (5.32) and (5.33) and making the substitutions shown in (5.10), show that one obtains Fock's solutions, equation (5.30) for $(n, l, m) = (1, 0, 0)$ and $(n, l, m) = (2, 0, 0)$.

(5) Derive equation (5.5). Use a pencil and paper rather than Mathematica.

Chapter 6

D-DIMENSIONAL HYDROGENLIKE ORBITALS IN DIRECT SPACE

6.1 Generalization of the 3-dimensional solution

In the previous chapter, we discussed V. Fock's method for finding hydrogenlike wave functions in reciprocal space, and its generalization to the d-dimensional problem:

$$\left[-\frac{1}{2}\Delta - \frac{Z}{r} - E\right]\psi(\mathbf{x}) = 0 \tag{6.1}$$

where r is the hyperradius and where Δ is the generalized Laplacian operator of equation (2.1): [Avery, 1989], [Avery, 2000], [Avery and Avery, 2006], [Avery, 2009], [Avery, Rettrup and Avery, 2012], [Vilinken, 1968], [Wen and Avery, 1985]

$$\Delta \equiv \sum_{j=1}^{d} \frac{\partial^2}{\partial x_j^2} = \frac{1}{r^{d-1}}\frac{\partial}{\partial r} r^{d-1}\frac{\partial}{\partial r} - \frac{\Lambda^2}{r^2} \tag{6.2}$$

Equation (6.1) can also be solved in direct space [Avery, 1989] (by a slight modification of the method normally used to treat the hydrogen atom), and it is interesting to compare the direct-space solutions with the reciprocal-space solutions discussed above. If we let

$$\rho \equiv Zr \tag{6.3}$$

and

$$\epsilon = \frac{2E}{Z^2} \tag{6.4}$$

then (6.1) becomes

$$\left[\frac{1}{\rho^{d-1}}\frac{\partial}{\partial \rho}\rho^{d-1}\frac{\partial}{\partial \rho} - \frac{\Lambda^2}{\rho^2} + \frac{2}{\rho} + \epsilon\right]\psi(\mathbf{x}) = 0 \tag{6.5}$$

Let us try to find solutions of the form

$$\psi_{n,\lambda,\mu}(\mathbf{x}) = R_{n,\lambda}(\rho) Y_{\lambda,\mu}(\mathbf{u}) \tag{6.6}$$

where $Y_{\lambda,\mu}(\mathbf{u})$ is a hyperspherical harmonic in the d-dimensional space. so that

$$\left[\Lambda^2 - \lambda(\lambda + d - 2)\right] Y_{\lambda,\mu}(\mathbf{u}) = 0 \tag{6.7}$$

Combining equations (6.5)–(6.7), we can see that the radial wave functions $R_{n,\lambda}(\rho)$ must be solutions of the differential equation

$$\left[\frac{1}{\rho^{d-1}}\frac{\partial}{\partial\rho}\rho^{d-1}\frac{\partial}{\partial\rho} - \frac{\lambda(\lambda+d-2)}{\rho^2} + \frac{2}{\rho} + \epsilon\right] R_{n,\lambda}(\rho) = 0 \tag{6.8}$$

Equation (6.8) has solutions of the form

$$\begin{aligned} &R_{n,\lambda}(\rho) \\ &= N_{n,\lambda}\ \rho^{\lambda}\ e^{-2\rho/(d+2n-3)} F\left[\ \lambda + 1 - n \mid 2\lambda + d - 1 \mid 4\rho/(d+2n-3)\ \right] \end{aligned} \tag{6.9}$$

where

$$N_{n,\lambda} = \frac{Z^{d/2}}{2(2\lambda+d-2)!}\left(\frac{(\lambda+n+d-3)!}{(n-1-\lambda)!}\right)^{1/2}\left(\frac{4}{d+2n-3}\right)^{\lambda+d/2+1/2} \tag{6.10}$$

and where $F[a|b|x]$ is a confluent hypergeometric function:

$$F[a|b|x] \equiv 1 + \frac{ax}{b} + \frac{a(a+1)x^2}{b(b+1)2!} + \frac{a(a+1)(a+2)x^3}{b(b+1)(b+2)3!} + \cdots \tag{6.11}$$

The solutions $R_{n,\lambda}(\rho)$, shown in equation (6.9), correspond to the energies:

$$E = -\frac{2Z^2}{(d+2n-3)^2} \tag{6.12}$$

In order for the radial functions $R_{n,\lambda}(\rho)$ to tend to zero as $r \to \infty$, the series defined by (6.9) and (6.11) must terminate. This boundary condition requires that $\lambda + 1 - n$ be zero or a negative integer, and it will be satisfied if

$$\begin{aligned} n &= 1, 2, 3, \ldots, \infty \\ \lambda &= 0, 1, 2, \ldots, n-1 \end{aligned} \tag{6.13}$$

The first few d-dimensional hydrogenlike radial wave functions are

$$\begin{aligned}
\frac{1}{N_{1,0}}R_{1,0}(\rho) &= e^{-2\rho/(d-1)} \\
\frac{1}{N_{2,0}}R_{2,0}(\rho) &= e^{-2\rho/(d+1)}\left[1-\frac{4\rho}{(d-1)(d+1)}\right] \\
\frac{1}{N_{2,1}}R_{2,1}(\rho) &= \rho e^{-2\rho/(d+1)} \\
\frac{1}{N_{3,0}}R_{3,0}(\rho) &= e^{-2\rho/(d+3)}\left[1-\frac{8\rho}{(d-1)(d+3)}+\frac{16\rho^2}{(d-1)d(d+3)^2}\right] 6-14a \\
\frac{1}{N_{3,1}}R_{3,1}(\rho) &= \rho e^{-2\rho/(d+3)}\left[1-\frac{4\rho}{(d+1)(d+3)}\right] \\
\frac{1}{N_{3,2}}R_{3,2}(\rho) &= \rho^2 e^{-2\rho/(d+3)}
\end{aligned} \tag{6.14}$$

For $d = 3$ and $\lambda = l$, equations (6.9)–(6.14) reduce to the familiar wave functions and energies of hydrogenlike atoms:

$$\begin{aligned}
\frac{1}{N_{1,0}}R_{1,0}(\rho) &= e^{-\rho} \\
\frac{1}{N_{2,0}}R_{2,0}(\rho) &= e^{-\rho/2}\left[1-\frac{\rho}{2}\right] \\
\frac{1}{N_{2,1}}R_{2,1}(\rho) &= \rho e^{-\rho/2} \\
\frac{1}{N_{3,0}}R_{3,0}(\rho) &= e^{-\rho/3}\left[1-\frac{2\rho}{3}+\frac{2\rho^2}{27}\right] \\
\frac{1}{N_{3,1}}R_{3,1}(\rho) &= \rho e^{-\rho/3}\left[1-\frac{\rho}{6}\right] \\
\frac{1}{N_{3,2}}R_{3,2}(\rho) &= \rho^2 e^{-\rho/3}
\end{aligned} \tag{6.15}$$

Substituting (6.12) into (6.1), we find that the functions (6.6) obey the wave equation

$$\left[-\frac{1}{2}\Delta-\frac{Z}{r}+\frac{2Z^2}{(d+2n-3)^2}\right]\psi_{n,\lambda,\mu}(\mathbf{x}) = 0 \tag{6.16}$$

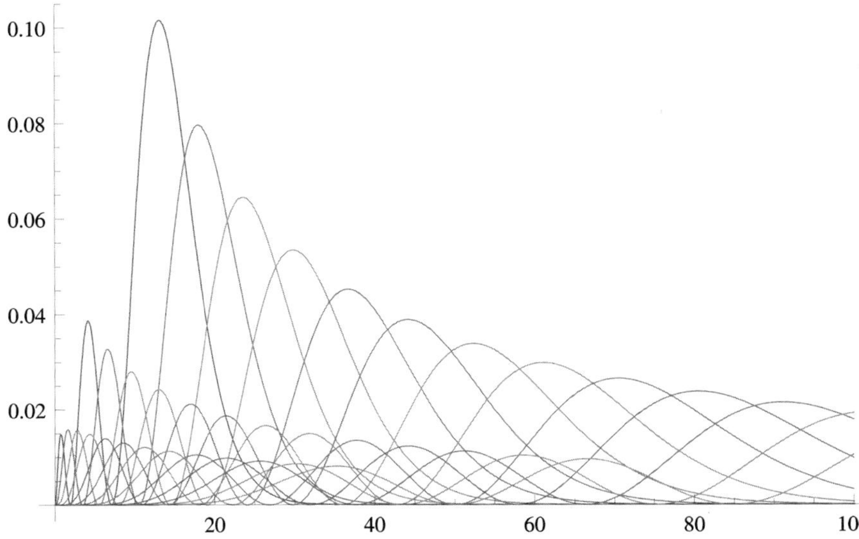

Fig. 6.1: This figure shows the radial distribution function $|R_{3,0}(r)|^2\ r^{d-1}$ of equation (6.9) plotted as a function of the hyperradius for various values of d. The curves for $d = 3, 4, 5, \ldots, 15$ are shown the figure. The highest maximum corresponds to $d = 3$. As d increases, the maxima and zeros occur at larger and larger values of r. The maxima occur on the third peak of each function.

6.2 Orthonormality

The orthonormality relation for d-dimensional hydrogenlike orbitals in direct space is given by

$$\int d\mathbf{x}\ \psi^*_{n',\lambda',\mu'}(\mathbf{x})\ \psi_{n,\lambda,\mu}(\mathbf{x}) = \delta_{n',n}\delta_{\lambda',\lambda}\delta_{\mu',\mu} \tag{6.17}$$

Written out in more detail, this becomes

$$\int_0^\infty dr\ r^{d-1}\ R_{n',\lambda'}(r)R_{n,\lambda}(r)\ \int d\Omega_d\ Y^*_{\lambda',\mu'}(\mathbf{u})Y_{\lambda,\mu}(\mathbf{u}) = \delta_{n',n}\delta_{\lambda',\lambda}\delta_{\mu',\mu} \tag{6.18}$$

Since the hyperspherical harmonics are constructed in such a way that they obey the orthonormality relation

$$\int d\Omega_d\ Y^*_{\lambda',\mu'}(\mathbf{u})Y_{\lambda,\mu}(\mathbf{u}) = \delta_{\lambda',\lambda}\delta_{\mu',\mu} \tag{6.19}$$

equation (6.18) implies that

$$\int_0^\infty dr\ r^{d-1}\ R_{n',\lambda}(r)R_{n,\lambda}(r)\ = \delta_{n',n} \tag{6.20}$$

6.3 *d*-dimensional Coulomb Sturmians

It was mentioned above that the interest in d-dimensional hydrogenlike orbitals stems from the fact that they can be used as basis functions for $d = 3N$-dimensional problems, where N is the number of particles. However, it is well known that the 3-dimensional hydrogenlike orbitals do not form a complete set for the representation of bound-state solutions to the 1-particle Schrödinger equation, unless the continuum is included. To remedy this situation, Shull and Löwdin introduced a type of basis function that has become known as *Coulomb Sturmians.* They have precisely the same form as the 3-dimensional hydrogenlike atomic orbitals except that Z/n is everywhere replaced by a constant k, which is the same for all members of the basis set. The first few of the familiar 3-dimensional hydrogenlike radial wave functions are shown here:

$$\begin{aligned}
R_{1,0}(r) &= 2Z^{3/2}e^{-Zr} \\
R_{2,0}(r) &= 2\left(\frac{Z}{2}\right)^{3/2} e^{-Zr/2}\left[1-\frac{Zr}{2}\right] \\
R_{2,1}(r) &= \frac{2}{\sqrt{3}}\left(\frac{Z}{2}\right)^{3/2} e^{-Zr/2}\,\frac{Zr}{2} \\
\vdots \quad & \quad \vdots \qquad\qquad \vdots
\end{aligned} \tag{6.21}$$

and so on. As we mentioned in Chapter 5, if we replace Z/n everywhere by a constant k, we obtain the Coulomb Sturmian radial wave functions

$$\begin{aligned}
R'_{1,0}(r) &= 2k^{3/2}e^{-kr} \\
R'_{2,0}(r) &= 2k^{3/2}e^{-kr}\left[1-kr\right] \\
R'_{2,1}(r) &= \frac{2}{\sqrt{3}}k^{3/2}e^{-kr}\,kr \\
\vdots \quad & \quad \vdots \qquad\qquad \vdots
\end{aligned} \tag{6.22}$$

Table 6.1: One-electron 3-dimensional Coulomb Sturmian radial functions. If k is replaced by Z/n they are identical to the familiar hydrogenlike radial wave functions.

n	l	$R'_{n,l}(r)$
1	0	$2k^{3/2}e^{-kr}$
2	0	$2k^{3/2}(1-kr)e^{-kr}$
2	1	$\frac{2k^{3/2}}{\sqrt{3}}\ kr\ e^{-kr}$
3	0	$2k^{3/2}\left(1-2kr+\frac{2(kr)^2}{3}\right)e^{-kr}$
3	1	$2k^{3/2}\frac{2\sqrt{2}}{3}\ kr\ \left(1-\frac{kr}{2}\right)e^{-kr}$
3	2	$2k^{3/2}\frac{\sqrt{2}}{3\sqrt{5}}\ (kr)^2\ e^{-kr}$
4	0	$2k^{3/2}\left(1-3kr+2(kr)^2-\frac{(kr)^3}{3}\right)e^{-kr}$
4	1	$2k^{3/2}\sqrt{\frac{5}{3}}\ kr\left(1-kr+\frac{(kr)^2}{5}\right)\ e^{-kr}$
4	2	$2k^{3/2}\frac{1}{\sqrt{5}}\ (kr)^2\left(1-\frac{kr}{3}\right)\ e^{-kr}$
4	3	$2k^{3/2}\frac{1}{3\sqrt{35}}\ (kr)^3\ e^{-kr}$

Notice that apart from the factor $k^{3/2}$, the Coulomb Sturmian radial wave functions are pure functions of the parameter $kr \equiv s$. Like the hydrogenlike orbitals, the Coulomb Sturmians use the familiar 3-dimensional spherical harmonics for their angular part:

$$\chi_{n,l,m}(\mathbf{x}) = R'_{n,l}(r) Y_{l,m}(\mathbf{u}) \tag{6.23}$$

The 3-dimensional Coulomb Sturmians are complete in the sense that any bound-state solution of the 1-particle Schrödinger equation may be represented in terms of them. Because of these properties, Coulomb Sturmians are very widely used in atomic physics and other branches of physics.

It is interesting to ask whether we can construct d-dimensional Coulomb Sturmians with analogous completeness and convergence properties. In going from the hydrogenlike radial wave functions of equation (6.21) to the Coulomb Sturmian functions of (6.22), we replaced Z/n with a constant k. Comparison of the 3-dimensional energies

$$E = -\frac{Z^2}{2n^2} \to -\frac{k^2}{2} \tag{6.24}$$

with the d-dimensional energies

$$E = -\frac{2Z^2}{(d+2n-3)^2} \to -\frac{k^2}{2} \tag{6.25}$$

suggests that we ought to make the replacement

$$\frac{2Z}{(d+2n-3)} \to k \tag{6.26}$$

If we make this substitution in equations (6.9) and (6.10), then we have instead of (6.14):

$$\begin{aligned}
\frac{1}{N'_{1,0}} R'_{1,0}(r) &= k^{d/2} e^{-kr} \\
\frac{1}{N'_{2,0}} R'_{2,0}(r) &= k^{d/2} e^{-kr} \left[1 - \frac{2kr}{(d-1)}\right] \\
\frac{1}{N'_{2,1}} R'_{2,1}(r) &= k^{d/2} (kr) e^{-kr} \\
\frac{1}{N'_{3,0}} R'_{3,0}(r) &= k^{d/2} e^{-kr} \left[1 - \frac{4kr}{(d-1)} + \frac{4(kr)^2}{(d-1)d}\right] \\
\frac{1}{N'_{3,1}} R'_{3,1}(r) &= k^{d/2} (kr) e^{-kr} \left[1 - \frac{2kr}{(d+1)}\right] \\
\frac{1}{N'_{3,2}} R'_{3,2}(r) &= k^{d/2} (kr)^2 e^{-kr}
\end{aligned} \tag{6.27}$$

Notice that, apart from the factor $k^{d/2}$, the d-dimensional Coulomb Sturmian radial functions are pure functions of $kr \equiv s$. The equation analogous to (6.23) is

$$\chi_{n,\lambda,\mu}(\mathbf{x}) = R'_{n,\lambda}(r) Y_{\lambda,\mu}(\mathbf{u}) \tag{6.28}$$

To find the wave equation obeyed by the d-dimensional Coulomb Sturmians, we make the substitution of equation (6.26) into (6.15). This gives us

$$\left[-\frac{1}{2}\Delta - \frac{(d+2n-3)k}{2r} + \frac{k^2}{2} \right] \chi_{n,\lambda,\mu}(\mathbf{x}) = 0 \tag{6.29}$$

6.4 Potential-weighted orthonormality

Coulomb Sturmians are normalized in the usual way, so that

$$\int d\mathbf{x}\ |\chi_{n,\lambda,\mu}(\mathbf{x})|^2 = 1 \tag{6.30}$$

However, instead of obeying an orthonormality relation like (6.17), they obey a *potential-weighted orthonormality relation.* To understand why this is the case, let us think of the 3-dimensional Coulomb Sturmians. They were derived from the hydrogenlike orbitals by making the substitution $Z/n \to k$. The hydrogenlike orbitals obey the 1-electron Schrödinger equation:

$$\left[-\frac{1}{2}\nabla^2 - \frac{Z}{r} + \frac{Z^2}{2n^2} \right] \psi_{n,l,m}(\mathbf{x}) = 0 \tag{6.31}$$

Making the substitution $Z/n \to k$ both in the wave equation and in the solution yields:

$$\left[-\frac{1}{2}\nabla^2 - \frac{nk}{r} + \frac{k^2}{2} \right] \chi_{n,l,m}(\mathbf{x}) = 0 \tag{6.32}$$

We can see that the set of Coulomb Sturmians corresponding to a given value of k are *isoenergetic.* All of the basis functions in the set correspond to the energy

$$E = -\frac{k^2}{2} \tag{6.33}$$

The completeness property of a Coulomb Sturmian basis set depends on the fact that all of the members of the set correspond to the same value of k, and hence the same energy. Equation (6.32) can be written in the form:

$$\left[-\frac{1}{2}\nabla^2 + \frac{k^2}{2} \right] \chi_{n,l,m}(\mathbf{x}) = \beta_n \frac{1}{r} \chi_{n,l,m}(\mathbf{x}) \tag{6.34}$$

where $\beta_n = kn$. We can see that this is still an eigenvalue equation, but the eigenvalue is not the energy. It is instead a weighting factor attached to the potential. This weighting factor is especially adjusted in such a way that all of the solutions to (6.34) will correspond to the same energy. Multiplying (6.34) on the left by $\chi^*_{n',l,m}(\mathbf{x})$ and integrating over the coordinates, we have

$$\int d\mathbf{x}\ \chi^*_{n',l,m}(\mathbf{x}) \left[-\frac{1}{2}\nabla^2 + \frac{k^2}{2}\right] \chi_{n,l,m}(\mathbf{x}) = \beta_n \int d\mathbf{x}\ \chi^*_{n',l,m}(\mathbf{x}) \frac{1}{r} \chi_{n,l,m}(\mathbf{x}) \tag{6.35}$$

and similarly

$$\int d\mathbf{x}\ \chi^*_{n,l,m}(\mathbf{x}) \left[-\frac{1}{2}\nabla^2 + \frac{k^2}{2}\right] \chi_{n',l,m}(\mathbf{x}) = \beta_{n'} \int d\mathbf{x}\ \chi_{n',l,m}(\mathbf{x}) \frac{1}{r} \chi^*_{n,l,m}(\mathbf{x}) \tag{6.36}$$

If we take the complex conjugate of equation (6.36) and subtract it from (6.35), we have

$$0 = (\beta_n - \beta_{n'}) \int d\mathbf{x}\ \chi^*_{n',l,m}(\mathbf{x}) \frac{1}{r} \chi_{n,l,m}(\mathbf{x}) \tag{6.37}$$

where the left-hand side of (6.37) vanishes because of Hermiticity. Thus, when $(\beta_n - \beta_{n'}) \neq 0$, i.e when $n \neq n'$,

$$\int d\mathbf{x}\ \chi^*_{n',l,m}(\mathbf{x}) \frac{1}{r} \chi_{n,l,m}(\mathbf{x}) = 0 \qquad \text{if } n \neq n' \tag{6.38}$$

Then, making use of the normalization condition (6.30) and the orthonormality of the spherical harmonics, we obtain the potential-weighted orthonormality relation for 3-dimensional Coulomb Sturmians:

$$\int d\mathbf{x}\ \chi^*_{n',l',m'}(\mathbf{x}) \frac{n}{kr} \chi_{n,l,m}(\mathbf{x}) = \delta_{n',n}\delta_{l',l}\delta_{m',m} \tag{6.39}$$

In a similar way, one can show that the d-dimensional Coulomb Sturmians obey the potential-weighted orthonormality relation,

$$\int d\mathbf{x}\ \chi^*_{n',\lambda',\mu'}(\mathbf{x}) \frac{(2n+d-3)}{2kr} \chi_{n,\lambda,\mu}(\mathbf{x}) = \delta_{n',n}\delta_{\lambda',\lambda}\delta_{\mu',\mu} \tag{6.40}$$

where $\delta_{\mu',\mu} = 1$ if and only if each member of the set of indices μ' is equal to the corresponding member of the set of indices μ. You can verify that equations (6.39) and (6.40) hold by carrying out the radial integration for particular values of n and λ.

6.5 Fourier transforms of Coulomb Sturmians

We mentioned above that with the substitution

$$\frac{2Z}{(d+2n-3)} \to k$$
$$\rho = Zr \to kr\ (d+2n-3)/2 \tag{6.41}$$

the d-dimensional hydrogenlike orbitals of equations (6.9) and (6.10) are changed into d-dimensional Coulomb Sturmians. The Fourier transforms of d-dimensional Coulomb Sturmians are given by

$$\begin{aligned} \chi^t_{n,\lambda,\mu}(\mathbf{p}) &= \frac{1}{(2\pi)^{d/2}} \int d\mathbf{x}\ e^{-i\mathbf{p}\cdot\mathbf{x}} \chi_{n,\lambda,\mu}(\mathbf{x}) \\ &= Y_{\lambda,\mu}(\mathbf{u}_p) \frac{(-i)^\lambda (d-2)!!I(0)}{(2\pi)^{d/2}} \int_0^\infty dr\ r^{d-1} j^d_\lambda(pr) R'_{n,\lambda}(r) \end{aligned} \tag{6.42}$$

where we have made use of equation (4.40). In the special case where $d = 3$, this becomes

$$\begin{aligned} \chi^t_{n,\lambda,\mu}(\mathbf{p}) &= \frac{1}{(2\pi)^{d/2}} \int d\mathbf{x}\ e^{-i\mathbf{p}\cdot\mathbf{x}} \chi_{n,\lambda,\mu}(\mathbf{x}) \\ &= Y_{l,m}(\mathbf{u}_p)(-i)^l \sqrt{\frac{2}{\pi}} \int_0^\infty dr\ r^2 j_l(pr) R'_{n,l}(r) \end{aligned} \tag{6.43}$$

If we compare the results obtained by by carrying out the hyperradial integral in (6.42) with equation (5.75), taking into account the projection (5.47), we find that they are identical. The reader may verify that this is the case by evaluating the integrals for particular values of n and λ. In other words, the results of the d-dimensional generalization of V. Fock's method for treating hydrogenlike atoms do not give us sets of Fourier-transformed d-dimensional hydrogenlike orbitals, but rather sets of Fourier-transformed d-dimensional Coulomb Sturmians. Like the Coulomb Sturmians in direct space, the Coulomb Sturmians in reciprocal space are characterized by a parameter k, whose significance we will discuss in detail later.

6.6 Use of d-dimensional Coulomb Sturmians as a basis

Suppose that we wish to solve the d-dimensional Schrödinger equation

$$\left[-\frac{1}{2}\Delta + V(\mathbf{x}) - E_\kappa\right] \Psi_\kappa(\mathbf{x}) = 0 \tag{6.44}$$

where Δ is the d-dimensional generalized Laplacian operator and

$$\mathbf{x} \equiv (x_1, x_2, \ldots, x_d) \tag{6.45}$$

We can use d-dimensional Coulomb Sturmians as a basis for solving (6.44), letting

$$\Psi_\kappa(\mathbf{x}) = \sum_{n,\lambda,\mu} \chi_{n,\lambda,\mu}(\mathbf{x}) B_{n,\lambda,\mu;\kappa} \tag{6.46}$$

Substituting this superposition into (6.44), multiplying from the left by $\chi^*_{n',\lambda',\mu'}(\mathbf{x})$, and integrating over the coordinates, we obtain the set of secular equations:

$$\sum_{n,\lambda,\mu} \int d\mathbf{x}\; \chi^*_{n',\lambda',\mu'}(\mathbf{x}) \left[-\frac{1}{2}\Delta + V(\mathbf{x}) - E_\kappa\right] \chi_{n,\lambda,\mu}(\mathbf{x}) B_{n,\lambda,\mu;\kappa} = 0 \tag{6.47}$$

Since all of the members of our basis set obey equation (6.29), we can rewrite the secular equation in the form

$$\sum_{n,\lambda,\mu} \int d\mathbf{x}\; \chi^*_{n',\lambda',\mu'}(\mathbf{x}) \times \left[\frac{(d+2n-3)k}{2r} + V(\mathbf{x}) - E_\kappa - \frac{k^2}{2}\right] \chi_{n,\lambda,\mu}(\mathbf{x}) B_{n,\lambda,\mu;\kappa} = 0 \tag{6.48}$$

We start the calculation by choosing a value of k and evaluating the integrals in (6.48). Then, by solving the secular equations (6.48), we obtain the bound-state eigenvalues and eigenvectors. Because of the completeness properties of Coulomb Sturmian basis sets, if we use a large basis set, the results will only weakly depend on the value of k chosen. We can, of course, repeat the calculation and optimize the value of k.

In later chapters, we will discuss systems which interact entirely through Coulomb potentials. In such systems we can take advantage of special properties of Coulomb Sturmians. However here, we shall suppose the potential to be a general one. The most simple such case is the one where $V(\mathbf{x})$ depends only on the hyperradius. In that case, we have

$$\sum_{n} \int_0^\infty dr\; r^{d-1}\; R'_{n',\lambda}(r) \times \left[\frac{(d+2n-3)k}{2r} + V(r) - E_\kappa - \frac{k^2}{2}\right] R'_{n,\lambda}(r) B_{n,\lambda;\kappa} = 0 \tag{6.49}$$

where we have carried out the hyperangular integration, making use of the orthonormality of the hyperspherical harmonics. Solving the secular equations (6.49) for each value of λ gives us the solutions.

$$\Psi_{\kappa,\lambda,\mu}(\mathbf{x}) = Y_{\lambda,\mu}(\mathbf{u}) \sum_n R'_{n,\lambda}(r) B_{n,\lambda;\kappa} \tag{6.50}$$

Notice that when the potential depends only on the hyperradius, the expansion coefficients $B_{n,\lambda;\kappa}$ no longer depend on μ. Notice also that the radial integral for the first term in (6.49) can be carried out exactly by means of equation (6.40).

6.7 Cases where the potential may have angular dependence

For the more general case, where $V(\mathbf{x})$ may depend on angles, the integrals in equation (6.48) become:

$$\begin{aligned}
&\int d\mathbf{x}\ \chi^*_{n',\lambda',\mu'}(\mathbf{x}) \left[\frac{(d+2n-3)k}{2r} + V(\mathbf{x}) - E_\kappa - \frac{k^2}{2}\right] \chi_{n,\lambda,\mu}(\mathbf{x}) \\
&= \delta_{\lambda',\lambda}\delta_{\mu',\mu} \int_0^\infty dr\ r^{d-1} R'_{n',\lambda}(r) \left[\frac{(d+2n-3)k}{2r} - E_\kappa - \frac{k^2}{2}\right] R'_{n,\lambda}(r) \\
&+ \int d\mathbf{x}\ \chi^*_{n',\lambda',\mu'}(\mathbf{x}) V(\mathbf{x}) \chi_{n,\lambda,\mu}(\mathbf{x})
\end{aligned} \tag{6.51}$$

Let us now consider the problem of evaluating the matrix elements

$$V_{\nu',\nu} \equiv \int d\mathbf{x}\ \chi^*_{\nu'}(\mathbf{x})\chi_\nu(\mathbf{x})\ V(\mathbf{x}) \tag{6.52}$$

where we have introduced the notation $\nu \equiv (n, \lambda, \mu)$. This problem becomes simpler if we let

$$\chi^*_{\nu'}(\mathbf{x})\chi_\nu(\mathbf{x}) = \sum_{\nu''} \chi_{\nu''}(2k, \mathbf{x}) \mathcal{C}^{\nu''}_{\nu',\nu} \tag{6.53}$$

In other words, we expand a product of two Coulomb Sturmians as a series of single Coulomb Sturmians with double the k value. The coefficients $\mathcal{C}^{\nu''}_{\mu',\mu}$ are easy to evaluate: We start by letting

$$Y^*_{\lambda',\mu'}(\mathbf{u}) Y_{\lambda,\mu}(\mathbf{u}) = \sum_{\lambda''=|\lambda-\lambda'|}^{\lambda+\lambda'} \sum_{\mu''\leq\lambda''} Y_{\lambda'',\mu''}(\mathbf{u})\ I^{\lambda'',\mu''}_{\lambda',\mu';\lambda,\mu} \tag{6.54}$$

Then, using the orthonormality of the hyperspherical harmonics, we obtain

$$I^{\lambda'',\mu''}_{\lambda',\mu';\lambda,\mu} = \int d\Omega_d\ Y^*_{\lambda'',\mu''}(\mathbf{u}) Y^*_{\lambda',\mu'}(\mathbf{u}) Y_{\lambda,\mu}(\mathbf{u}) \tag{6.55}$$

The integral in equation (6.55) can be evaluated rapidly and exactly by means of our general hyperangular integration theorem (2.28). When we are in possession of the coefficients $I^{\lambda'',\mu''}_{\lambda',\mu';\lambda,\mu}$, we can rewrite equation (6.53) as

$$\begin{aligned} &R'_{n',\lambda'}(r)R'_{n,\lambda}(r)Y^*_{\lambda',\mu'}(\mathbf{u})Y_{\lambda,\mu}(\mathbf{u}) \\ &= \sum_{n'',\lambda'',\mu''} R'_{n'',\lambda''}(2k,r)Y_{\lambda''\mu''}(\mathbf{u})\; b^{n''}_{n',n} I^{\lambda'',\mu''}_{\lambda',\mu';\lambda,\mu} \end{aligned} \tag{6.56}$$

where

$$\mathcal{C}^{\mu''}_{\mu',\mu} \equiv b^{n''}_{n',n} I^{\lambda'',\mu''}_{\lambda',\mu';\lambda,\mu} \tag{6.57}$$

and

$$R'_{n',\lambda'}(r)R'_{n,\lambda}(r) = \sum_{n''} R'_{n'',\lambda''}(2k,r)\; b^{n''}_{n',n} \tag{6.58}$$

In order to evaluate the coefficients $b^{n''}_{n',n}$, we make use of the potential-weighted orthonormality relations (6.40) obeyed by d-dimensional Coulomb Sturmians. From (6.40) and from the orthonormality of the hyperspherical harmonics, it follows that

$$\frac{2n+d-3}{4k}\int_0^\infty dr\; r^{d-2} R'_{n'',\lambda''}(2k,r)R'_{n''',\lambda''}(2k,r) = \delta_{n''',n''} \tag{6.59}$$

Equation (6.59) allows us to solve (6.58) for the coefficients $b^{n'''}_{n',n}$:

$$\begin{aligned} &\frac{2n+d-3}{4k}\int_0^\infty dr\; r^{d-2} R'_{n''',\lambda''}(2k,r)R'_{n',\lambda'}(r)R'_{n,\lambda}(r) \\ &= \sum_{n''} \frac{2n+d-3}{4k}\int_0^\infty dr\; r^{d-2} R'_{n''',\lambda''}(2k,r)R'_{n'',\lambda''}(2k,r)\; b^{n''}_{n',n} \\ &= \sum_{n''} \delta_{n''',n''} b^{n''}_{n',n} = b^{n'''}_{n',n} \end{aligned} \tag{6.60}$$

The series in equation (6.58) has a limited number of terms and is exact. Having calculated the coefficients $\mathcal{C}^{\nu''}_{\nu',\nu}$ in this way, we can rewrite (6.52) in the form

$$\begin{aligned} V_{\nu',\nu} &= \sum_{\nu''} \mathcal{C}^{\nu''}_{\nu',\nu} \int d\mathbf{x}\; \chi_{\nu''}(\mathbf{x})V(\mathbf{x}) \\ &= \sum_{\nu''} \mathcal{C}^{\nu''}_{\nu',\nu} \int d\mathbf{p}\; \chi^t_{\nu''}(\mathbf{p})V^t(-\mathbf{p}) \end{aligned} \tag{6.61}$$

where the second step has been calculated using equation (4.8), and where

$$V^t(-\mathbf{p}) = \frac{1}{(2\pi)^{d/2}}\int d\mathbf{x}\; e^{i\mathbf{p}\cdot\mathbf{x}}\; V(\mathbf{x}) \tag{6.62}$$

and where $\chi^t_{\mu''}(\mathbf{p})$ is calculated using the methods discussed in Chapter 5. Thus we can evaluate the matrix elements either in direct space or in reciprocal space, whichever is more convenient. The coefficients $\mathcal{C}_{\nu'';\nu',\nu}$ are universals which are characteristic of the Coulomb Sturmian basis set and independent of the potentials. They form a large but sparse matrix, which may be precalculated and stored. If $V(\mathbf{x})$ is expanded in terms of functions of the hyperradius multiplied by d-dimensional hyperspherical harmonics, the orthonormality of the harmonics can be used to evaluate all the integrals except the hyperradial ones.

6.8 Exercises

(1) Type `?radwf` to obtain the definitions of this function. It generates the radial function of the d-dimensional hydrogenlike orbitals defined in equation (6.9). For $d = 3$, make a table of these orbitals up to $n = 2$ and check that they are properly normalized. Multiply the orbitals in the table by `y3[l,0]` and integrate over both solid angle and r^2dr to check that the set of functions is orthonormal.

(2) Repeat this procedure for d equal to 4, 5, and 6, taking into account the change in the Jacobian when going to a space of higher dimension.

(3) Type `?strwf` to obtain the definitions of this function. It generates the radial function of the d-dimensional Coulomb Sturmians obtained from hydrogenlike radial functions by the substitution of equation (6.26). For $d = 3$, make a table of these orbitals up to $n = 2$ and check that they are properly normalized. Compare to the hydrogenlike wave functions.

(4) Check that the d-dimensional Coulomb Sturmians are properly normalized for d equal to 4, 5, and 6, taking into account the change in the Jacobian when going to a space of higher dimension.

(5) Check that the d-dimensional Coulomb Sturmians obey the potential-weighted orthonormality relations shown in equations (6.39) and (6.40).

(6) Make a plot of the radial part of the 3-dimensional hydrogenlike wave functions with the charge $Z = 1$, $l = 0$ and $n = 1, 2, ..., 15$. Compare this with a plot of the radial part of the 3-dimensional Coulomb Sturmian wave functions with k equal to 1. Notice especially where the nodes occur.

(7) Use equations (6.54) and (6.55) to represent the products of 4-dimensional hyperspherical harmonics as sums of single 4-dimensional hyperspherical harmonics.

Chapter 7

GENERALIZED STURMIANS

7.1 Generalized Sturmians and many-particle bound state problems

In a pioneering 1968 paper [Goscinski, 2003], Osvaldo Goscinski generalized the concept of Sturmians (see also [Avery, 1989; Avery, 2000; Avery and Avery, 2006; Avery, 2009; Avery, Rettrup and Avery, 2012; Vilinken, 1968; Wen and Avery, 1985]). He regarded Sturmians as isoenergetic solutions to a general d-dimensional Schrödinger-like equation of the form

$$\left[-\frac{1}{2}\Delta + \beta_\nu V_0(\mathbf{x}) + \frac{p_\kappa^2}{2}\right] \Phi_\nu(\mathbf{x}) = 0 \tag{7.1}$$

with a weighted potential $\beta_\nu V_0(\mathbf{x})$, the weighting constants β_ν being chosen in such a way as to make all of the solutions correspond to the same energy, $-p_\kappa^2/2$, where p_κ is real. Equation (7.1) could be applied to N particles of different masses m_j with

$$\Delta \equiv \sum_{j=1}^{N} \frac{1}{m_j}\nabla_j^2 \tag{7.2}$$

For a collection of N electrons, one can let

$$\Delta \equiv \sum_{j=1}^{d} \frac{\partial^2}{\partial x_j^2} \tag{7.3}$$

where $d = 3N$ and

$$\mathbf{x} = (x_1, x_2, \ldots, x_d) \tag{7.4}$$

since, in atomic units, $m_j = 1$ for all the electrons. Goscinski was able to show that generalized Sturmians obey a potential-weighted orthogonality relation of the form

$$\int d\mathbf{x}\ \Phi_{\nu'}^*(\mathbf{x}) V_0(\mathbf{x}) \Phi_\nu(\mathbf{x}) = 0 \qquad \text{if } \beta_{\nu'}^* \neq \beta_\nu \tag{7.5}$$

We can see why this is true by the following argument, which is a d-dimensional generalization of the method used in equations (6.35)–(6.39): The set of functions $\Phi_\nu(\mathbf{x})$ all obey equation (7.1). Therefore we can write:

$$\begin{aligned}\int d\mathbf{x}\ \Phi^*_{\nu'}\left[-\frac{1}{2}\Delta+\frac{p_\kappa^2}{2}\right]\Phi_\nu(\mathbf{x}) &= -\beta_\nu\int d\mathbf{x}\ \Phi^*_{\nu'}(\mathbf{x})V_0(\mathbf{x})\Phi_\nu(\mathbf{x})\\ \int d\mathbf{x}\ \Phi^*_{\nu}\left[-\frac{1}{2}\Delta+\frac{p_\kappa^2}{2}\right]\Phi_{\nu'}(\mathbf{x}) &= -\beta_{\nu'}\int d\mathbf{x}\ \Phi^*_{\nu}(\mathbf{x})V_0(\mathbf{x})\Phi_{\nu'}(\mathbf{x})\end{aligned}\tag{7.6}$$

Taking the conjugate transpose of the matrices in the lower equation in (7.6), and subtracting this from the top equation, we obtain:

$$0=(\beta_\nu-\beta^*_{\nu'})\int d\mathbf{x}\ \Phi^*_{\nu'}(\mathbf{x})V_0(\mathbf{x})\Phi_\nu(\mathbf{x})\tag{7.7}$$

and hence the orthogonality relationship (7.5). In setting the left-hand side of equation (7.7) equal to zero, we have made use of the Hermiticity of the operator Δ. We should notice that in equations (7.5) and (7.6), ν stands for a set of quantum numbers, some of which influence the value of β_ν, while others do not. We can call these respectively *major* and *minor* quantum numbers. Orthogonality with respect to the minor quantum numbers can often be established by means of symmetry properties. When this is not possible, the members of a Sturmian basis set can be made orthogonal with respect to the remaining minor quantum numbers by means of for example Gram-Schmidt or Löwdin orthogonalization. We shall normalize our generalized Sturmian basis sets in such a way that the potential-weighted orthonormality relations take the form

$$\int d\mathbf{x}\ \Phi^*_{\nu'}(\mathbf{x})V_0(\mathbf{x})\Phi_\nu(\mathbf{x})=-\delta_{\nu',\nu}\frac{p_\kappa^2}{\beta_\nu}\tag{7.8}$$

This type of normalization is convenient because

(1) It makes (7.8) reduce to the usual potential-weighted orthonormality relations for the case of Coulomb Sturmians, when $d=3$, $p_\kappa=k$ and $\beta_\nu=nk$, equation (6.39).
(2) It makes the generalized Sturmians fit more naturally into the theory of Sobolev spaces.
(3) The special Goscinskian configurations, which will be introduced later, are already properly normalized.

7.2 Sturmians and the many-particle Schrödinger equation

The non-relativistic many-particle Schrödinger equation can be written in the form

$$\left[-\frac{1}{2}\Delta + V(\mathbf{x}) - E_\kappa\right]\Psi_\kappa(\mathbf{x}) = 0 \tag{7.9}$$

where atomic units are used and where Δ is defined by (7.2) or (7.3). We now expand the wave function $\Psi_\kappa(\mathbf{x})$ in terms of a generalized Sturmian basis set.

$$\Psi_\kappa(\mathbf{x}) = \sum_\nu \Phi_\nu(\mathbf{x}) B_{\nu,\kappa} \tag{7.10}$$

We need not do so, but if $E_\kappa < 0$, we can let

$$p_\kappa^2 \equiv -2E_\kappa \tag{7.11}$$

In the analysis of scattering experiments, where $E_\kappa > 0$, we can still use the relation (7.11), provided that we allow p_κ to be complex. Substituting the expansion of (7.10) into the Schrödinger equation, and using the fact that all the members of our basis set obey equation (7.1), we obtain

$$\sum_\nu \left[-\frac{1}{2}\Delta + V(\mathbf{x}) - E_\kappa\right]\Phi_\nu(\mathbf{x}) B_{\nu,\kappa}$$

$$= \sum_\nu \left[V(\mathbf{x}) - \beta_\nu V_0(\mathbf{x})\right]\Phi_\nu(\mathbf{x}) B_{\nu,\kappa} = 0 \tag{7.12}$$

If we multiply (7.12) on the left by a conjugate function from our basis set and integrate over all the coordinates, we obtain

$$\sum_\nu \int d\mathbf{x}\ \Phi_{\nu'}^*(\mathbf{x}) \left[V(\mathbf{x}) - \beta_\nu V_0(\mathbf{x})\right]\Phi_\nu(\mathbf{x}) B_{\nu,\kappa} = 0 \tag{7.13}$$

We next introduce the notation

$$T_{\nu',\nu} \equiv -\frac{1}{p_\kappa}\int d\mathbf{x}\ \Phi_{\nu'}^*(\mathbf{x}) V(\mathbf{x}) \Phi_\nu(\mathbf{x}) \tag{7.14}$$

For systems interacting through Coulomb potentials, $T_{\nu',\nu}$ is found to be independent of p_κ. This is important because it means that for systems interacting through Coulomb potentials, the spectrum of p_κ values can be found immediately by solving the secular equations. Finally, making use of the potential-weighted orthonormality relations (7.8), we obtain the Sturmian secular equations.

$$\sum_\nu \left[T_{\nu',\nu} - p_\kappa \delta_{\nu',\nu}\right] B_{\nu,\kappa} = 0 \tag{7.15}$$

These secular equations have several remarkable features that will be discussed in more detail in later sections. For the moment, it is interesting to notice that the kinetic energy term has disappeared. Furthermore, the eigenvalues are not energies, but values of the "scaling parameter" p_κ, which is related to the energy spectrum through equation (7.11).

7.3 Momentum-space orthonormality of generalized Sturmian basis sets

We would now like to find the momentum-space orthonormality relations obeyed by Fourier transforms of the generalized Sturmian basis set. Because the Fourier transform is unitary, the inner product of any two functions in L_2 is preserved under the operation of taking their Fourier transforms, i.e.

$$\int d\mathbf{x}\ f(\mathbf{x})^* g(\mathbf{x}) = \int d\mathbf{p}\ f^t(\mathbf{p})^* g^t(\mathbf{p}) \tag{7.16}$$

Using this well-known relationship with $f^*(\mathbf{x}) = \Phi^*_{\nu'}(\mathbf{x})$ and $g(\mathbf{x}) = V_0(\mathbf{x})\Phi_\nu(\mathbf{x})$, we have

$$\int d\mathbf{x}\ \Phi^*_{\nu'}(\mathbf{x})V_0(\mathbf{x})\Phi_\nu(\mathbf{x}) = \int d\mathbf{p}\ \Phi^{t*}_{\nu'}(\mathbf{p}) \left[V_0\Phi_\nu\right]^t(\mathbf{p}) \tag{7.17}$$

In order to evaluate $\left[V_0\Phi_\nu\right]^t(\mathbf{p})$, we remember the Fourier convolution theorem, which states that the Fourier transform of the product of two functions is the convolution of their Fourier transforms. Thus if a and b are any two functions in L_2,

$$\left[ab\right]^t(\mathbf{p}') \equiv \frac{1}{(2\pi)^{d/2}} \int d\mathbf{x}\ e^{-i\mathbf{p}'\cdot\mathbf{x}} a(\mathbf{x})b(\mathbf{x}) = \frac{1}{(2\pi)^{d/2}} \int d\mathbf{p}\ a^t(\mathbf{p}'-\mathbf{p})b^t(\mathbf{p}) \tag{7.18}$$

Letting $a(\mathbf{x}) = V_0(\mathbf{x})$ and $b(\mathbf{x}) = \Phi_\nu(\mathbf{x})$ we have

$$\left[V_0\Phi_\nu\right]^t(\mathbf{p}') = \frac{1}{(2\pi)^{d/2}} \int d\mathbf{p}\ V_0^t(\mathbf{p}'-\mathbf{p})\Phi^t_\nu(\mathbf{p}) \tag{7.19}$$

Since, using a method similar to that used in deriving equation (5.40), we find that the momentum-space integral equation corresponding to (7.1) has the form

$$(p'^2 + p_\kappa^2)\Phi^t_\nu(\mathbf{p}') = -\frac{2\beta_\nu}{(2\pi)^{d/2}} \int d\mathbf{p}\ V_0^t(\mathbf{p}'-\mathbf{p})\Phi^t_\nu(\mathbf{p}) \tag{7.20}$$

it follows that

$$\left[V_0\Phi_\nu\right]^t(\mathbf{p}) = -\frac{(p^2+p_\kappa^2)}{2\beta_\nu}\Phi^t_\nu(\mathbf{p}) \tag{7.21}$$

Finally, substituting (7.21) into (7.17) and using (7.8), we obtain the momentum-space orthonormality relations for a set of generalized Sturmian basis functions:

$$\int d\mathbf{p}\ \Phi^{t*}_{\nu'}(\mathbf{p}) \left(\frac{p^2+p_\kappa^2}{2p_\kappa^2}\right) \Phi^t_\nu(\mathbf{p}) = \delta_{\nu',\nu} \tag{7.22}$$

Because all of the functions $\Phi_\nu(\mathbf{x})$ in the generalized Sturmian basis set obey equation (7.1), the potential-weighted direct space orthonormality relations shown in equation (7.8) can be rewritten in the form

$$\int d\mathbf{x}\; \Phi^*_{\nu'}(\mathbf{x}) \left(\frac{-\Delta + p_\kappa^2}{2p_\kappa^2} \right) \Phi_\nu(\mathbf{x}) = \delta_{\nu',\nu} \tag{7.23}$$

so that the momentum-space and direct-space orthonormality relations can be seen to be related to each other in a symmetrical way. These weighted orthonormality relations in $L_2(\mathbb{R}^d)$ are the usual orthonormality relations in the Sobolev space $W_2^{(1)}(\mathbb{R}^d)$.

7.4 Sturmian expansions of *d*-dimensional plane waves

If the set of generalized Sturmian basis functions is complete in the sense of spanning the Sobolev space $W_2^{(1)}(\mathbb{R}^d)$, we can use it to construct a weakly convergent expansion of a d-dimensional plane wave (valid only in the sense of distributions). Suppose that we let

$$e^{i\mathbf{p}\cdot\mathbf{x}} = \left(\frac{p_\kappa^2 + p^2}{2p_\kappa^2} \right) \sum_\nu \Phi_\nu^{*t}(\mathbf{p}) a_\nu \tag{7.24}$$

We can then determine the coefficients a_ν by means of the orthonormality relations (7.22). Multiplying (7.24) on the left by $\Phi^t_{\nu'}(\mathbf{p})$ and integrating over $d\mathbf{p}$ making use of (7.23), we obtain

$$\begin{aligned}\int d\mathbf{p}\; \Phi^t_{\nu'}(\mathbf{p}) e^{i\mathbf{p}\cdot\mathbf{x}} &= \int d\mathbf{p}\; \Phi^t_{\nu'}(\mathbf{p}) \left(\frac{p_\kappa^2 + p^2}{2p_\kappa^2} \right) \sum_\nu \Phi_\nu^{*t}(\mathbf{p}) a_\nu \\ &= \sum_\nu \delta_{\nu',\nu} a_\nu = a_{\nu'} \end{aligned} \tag{7.25}$$

so that

$$a_\nu = \int d\mathbf{p}\; e^{i\mathbf{p}\cdot\mathbf{x}} \Phi^t_\nu(\mathbf{p}) = (2\pi)^{d/2} \Phi_\nu(\mathbf{x}) \tag{7.26}$$

Thus finally we obtain an expansion of the form

$$e^{i\mathbf{p}\cdot\mathbf{x}} = (2\pi)^{d/2} \left(\frac{p_\kappa^2 + p^2}{2p_\kappa^2} \right) \sum_\nu \Phi_\nu^{t*}(\mathbf{p}) \Phi_\nu(\mathbf{x}) \tag{7.27}$$

If the set of generalized Sturmians $\Phi_\nu(\mathbf{x})$ does not span $W_2^{(1)}(\mathbb{R}^d)$, equation (7.27) becomes

$$P\left[e^{i\mathbf{p}\cdot\mathbf{x}} \right] = (2\pi)^{d/2} \left(\frac{p_\kappa^2 + p^2}{2p_\kappa^2} \right) \sum_\nu \Phi_\nu^{t*}(\mathbf{p}) \Phi_\nu(\mathbf{x}) \tag{7.28}$$

where $P\left[e^{i\mathbf{p}\cdot\mathbf{x}}\right]$ is the projection of the d-dimensional plane wave onto the subspace spanned by the set $\{\Phi_\nu(\mathbf{x})\}$. For example, if we are considering a system of N electrons, with $d = 3N$, the generalized Sturmian basis set might be antisymmetric with respect to exchange of the N electron coordinates but otherwise complete. In that case, $P\left[e^{i\mathbf{p}\cdot\mathbf{x}}\right]$ would represent the projection of the plane wave onto that part of Hilbert space corresponding to functions of $\mathbf{x}$ that are antisymmetric with respect to exchange of the N electron coordinates. Neither the expansion shown in equation (7.28) nor that shown in equation (7.27) is point-wise convergent. In other words, we cannot perform the sums shown on the right-hand sides of these equations and expect them to give point-wise convergent representations of the plane wave or its projection. However, the expansions are valid in the sense of distributions. Thus, for example, if

$$f(\mathbf{p}) = \sum_{\nu'} \Phi^t_{\nu'}(\mathbf{p}) c_{\nu'} \tag{7.29}$$

then

$$\frac{1}{(2\pi)^{d/2}} \int d\mathbf{p}\ e^{i\mathbf{p}\cdot\mathbf{x}} f(\mathbf{p}) = \frac{1}{(2\pi)^{d/2}} \int d\mathbf{p}\ e^{i\mathbf{p}\cdot\mathbf{x}} \sum_{\nu'} \Phi^t_{\nu'}(\mathbf{p}) c_{\nu'} \tag{7.30}$$

Replacing the plane wave in (7.30) by the expression given in (7.27), and making use of the reciprocal-space orthonormality relations (7.22), we obtain:

$$\begin{aligned}\frac{1}{(2\pi)^{d/2}} \int d\mathbf{p}\ e^{i\mathbf{p}\cdot\mathbf{x}} f(\mathbf{p}) &= \sum_{\nu',\nu} c_{\nu'} \Phi_\nu(\mathbf{x}) \int d\mathbf{p}\ \Phi^{t*}_\nu(\mathbf{p}) \left(\frac{p_\kappa^2 + p^2}{2p_\kappa^2}\right) \Phi^t_{\nu'}(\mathbf{p}) \\ &= \sum_{\nu',\nu} \Phi_\nu(\mathbf{x}) c_{\nu'} \delta_{\nu',\nu} = \sum_\nu \Phi_\nu(\mathbf{x}) c_\nu \end{aligned} \tag{7.31}$$

This result can be seen to be valid if the Fourier transforms of both sides are compared with equation (7.29).

7.5 Iteration of the Schrödinger equation

We are now in a position to iterate various integral forms of the Schrödinger equation. Using the Fourier convolution theorem (4.61), we can rewrite the momentum-space wave equation in the form

$$(p_\kappa^2 + p^2)\Psi^t_\kappa(\mathbf{p}) = -\frac{2}{(2\pi)^{d/2}} \int d\mathbf{x}\ e^{-i\mathbf{p}\cdot\mathbf{x}} V(\mathbf{x}) \Psi_\kappa(\mathbf{x}) \tag{7.32}$$

Dividing both sides of (7.32) by $(p_\kappa^2 + p^2)$, multiplying by $e^{i\mathbf{p}\cdot\mathbf{x}'}$ and integrating over $d\mathbf{p}$, we have

$$\int d\mathbf{p}\ e^{i\mathbf{p}\cdot\mathbf{x}'}\Psi_\kappa^t(\mathbf{p}) = -\frac{2}{(2\pi)^{d/2}}\int d\mathbf{x} \int d\mathbf{p}\ \frac{e^{i\mathbf{p}\cdot(\mathbf{x}'-\mathbf{x})}}{(p_\kappa^2 + p^2)} V(\mathbf{x})\Psi_\kappa(\mathbf{x}) \quad (7.33)$$

This can be rewritten in the form

$$\Psi_\kappa(\mathbf{x}') = -2\int d\mathbf{x}\ G(\mathbf{x}' - \mathbf{x})V(\mathbf{x})\Psi_\kappa(\mathbf{x}) \quad (7.34)$$

where

$$G(\mathbf{x}' - \mathbf{x}) = \frac{1}{(2\pi)^d}\int d\mathbf{p}\ \frac{e^{i\mathbf{p}\cdot(\mathbf{x}'-\mathbf{x})}}{p_\kappa^2 + p^2} \quad (7.35)$$

If the set of generalized Sturmian basis functions $\Phi_\nu(\mathbf{x})$ is complete, we can substitute the complex conjugate of (7.27) into (7.32). This gives us

$$\Psi_\kappa^t(\mathbf{p}) = -\frac{1}{p_\kappa^2}\sum_\nu \Phi_\nu^t(\mathbf{p})\int d\mathbf{x}\ \Phi_\nu^*(\mathbf{x})V(\mathbf{x})\Psi_\kappa(\mathbf{x}) \quad (7.36)$$

Taking the Fourier transform on both sides of (7.36), we have

$$\Psi_\kappa(\mathbf{x}') = -\frac{1}{p_\kappa^2}\sum_\nu \Phi_\nu(\mathbf{x}')\int d\mathbf{x}\ \Phi_\nu^*(\mathbf{x})V(\mathbf{x})\Psi_\kappa(\mathbf{x}) \quad (7.37)$$

Finally, if we compare (7.37) with (7.34), we can make the identification (in the sense of distributions)

$$G(\mathbf{x}' - \mathbf{x}) = \frac{1}{2p_\kappa^2}\sum_\nu \Phi_\nu(\mathbf{x}')\Phi_\nu^*(\mathbf{x}) \quad (7.38)$$

In other words, for any complete set of generalized Sturmian basis functions, we can make the identification

$$\frac{1}{2p_\kappa^2}\sum_\nu \Phi_\nu(\mathbf{x}')\Phi_\nu^*(\mathbf{x}) = \frac{1}{(2\pi)^d}\int d\mathbf{p}\ \frac{e^{i\mathbf{p}\cdot(\mathbf{x}'-\mathbf{x})}}{p_\kappa^2 + p^2} \quad (7.39)$$

which is valid in the sense of distributions.

Suppose that we now wish to iterate the Schrödinger equation in the form shown in equation (7.37). Substituting an initial solution into the integral on the right-hand side, we obtain a first-iterated solution. This can in turn be substituted into the integral on the right-hand side, yielding a second-iterated solution, and so on.

$$\begin{aligned}
\Psi_\kappa^{(1)}(\mathbf{x}') &= -\frac{1}{p_\kappa^2}\sum_\nu \Phi_\nu(\mathbf{x}')\int d\mathbf{x}\ \Phi_\nu^*(\mathbf{x})V(\mathbf{x})\Psi_\kappa^{(0)}(\mathbf{x}) \\
\Psi_\kappa^{(2)}(\mathbf{x}') &= -\frac{1}{p_\kappa^2}\sum_\nu \Phi_\nu(\mathbf{x}')\int d\mathbf{x}\ \Phi_\nu^*(\mathbf{x})V(\mathbf{x})\Psi_\kappa^{(1)}(\mathbf{x}) \\
\vdots \quad & \quad \vdots \qquad\qquad \vdots
\end{aligned} \quad (7.40)$$

From equation (7.14) we have

$$T_{\nu',\nu} \equiv -\frac{1}{p_\kappa} \int d\mathbf{x}\; \Phi^*_{\nu'}(\mathbf{x}) V(\mathbf{x}) \Phi_\nu(\mathbf{x}) \tag{7.41}$$

If we now let

$$\Psi^{(i)}_\kappa(\mathbf{x}) = \sum_{\nu \in I_i} \Phi_\nu(\mathbf{x}) B^{(i)}_{\nu,\kappa} \tag{7.42}$$

it follows that

$$B^{(i+1)}_{\nu',\kappa} = \frac{1}{p_\kappa} \sum_{\nu \in I_i} T_{\nu',\nu} B^{(i)}_{\nu,\kappa} \qquad \nu' \in I_{i+1} \tag{7.43}$$

Thus the iteration can easily be performed in practice provided that we are able to evaluate the matrix elements $T_{\nu',\nu}$. An initial solution can be obtained by solving the Sturmian secular equations

$$\sum_{\nu \in I_0} \left[T_{\nu',\nu} - p_\kappa \delta_{\nu',\nu}\right] B^{(0)}_{\nu,\kappa} = 0 \tag{7.44}$$

with a truncated basis set contained in the domain I_0. Equation (7.43) can then be used as a criterion for automatic selection of a larger basis set to be used in a more accurate version of the secular equations. If the initial solution has a symmetry corresponding to one of the irreducible representations of the symmetry group of $V(\mathbf{x})$, the first-iterated solution will be of the same symmetry, provided that the domain I_1 is appropriately chosen. Thus iteration can be used to construct symmetry-adapted basis sets. Having found symmetry-adapted coefficients $B^{(i)}_{\nu,\kappa}$ we have the approximate equation:

$$\sum_{\nu \in I_i} \left[T_{\nu',\nu} - \delta_{\nu',\nu} p_\kappa\right] B^{(i)}_{\nu,\kappa} \approx 0 \tag{7.45}$$

or

$$\sum_{\nu \in I_i} T_{\nu',\nu} B^{(i)}_{\nu,\kappa} \approx p_\kappa B^{(i)}_{\nu',\kappa} \tag{7.46}$$

from which it follows that

$$p_\kappa \approx \frac{1}{B^{(i)}_{\nu',\kappa}} \sum_{\nu \in I_i} T_{\nu',\nu} B^{(i)}_{\nu,\kappa} \tag{7.47}$$

The values of p_κ thus obtained will depend slightly on the values of ν'. The degree of variation gives us a measure of the accuracy or inaccuracy of the approximation.

If we substitute (7.28) into (7.32), we can see that equations (7.42)–(7.44) can be used to iterate the N-electron Schrödinger equation even when the generalized Sturmian basis set $\Phi_\nu(\mathbf{x})$ is not complete, but of course the iterated solutions will never leave the part of Sobolev space spanned by the basis set.

7.6 Molecular spectra

It might be asked whether solutions to the many-center one-electron wave equation can be used to study the spectra of molecules and molecular ions with several electrons. To do this, we can try to construct generalized Sturmians analogous to those used to study the spectra of atoms and atomic ions. Let us suppose that we have solved

$$\left[-\frac{1}{2}\nabla_i^2 + \beta_\nu v(\mathbf{x}_i) - \epsilon_j\right]\varphi_j(\mathbf{x}_i) = 0 \qquad i = 1, 2, \ldots, N \tag{7.48}$$

Here i is the index of the electron, while j is an index labeling 1-electron energy parameters ϵ_j and the orbitals corresponding to them. The index ν labels N-electron configurations, and $v(\mathbf{x}_i)$ is the zeroth-order potential experienced by the ith electron. For an N-electron system, we choose the weighting factors β_ν in such a way that

$$\epsilon_1 + \epsilon_2 + \cdots + \epsilon_N = E_\kappa \tag{7.49}$$

Then, if we construct Slater determinants of the form

$$\Phi_\nu(\mathbf{x}) = |\varphi_1\varphi_2\varphi_3\cdots| \tag{7.50}$$

they will be solutions to

$$\left[-\frac{1}{2}\Delta + \beta_\nu V_0(\mathbf{x}) - E_\kappa\right]\Phi_\nu(\mathbf{x}) = 0 \tag{7.51}$$

where

$$\Delta \equiv \nabla_1^2 + \nabla_2^2 + \cdots + \nabla_N^2 \tag{7.52}$$

and

$$V_0(\mathbf{x}) \equiv \sum_{i=1}^{N} v(\mathbf{x}_i) \tag{7.53}$$

The generalized Sturmian basis functions $\Phi_\nu(\mathbf{x})$ will obey a potential-weighted orthonormality relation, and we can normalize them in such a way that they obey

$$\int d\mathbf{x}\ \Phi_{\nu'}^*(\mathbf{x})V_0(\mathbf{x})\Phi_\nu(\mathbf{x}) = \delta_{\nu',\nu}\frac{2E_\kappa}{\beta_\nu} \tag{7.54}$$

We can then use the generalized Sturmian basis set to try to solve the true Schrödinger equation

$$\left[-\frac{1}{2}\Delta + V(\mathbf{x}) - E_\kappa\right]\Psi_\kappa(\mathbf{x}) = 0 \tag{7.55}$$

where

$$V(\mathbf{x}) = V_0(\mathbf{x}) + V'(\mathbf{x}) \tag{7.56}$$

and where V' is the interelectron repulsion term

$$V'(\mathbf{x}) = \sum_{i'>i}^{N} \frac{1}{|\mathbf{x}_i - \mathbf{x}_{i'}|} \tag{7.57}$$

If we let

$$\Psi_\kappa(\mathbf{x}) = \sum_\nu \Phi_\nu(\mathbf{x}) B_{\nu,\kappa} \tag{7.58}$$

we obtain the set of secular equations

$$\sum_\nu \int d\mathbf{x}\ \Phi^*_{\nu'}(\mathbf{x}) \left[V(\mathbf{x}) - \beta_\nu V_0(\mathbf{x})\right] \Phi_\nu(\mathbf{x}) B_{\nu,\kappa} = 0 \tag{7.59}$$

With the help of the potential-weighted orthonormality relations and the definitions

$$T_{\nu',\nu} \equiv -\frac{1}{p_\kappa} \int d\mathbf{x}\ \Phi^*_{\nu'}(\mathbf{x}) V(\mathbf{x}) \Phi_\nu(\mathbf{x}) \tag{7.60}$$

and

$$p_\kappa \equiv \sqrt{-2E_\kappa} \tag{7.61}$$

these secular equations can be written in the familiar form

$$\sum_\nu \left[T_{\nu',\nu} - p_\kappa \delta_{\nu',\nu}\right] B_{\nu,\kappa} = 0 \tag{7.62}$$

Just as in the case of atoms, the nuclear attraction term is diagonal:

$$T^{(0)}_{\nu',\nu} \equiv -\frac{1}{p_\kappa} \int d\mathbf{x}\ \Phi^*_{\nu'}(\mathbf{x}) V_0(\mathbf{x}) \Phi_\nu(\mathbf{x}) = \delta_{\nu',\nu} \frac{p_\kappa}{\beta_\nu} \tag{7.63}$$

For a given configuration Φ_ν and a given energy E_κ, we can find the appropriate weighting factors β_ν in the following way:

(1) Solve equation (7.48) for many values of β_ν and make a table of the ϵ_j values corresponding to the solutions $\varphi_j(\mathbf{x}_i)$

(2) For each configuration $\Phi_\nu(\mathbf{x})$, calculate the value of E_κ for every value of β_ν.

(3) Use interpolation to invert the relationship and find β_ν as a function of E_κ.

(4) Solve (7.62) using $p_\kappa = \sqrt{-2E_\kappa}$ for the nuclear attraction term shown in equation (7.63) and find an improved value of p_κ. If necessary, solve (7.62) again using the improved p_κ/β_ν value for the nuclear attraction term.

7.7 Goscinskian configurations in atomic physics

Atomic calculations using generalized Sturmians are particularly simple. If one chooses $V_0(\mathbf{x})$ to be the nuclear attraction potential,

$$V_0(\mathbf{x}) = -\sum_{j=1}^{N} \frac{Z}{r_j} \tag{7.64}$$

then (as was shown by Goscinski in a pioneering paper [Goscinski, 2003]) an exact solution to equation (7.1) is given by a Slater determinant of one-electron hydrogenlike spin-orbitals:

$$\Phi_\nu(\mathbf{x}) = |\chi_{\mu_1}\chi_{\mu_2}\chi_{\mu_3}\cdots| \tag{7.65}$$

where we set $\nu \equiv (\mu_1, \ldots, \mu_N)$ and $\mu \equiv (n, l, m, m_s)$. The one-electron hydrogenlike spin-orbitals have the familiar form

$$\begin{aligned}\chi_{nlm,+1/2}(\mathbf{x}_j) &= R_{nl}(r_j)Y_{lm}(\theta_j, \phi_j)\alpha(j)\\ \chi_{nlm,-1/2}(\mathbf{x}_j) &= R_{nl}(r_j)Y_{lm}(\theta_j, \phi_j)\beta(j)\end{aligned} \tag{7.66}$$

but they correspond to a weighted nuclear charge $Q_\nu = \beta_\nu Z$ that is characteristic for the configuration. In equation (7.66), Y_{lm} is a spherical harmonic and the radial part of the orbital is defined as

$$R_{nl}(r_j) = \mathcal{N}_{nl}\left(\frac{2Q_\nu r_j}{n}\right)^l e^{-Q_\nu \frac{r_j}{n}} F\left(l+1-n\middle|2l+2\middle|2Q_\nu \frac{r_j}{n}\right) \tag{7.67}$$

with

$$\mathcal{N}_{nl} = \frac{2}{(2l+1)!}\left(\frac{Q_\nu}{n}\right)^{3/2}\sqrt{\frac{(l+n)!}{n(n-l-1)!}} \tag{7.68}$$

and where F is the confluent hypergeometric function ${}_1F_1$:

$$F\left(a|b|x\right) \equiv \sum_{k=0}^{\infty} \frac{a^{\overline{k}}}{b^{\overline{k}}k!}x^k = 1 + \frac{a}{b}\frac{x}{1!} + \frac{a(a+1)}{b(b+1)}\frac{x^2}{2!} + \cdots \tag{7.69}$$

When $a = 1+l-n$ is zero or a negative integer, the series terminates (thus fulfilling the boundary conditions of the hydrogenlike Schrödinger equation) and we have the polynomial

$$F\left(1+l-n\middle|2l+2\middle|2Q_\nu \frac{r_j}{n}\right) = \sum_{k=0}^{n-l-1} \frac{(1+l-n)^{\overline{k}}}{(2l+2)^{\overline{k}}k!}\left(\frac{2Q_\nu}{n}\right)^k r_j^k \tag{7.70}$$

Table 7.1: The first few radial functions of equation (7.67).

$R_{1,0}(r_j) = 2Q_\nu^{3/2} e^{-Q_\nu r_j}$
$R_{2,0}(r_j) = \frac{Q_\nu^{3/2}}{\sqrt{2}} e^{-Q_\nu r_j/2} \left(1 - \frac{Q_\nu r_j}{2}\right)$
$R_{2,1}(r_j) = \frac{Q_\nu^{3/2}}{2\sqrt{6}} e^{-Q_\nu r_j/2} Q_\nu r_j$
$\vdots \quad \vdots \quad \vdots$

In order for the Goscinskian configurations shown in equation (7.65) to fulfill equation (7.1), with $p_\kappa^2 = -2E_\kappa$, one needs to choose the effective charge to be

$$Q_\nu = \beta_\nu Z = \left(\frac{-2E_\kappa}{\frac{1}{n^2} + \frac{1}{n'^2} + \cdots}\right)^{1/2} \tag{7.71}$$

The hydrogenlike one-electron spin-orbitals have the following properties: They obey the hydrogenlike wave equation

$$\left[-\frac{1}{2}\nabla_j^2 + \frac{Q_\nu^2}{2n^2} - \frac{Q_\nu}{r_j}\right] \chi_\mu(\mathbf{x_j}) = 0 \tag{7.72}$$

They are orthonormal

$$\int d\tau_j \ \chi_{\mu'}^*(\mathbf{x}_j)\chi_\mu(\mathbf{x}_j) = \delta_{\mu',\mu} \tag{7.73}$$

and they also obey the virial theorem, which in our case can be written in the form:

$$-\int d\tau_j \ |\chi_\mu(\mathbf{x}_j)|^2 \frac{Q_\nu}{r_j} = -\frac{Q_\nu^2}{n^2} \tag{7.74}$$

Applying the kinetic energy operator to Φ_ν and using equation (7.72) for each χ_μ we have

$$\begin{aligned} &\left[-\frac{1}{2}\sum_{j=1}^{N} \nabla_j^2\right] \Phi_\nu(\mathbf{x}) \\ &= \left[-\left(\frac{Q_\nu^2}{2n^2} + \frac{Q_\nu^2}{2n'^2} + \cdots\right) + \left(\frac{Q_\nu}{r_1} + \frac{Q_\nu}{r_2} + \cdots\right)\right] \Phi_\nu(\mathbf{x}) \\ &= \left[E_\kappa - \beta_\nu V_0(\mathbf{x})\right] \Phi_\nu(\mathbf{x}) \end{aligned} \tag{7.75}$$

showing that Φ_ν is indeed a solution to equation (7.1).

The normalization of the Goscinskian configurations can be found by making use of equations (7.71), (7.73) and (7.74). In this way we obtain

$$\begin{aligned}\int d\tau\ V_0(\mathbf{x})|\Phi_\nu(\mathbf{x})|^2 &= -\sum_{\mu\in\nu}\int d\tau_j\ |\chi_\mu(\mathbf{x}_j)|^2\frac{Z}{r_j}\\ &= -\frac{1}{\beta_\nu}\sum_{\mu\in\nu}\int d\tau_j\ |\chi_\mu(\mathbf{x}_j)|^2\frac{Q_\nu}{r_j}\\ &= -\frac{Q_\nu^2}{\beta_\nu}\sum_{\mu\in\nu}\frac{1}{n^2} = \frac{2E_\kappa}{\beta_\nu}\end{aligned} \tag{7.76}$$

Combining equations (7.76) and (7.5), we obtain a potential-weighted orthonormality relation for the Goscinskian configurations:

$$\int d\tau\ \Phi_{\nu'}^*(\mathbf{x})V_0(\mathbf{x})\Phi_\nu(\mathbf{x}) = \delta_{\nu',\nu}\frac{2E_\kappa}{\beta_\nu} = -\delta_{\nu',\nu}\frac{p_\kappa^2}{\beta_\nu} \tag{7.77}$$

in agreement with equation (7.8). Thus the Goscinskian configurations, constructed in the way that we have described, are already properly normalized.

7.8 Derivation of the secular equations

We are now in a position to derive the generalized Sturmian secular equations appropriate for atoms and ions. In this derivation, it will be convenient to introduce the variable

$$\mathcal{R}_\nu \equiv \sqrt{\frac{1}{n^2}+\frac{1}{n'^2}+\cdots} \tag{7.78}$$

In equation (7.78), the n's are the principal quantum numbers of the hydrogenlike spin-orbitals occurring in the Goscinskian configurations Φ_ν (7.65). In terms of these two new variables, equation (7.71) takes on the simple form:

$$Q_\nu = \beta_\nu Z = \frac{p_\kappa}{\mathcal{R}_\nu} \tag{7.79}$$

If we neglect the motion of the nucleus, spin-orbit coupling and spin-spin coupling, the potential that appears in the non-relativistic Schrödinger equation of an atom or ion without external fields is

$$V(\mathbf{x}) = V_0(\mathbf{x}) + V'(\mathbf{x}) = -\sum_{j=1}^{N}\frac{Z}{r_j} + \sum_{j>i}^{N}\sum_{i=1}^{N}\frac{1}{r_{ij}} \tag{7.80}$$

Table 7.2: ^1S excited state energies (in Hartrees) for the 2-electron isoelectronic series. The basis set used consisted of 40 generalized Sturmians of the Goscinski type. Experimental values are taken from the NIST tables [NIST Atomic Spectra Database] (http://physics.nist.gov/asd). Discrepancies between calculated and experimental energies for the heavier ions are due mainly to relativistic effects.

	He	Li^+	Be^{2+}	B^{3+}	C^{4+}	N^{5+}
$1s2s\ ^1S$	−2.1429	−5.0329	−9.1730	−14.564	−21.206	−29.098
expt.	−2.1458	−5.0410	−9.1860	−14.582	−21.230	−29.131
$1s3s\ ^1S$	−2.0603	−4.7297	−8.5099	−13.402	−19.406	−26.521
expt.	−2.0611	−4.7339	−8.5183	−13.415	−19.425	−26.548
$1s4s\ ^1S$	−2.0332	−4.6276	−8.2837	−13.003	−18.785	−25.629
expt.	−2.0334	−4.6299	−8.2891			−25.654
$1s5s\ ^1S$	−2.0210	−4.5811	−8.1806	−12.820	−18.500	−25.220
expt.	−2.0210	−4.5825				−25.241
$1s6s\ ^1S$	−2.0144	−4.5562	−8.1250	−12.721	−18.346	−24.998
expt.	−2.0144	−4.5571				
$1s7s\ ^1S$	−2.0105	−4.5412	−8.0917	−12.662	−18.253	−24.865
expt.	−2.0104	−4.5418				
$1s8s\ ^1S$	−2.0080	−4.5315	−8.0701	−12.624	−18.194	−24.779
expt.	−2.0079					
$1s9s\ ^1S$	−2.0063	−4.5248	−8.0554	−12.598	−18.153	−24.720
expt.	−2.0062					
$1s10s\ ^1S$	−2.0051	−4.5201	−8.0449	−12.579	−18.124	−24.678
expt.	−2.0050					
$1s11s\ ^1S$	−2.0042	−4.5166	−8.0371	−12.566	−18.102	−24.647
expt.	−2.0041					
$1s12s\ ^1S$	−2.0034	−4.5140	−8.0312	−12.555	−18.086	−24.624
expt.	−2.0034					

where V_0 and V' respectively represent nuclear attraction and interelectron repulsion. We next introduce the matrices

$$T^0_{\nu',\nu} \equiv -\frac{1}{p_\kappa}\int d\tau\ \Phi^*_{\nu'}(\mathbf{x})V_0(\mathbf{x})\Phi_\nu(\mathbf{x}) \tag{7.81}$$

and

$$T'_{\nu',\nu} \equiv -\frac{1}{p_\kappa}\int d\tau\ \Phi^*_{\nu'}(\mathbf{x})V'(\mathbf{x})\Phi_\nu(\mathbf{x}) \tag{7.82}$$

From the potential-weighted orthogonality relation and normalization (7.77) and from equation (7.79) it follows that

$$T^0_{\nu',\nu} = \delta_{\nu'\nu}Z\mathcal{R}_\nu \tag{7.83}$$

Thus the nuclear attraction matrix $T^0_{\nu',\nu}$ defined by equation (7.81) is independent of p_κ. One can show that $T'_{\nu',\nu}$ is also independent of p_κ.

We now substitute the superposition (7.10) into the Schrödinger equation:

$$\sum_\nu \left[-\frac{1}{2}\Delta + V(\mathbf{x}) - E_\kappa\right] \Phi_\nu(\mathbf{x}) B_{\nu,\kappa} = 0 \tag{7.84}$$

Since each of the basis functions Φ_ν obeys (7.1), equation (7.84) can be written in the form

$$\sum_\nu \left[V(\mathbf{x}) - \beta_\nu V_0(\mathbf{x})\right] \Phi_\nu(\mathbf{x}) B_{\nu,\kappa} = 0 \tag{7.85}$$

If we multiply (7.85) from the left by a conjugate function from the basis set and integrate over space and spin coordinates, we obtain

$$\sum_\nu \int d\tau \; \Phi^*_{\nu'}(\mathbf{x}) \left[V(\mathbf{x}) - \beta_\nu V_0(\mathbf{x})\right] \Phi_\nu(\mathbf{x}) B_{\nu,\kappa} = 0 \tag{7.86}$$

With the help of equations (7.80), (7.82) and (7.83), this becomes

$$\sum_\nu \left[-p_\kappa \delta_{\nu',\nu} Z\mathcal{R}_\nu - p_\kappa T'_{\nu',\nu} + \beta_\nu p_\kappa \delta_{\nu',\nu} Z\mathcal{R}_\nu\right] B_{\nu,\kappa} = 0 \tag{7.87}$$

Finally, making use of the fact that

$$\beta_\nu Z\mathcal{R}_\nu = p_\kappa \tag{7.88}$$

changing signs, and dividing by p_κ, we obtain:

$$\sum_\nu \left[\delta_{\nu',\nu} Z\mathcal{R}_\nu + T'_{\nu',\nu} - p_\kappa \delta_{\nu',\nu}\right] B_{\nu,\kappa} = 0 \tag{7.89}$$

Equation (7.89) differs in several remarkable respects from the conventional set of secular equations that we would obtain by diagonalizing the Hamiltonian of an atom or ion:

(1) The kinetic energy term has vanished.
(2) The nuclear attraction matrix is diagonal and energy-independent.
(3) The interelectron repulsion matrix is energy-independent.
(4) The roots are not energies but values of the parameter p_κ, which is proportional to the square roots of the binding energies (since $p_\kappa = \sqrt{-2E_\kappa}$).

Because the effective nuclear charges that characterize the basis functions depend on p_κ, the basis set is not completely known before the secular equations have been solved. Only the form of the basis functions is known, but not their scale. Solving the secular equations gives us at one stroke a spectrum of energies, and a near-optimal set of basis functions for representing the corresponding states. The Slater-exponents that appear in the hydrogenlike orbitals of the Goscinskian configurations turn out to be almost exactly those that would have been found by a variational optimization of the basis. In the generalized Sturmian method, we are able to dispense with the time-consuming variational optimization that would otherwise be needed to find the optimum basis for representing each excited state. This characteristic of the generalized Sturmian method makes it especially suitable for calculating large numbers of excited states of few-electron atoms.

Table 7.2 shows the results of a small calculation on the ^1S excited states of the 2-electron isoelectronic series, compared with experimental results. The calculation took only a few seconds of computational time.

7.9 The Large-Z approximation

When nuclear attraction dominates completely over interelectron repulsion, the term $T'_{\nu',\nu}$ can be neglected in (7.89), we obtain an explicit expression for the energies: [Avery and Avery, 2006]

$$\frac{E_\kappa}{Z^2} \to -\frac{1}{2}\mathcal{R}_\nu^2 = -\frac{1}{2n^2} - \frac{1}{2n'^2} - \frac{1}{2n''^2} - \cdots \tag{7.90}$$

In other words, if interelectron repulsion is totally neglected, the energy levels of an N-electron atom correspond to those of a set of N entirely independent electrons moving in the attractive potential of a nucleus with charge Z. We can introduce the name "$\mathcal{R}$-block" for the set of primitive configurations that are degenerate when interelectron repulsion is neglected. Even when Z is large, interelectron repulsion hybridizes the set of degenerate primitive configurations belonging to a particular value of $\mathcal{R}_\nu$ and very slightly removes the degeneracy, even though in the case of large Z nuclear attraction is a much larger effect than interelectron repulsion.

The large-Z approximation does not neglect interelectron repulsion, but in building up the wave function, it neglects all configurations outside the $\mathcal{R}$-block. In other words, for large Z, the interelectron repulsion matrix elements linking different $\mathcal{R}$-blocks are neglected, and equation (7.89) is solved one block at a time. For each $\mathcal{R}$-block, the term $\delta_{\nu',\nu}Z\mathcal{R}_\nu$ is a multiple of the unit matrix. From linearity, it follows that adding any multiple of the unit matrix to another matrix does not change its eigenvectors, while all its

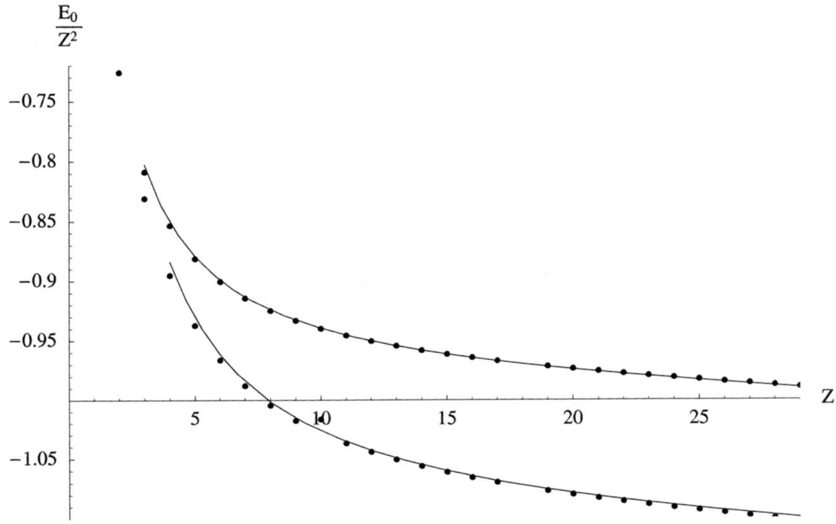

Fig. 7.1: This figure shows the large-Z approximation for the heliumlike and lithiumlike isoelectronic series ground states crudely corrected for relativistic effects. The smooth lines show $E_\kappa/Z^2 \simeq -f_\nu(Z)(Z\mathcal{R}_\nu - |\lambda_\kappa|)^2/(2Z^2)$. Here $f_\nu(Z)$ is the ratio between the energies with and without relativistic effects when one neglects interelectron repulsion. The dots represent experimental values.

roots are shifted by a constant amount. Therefore in the large-Z approximation we can find solutions for (7.89) by diagonalizing the interelectron repulsion matrix for a particular $\mathcal{R}$-block, i.e. by solving

$$\sum_{\nu}{}' \left[T'_{\nu',\nu} - \lambda_\kappa \delta_{\nu',\nu}\right] B_{\nu,\kappa} = 0 \tag{7.91}$$

The prime above the sum in equation (7.91) means that it is taken over all values of ν that correspond to a particular value of $\mathcal{R}_\nu$, i.e. to a particular set of principal quantum numbers $n, n', \ldots$. Thus in the large-Z approximation, the roots to the secular equations (7.89) are the same as those of (7.91), except that they are shifted by a constant amount, $Z\mathcal{R}_\nu$.

$$p_\kappa = Z\mathcal{R}_\nu + \lambda_\kappa \tag{7.92}$$

Then with the help of (7.11) we obtain

$$E_\kappa \approx -\frac{1}{2}(Z\mathcal{R}_\nu + \lambda_\kappa)^2 \tag{7.93}$$

Because of the minus sign in the definition of $T'_{\nu',\nu}$, the roots of the interelectron repulsion matrix are always negative. Thus, as we would expect, the effect of interelectron repulsion is to reduce the binding energy of a state. This point can be made more explicit by writing

$$E_\kappa \approx -\frac{1}{2}(Z\mathcal{R}_\nu - |\lambda_\kappa|)^2 \tag{7.94}$$

The true energy is only approximately equal to the expression on the right of (7.94) because of the restriction of the basis set to a single $\mathcal{R}$-block, but as the nuclear charge increases in an isoelectronic series, the approximation improves.

Equation (7.94) can be applied to any state whatever of an atom or ion in the large-Z approximation. To illustrate this point, we can think of the lowest carbon-like $\mathcal{R}$-block with

$$\mathcal{R}_\nu \equiv \sqrt{\frac{2}{1^2} + \frac{4}{2^2}} = \sqrt{3} \tag{7.95}$$

In other words, we consider the $\mathcal{R}$-block corresponding to a 6-electron configuration with 2 electrons in the $n = 1$ shell and 4 electrons in the $n = 2$ shell.

Because of the Pauli principle the full $\mathcal{R}$-block has a much smaller dimension than $2n^2 \times 2n'^2 \times 2n''^2 \times \cdots$. In fact, the carbon-like $\mathcal{R}$-block gives us a 70×70 matrix to diagonalize in equation (7.94). The dimension of the $\mathcal{R}$-block is given by the binomial coefficient

$$\binom{8}{4} = \frac{8!}{4!(8-4)!} = 70 \tag{7.96}$$

because this is the number of ways to select 4 different spin-orbitals from the 8 available one-electron hydrogenlike functions of the $n = 2$ shell, while there is only one way of putting the 2 remaining electrons into the $n = 1$ shell.

It is interesting to consider the lowest $\mathcal{R}$-blocks for the He-like, Li-like, Be-like and B-like isoelectronic series. These have $\mathcal{R}_\nu$ values and dimensions given respectively by

$$\text{He-like}: \quad \mathcal{R}_\nu = \sqrt{\frac{2}{1^2}} = \sqrt{2} \qquad \binom{8}{0} = \frac{8!}{0!(8-0)!} = 1$$

$$\text{Li-like}: \quad \mathcal{R}_\nu = \sqrt{\frac{2}{1^2} + \frac{1}{2^2}} = \frac{3}{2} \qquad \binom{8}{1} = \frac{8!}{1!(8-1)!} = 8$$

$$\text{Be-like}: \quad \mathcal{R}_\nu = \sqrt{\frac{2}{1^2} + \frac{2}{2^2}} = \frac{\sqrt{10}}{2} \qquad \binom{8}{2} = \frac{8!}{2!(8-2)!} = 28$$

$$\text{B-like}: \quad \mathcal{R}_\nu = \sqrt{\frac{2}{1^2} + \frac{3}{2^2}} = \frac{\sqrt{11}}{2} \qquad \binom{8}{3} = \frac{3!}{3!(8-3)!} = 56$$

The roots $|\lambda_\kappa|$ for these $\mathcal{R}$-blocks are shown in Tables 7.3 and 7.4. Similarly, the lowest $\mathcal{R}$-blocks of the interelectron repulsion matrix for the N-like, O-like, F-like and Ne-like isoelectronic series are characterized respectively by

$$\text{N-like}: \quad \mathcal{R}_\nu = \sqrt{\frac{2}{1^2} + \frac{5}{2^2}} = \frac{\sqrt{13}}{2} \qquad \binom{8}{5} = \frac{8!}{5!(8-5)!} = 56$$

$$\text{O-like}: \quad \mathcal{R}_\nu = \sqrt{\frac{2}{1^2} + \frac{6}{2^2}} = \frac{\sqrt{14}}{2} \qquad \binom{8}{6} = \frac{8!}{6!(8-6)!} = 28$$

$$\text{F-like}: \quad \mathcal{R}_\nu = \sqrt{\frac{2}{1^2} + \frac{7}{2^2}} = \frac{\sqrt{15}}{2} \qquad \binom{8}{7} = \frac{8!}{7!(8-7)!} = 8$$

$$\text{Ne-like}: \quad \mathcal{R}_\nu = \sqrt{\frac{2}{1^2} + \frac{8}{2^2}} = 2 \qquad \binom{8}{8} = \frac{8!}{8!(8-8)!} = 1$$

The roots and symmetries obtained by diagonalizing these $\mathcal{R}$-blocks are shown in Table 7.3. The solid lines in Figure 7.1 show the ground state energies of the heliumlike and lithiumlike isoelectronic series calculated in

Table 7.3: Roots of the lowest $\mathcal{R}$-block of the interelectron repulsion matrix for the He-like, Li-like, Be-like and B-like isoelectronic series.

He-like $\lvert\lambda_\kappa\rvert$	term	Li-like $\lvert\lambda_\kappa\rvert$	term	Be-like $\lvert\lambda_\kappa\rvert$	term	B-like $\lvert\lambda_\kappa\rvert$	term
0.441942	^{1}S	0.681870	^{2}S	0.986172	^{1}S	1.40355	^{2}P
		0.729017	^{2}P	1.02720	^{3}P	1.44095	^{4}P
				1.06426	^{1}P	1.47134	^{2}D
				1.09169	^{3}P	1.49042	^{2}S
				1.10503	^{1}D	1.49395	^{2}P
				1.13246	^{1}S	1.52129	^{4}S
						1.54037	^{2}D
						1.55726	^{2}P

the large-Z approximation, using the equations

$$\begin{aligned} &\text{He-like:} \quad E_\kappa \simeq -\tfrac{1}{2}\left(Z\sqrt{\frac{2}{1^2}} - 0.441942\right)^2 \\ &\text{Li-like:} \quad E_\kappa \simeq -\tfrac{1}{2}\left(Z\sqrt{\frac{2}{1^2}+\frac{1}{2^2}} - 0.68180\right)^2 \end{aligned} \tag{7.97}$$

The roots of the ground state $\mathcal{R}$-block of the interelectron repulsion matrix for these two series correspond respectively to ^{1}S and ^{2}S states. Figure 7.1 shows the experimental ground state energies compared with the expressions in equation (7.94) multiplied by a relativistic correction factor $f_\nu(Z)$. The factor used was the ratio between the relativistic energy of the configuration completely neglecting interelectron repulsion and the

Table 7.4: Roots of the lowest $\mathcal{R}$-block of $T'_{\nu',\nu}$ for the N-like, O-like, F-like and Ne-like isoelectronic series.

N-like $\|\lambda_\kappa\|$	term	O-like $\|\lambda_\kappa\|$	term	F-like $\|\lambda_\kappa\|$	term	Ne-like $\|\lambda_\kappa\|$	term
2.41491	^{4}S	3.02641	^{3}P	3.68415	^{2}P	4.38541	^{1}S
2.43246	^{2}D	3.03769	^{1}D	3.78926	^{2}S		
2.44111	^{2}P	3.05065	^{1}S				
2.49314	^{4}P	3.11850	^{3}P				
2.52109	^{2}D	3.14982	^{1}P				
2.53864	^{2}S	3.24065	^{1}S				
2.54189	^{2}P						
2.61775	^{2}P						

non-relativistic energy $-\frac{1}{2}(Z\mathcal{R}_\nu)^2$ also neglecting interelectron repulsion. The relativistic 1-electron energies, expressed in Hartrees, and not including the rest energy, are given by [Avery and Avery, 2006]

$$\epsilon_\mu = c^2 \left[1 + \left(\frac{Z}{c(\gamma + n - |j + 1/2|)}\right)^2\right]^{-1/2} - c^2 \qquad (7.98)$$

$$\gamma \equiv \sqrt{\left(j + \frac{1}{2}\right)^2 - \left(\frac{Z}{c}\right)^2} \qquad c = 137.036 \qquad (7.99)$$

where j is the total angular momentum (orbital plus spin) of a single electron, i.e. $l \pm \frac{1}{2}$. In the general case of an N-electron atom, it would be more complicated to construct symmetry-adapted basis functions of the Russell-Saunders type by means of group theoretical coupling coefficients.

Fortunately this is not necessary. We shall see that general symmetry-adapted basis sets can be derived by considering the large-Z approximation.

7.10 Exercises

(1) Type `?strwf` to obtain a definition of the function which generates the d-dimensional Coulomb Sturmian radial wave function. Letting $d = 3$ and $k = Q/n$, make a table of `strwf[3,n,l,Q/n]` for $n = 1, 2, 3, 4$, and $l = 0, 1, ..n - 1$, Check the functions in Table 7.1 and extend the table to $n = 4$.

(2) Use the values in Table 7.4 to generate the energies of the lowest ^{2}P and ^{2}S states of the Fluorine-like isoelectronic series in the large-Z approximation.

(3) Repeat for the ground states and lowest ^{1}S excited states of the Neon-like isoelectronic series.

(4) Use equations (7.98) and (7.99) to make plots of ϵ_μ/Z^2 for the case where $n = 1$, for $j = 1/2$, and for the case where $n = 2$ with $j = 1/2$ and $j = 3/2$. In the plots, show the non-relativistic energy for comparison.

(5) Use equation (7.97) to make a table of ground-state energies for the Helium-like isoelectronic series in the large-Z approximation. Show, for comparison, a table of the energies roughly corrected for relativistic effects.

Chapter 8

CHOOSING APPROPRIATE HYPERSPHERICAL REPRESENTATIONS

8.1 The alternative Coulomb Sturmians of Aquilanti and Coletti

The mathematical theory of alternative orthonormal sets of 4-dimensional hyperspherical harmonics and their associated alternative hydrogenlike Coulomb Sturmians has been developed by Professors V. Aquilanti and C. Coletti and their co-workers [Aquilanti et al., 1996]–[Aquilanti and Avery, 2001], [Caligiana, 2003]. We will now discuss how the alternative Coulomb Sturmians of Aquilanti and Coletti may be applied to calculations of molecular electronic structure, and especially, how the relevant integrals may be evaluated.

Let

$$\mathcal{Y}_{\lambda,\alpha,\beta}(\mathbf{u}) = \sum_{l,m} Y_{\lambda,l,m}(\mathbf{u})\ U^{(\lambda)}_{l,m;\alpha,\beta} \tag{8.1}$$

be a set of 4-dimensional hyperspherical harmonics related to the standard hyperspherical harmonics by a unitary transformation. Then

$$\begin{aligned}
\int d\Omega_4\ Y^*_{\lambda,l,m}(\mathbf{u})\mathcal{Y}_{\lambda,\alpha,\beta}(\mathbf{u}) &= \sum_{l',m'} \int d\Omega_4\ Y^*_{\lambda,l,m}(\mathbf{u}) Y_{\lambda,l',m'}(\mathbf{u})\ U^{(\lambda)}_{l',m';\alpha,\beta} \\
&= \sum_{l',m'} \delta_{l',l}\delta_{m',m}\ U^{(\lambda)}_{l',m';\alpha,\beta} \\
&= U^{(\lambda)}_{l,m;\alpha,\beta}
\end{aligned} \tag{8.2}$$

Because of the unitarity of the transformation, the alternative hyperspherical harmonics will satisfy the orthonormality relationship

$$\int d\Omega_4\ \mathcal{Y}^*_{\lambda',\alpha',\beta'}(\mathbf{u})\mathcal{Y}_{\lambda,\alpha,\beta}(\mathbf{u}) = \delta_{\lambda',\lambda}\delta_{\alpha',\alpha}\delta_{\beta',\beta} \tag{8.3}$$

The set of orbitals related to the transformed hyperspherical harmonics by

$$\begin{aligned}
\tilde{\chi}_{n,\alpha,\beta}(\mathbf{x}) &= \frac{1}{(2\pi)^{3/2}} \int d\mathbf{p}\; e^{i\mathbf{p}\cdot\mathbf{x}} M(p) \mathcal{Y}_{n-1,\alpha,\beta}(\mathbf{u}) \\
&= \sum_{l,m} \left(\frac{1}{(2\pi)^{3/2}} \int d\mathbf{p}\; e^{i\mathbf{p}\cdot\mathbf{x}} M(p) Y_{\lambda,l,m}(\mathbf{u}) \right) U^{(\lambda)}_{l,m;\alpha,\beta} \\
&= \sum_{l,m} \chi_{n,l,m}(\mathbf{x})\; U^{(\lambda)}_{l,m;\alpha,\beta} \\
\tilde{\chi}^t_{n,\alpha,\beta}(\mathbf{p}) &= \frac{1}{(2\pi)^{3/2}} \int d\mathbf{x}\; e^{-i\mathbf{p}\cdot\mathbf{x}}\; \tilde{\chi}_{n,\alpha,\beta}(\mathbf{x}) \\
&= \sum_{l,m} \left(\frac{1}{(2\pi)^{3/2}} \int d\mathbf{x}\; e^{-i\mathbf{p}\cdot\mathbf{x}}\; \chi_{n,l,m}(\mathbf{x}) \right) U^{(\lambda)}_{l,m;\alpha,\beta} \\
&= \sum_{l,m} \left(M(p) Y_{\lambda,l,m}(\mathbf{u}) \right) U^{(\lambda)}_{l,m;\alpha,\beta} \\
&= M(p) \mathcal{Y}_{n-1,\alpha,\beta}(\mathbf{u})
\end{aligned} \tag{8.4}$$

where $\lambda = n - 1$ and

$$M(p) \equiv \frac{4k^{5/2}}{(k^2 + p^2)^2} \tag{8.5}$$

will form an alternative Coulomb Sturmian basis set. Such a set obeys the hydrogenlike Sturmian wave equation

$$\left[-\frac{1}{2}\nabla^2 - \frac{nk}{r} + \frac{k^2}{2} \right] \tilde{\chi}_{n,\alpha,\beta}(\mathbf{x}) = 0 \tag{8.6}$$

since $\tilde{\chi}_{n,\alpha,\beta}(\mathbf{x})$ is a linear combination of terms, each of which obeys the Sturmian wave equation:

$$\begin{aligned}
&\left[-\frac{1}{2}\nabla^2 - \frac{nk}{r} + \frac{k^2}{2} \right] \tilde{\chi}_{n,\alpha,\beta}(\mathbf{x}) \\
&= \sum_{l,m} \left(\left[-\frac{1}{2}\nabla^2 - \frac{nk}{r} + \frac{k^2}{2} \right] \chi_{n,l,m}(\mathbf{x}) \right) U^{(\lambda)}_{l,m;\alpha,\beta} = 0
\end{aligned} \tag{8.7}$$

We now introduce a large, block-diagonal unitary transformation matrix of the form:

$$U_{n',l',m';n,\alpha,\beta} \equiv \delta_{n',n} U^{(n-1)}_{l,m;\alpha,\beta} \tag{8.8}$$

where we remember that $n - 1 = \lambda$. The alternative Coulomb Sturmian functions are then related to the standard Coulomb Sturmians by

$$\tilde{\chi}_{n,\alpha,\beta}(\mathbf{x}) = \sum_{n',l',m'} \chi_{n',l',m'}(\mathbf{x})\; U_{n',l',m';n,\alpha,\beta} \tag{8.9}$$

We would like to show that the alternative Coulomb Sturmians obey the same potential-weighted orthonormality relations as the standard Coulomb Sturmians. In other words, we would like to show that

$$\int d\mathbf{x}\ \tilde{\chi}^*_{n',\alpha',\beta'}(\mathbf{x})\frac{1}{r}\tilde{\chi}_{n,\alpha,\beta}(\mathbf{x}) = \frac{k}{n}\delta_{n',n}\delta_{\alpha',\alpha}\delta_{\beta',\beta} \tag{8.10}$$

To show this, we make use of equation (8.9) and write:

$$\begin{aligned}
&\int d\mathbf{x}\ \tilde{\chi}^*_{n',\alpha',\beta'}(\mathbf{x})\frac{1}{r}\tilde{\chi}_{n,\alpha,\beta}(\mathbf{x}) \\
&= \sum_{n'',l'',m''}\sum_{n,l,m} U^*_{n'',l'',m'';n',\alpha',\beta'}\ U_{n,l,m;n,\alpha,\beta} \\
&\times \int d\mathbf{x}\ \chi^*_{n'',l'',m''}(\mathbf{x})\frac{1}{r}\chi_{n,l,m}(\mathbf{x}) \\
&= \sum_{n'',l'',m''}\sum_{n,l,m} U^*_{n'',l'',m'';n',\alpha',\beta'}\ U_{n,l,m;n,\alpha,\beta}\ \frac{k}{n}\delta_{n'',n}\delta_{l'',l}\delta_{m'',m} \\
&= \sum_{n,l,m} U^*_{n,l,m;n',\alpha',\beta'}\ U_{n,l,m;n,\alpha,\beta}\ \frac{k}{n} \\
&= \sum_{l,m} U^{*(n-1)}_{l,m;\alpha',\beta'}\ U^{(n-1)}_{l,m;\alpha,\beta}\ \frac{k}{n} \\
&= \frac{k}{n}\delta_{\alpha',\alpha}\delta_{\beta',\beta}
\end{aligned} \tag{8.11}$$

Because they span the same Hilbert subspace as the standard Coulomb Sturmians, the alternative Coulomb Sturmians have the same completeness properties: Any function that can be expanded in terms of Coulomb Sturmians can also be expanded in terms of alternative Coulomb Sturmians. Thus if we can write

$$f(\mathbf{x}) = \sum_{n,l,m} \chi_{n,l,m}(\mathbf{x})\ a_{n,l,m} \tag{8.12}$$

then we can also write

$$f(\mathbf{x}) = \sum_{n,\alpha,\beta} \tilde{\chi}_{n,\alpha,\beta}(\mathbf{x})\ \tilde{a}_{n,\alpha,\beta} \tag{8.13}$$

From the potential weighted orthonormality relation, we have

$$a_{n,l,m} = \frac{n}{k}\int d\mathbf{x}\ \chi^*_{n,l,m}(\mathbf{x})\frac{1}{r}f(\mathbf{x}) \tag{8.14}$$

and similarly

$$\tilde{a}_{n,\alpha,\beta} = \frac{n}{k}\int d\mathbf{x}\ \tilde{\chi}^*_{n,\alpha,\beta}(\mathbf{x})\frac{1}{r}f(\mathbf{x}) \tag{8.15}$$

There is a relationship between $a_{n,l,m}$ and $\tilde{a}_{n,\alpha,\beta}$ because

$$\tilde{\chi}^*_{n,\alpha,\beta}(\mathbf{x}) = \sum_{l,m} \chi^*_{n,l,m}(\mathbf{x})\; U^{*(\lambda)}_{l,m;\alpha,\beta} \tag{8.16}$$

and therefore

$$\tilde{a}_{n,\alpha,\beta} = \sum_{l,m} a_{n,l,m}\; U^{*(\lambda)}_{l,m;\alpha,\beta} \tag{8.17}$$

The momentum-space orthonormality relation obeyed by Coulomb Sturmians is

$$\int d\mathbf{p}\; \chi^{*t}_{n',l',m'}(\mathbf{p}) \left(\frac{p^2+k^2}{2k^2}\right) \chi^t_{n,l,m}(\mathbf{p}) = \delta_{n',n}\delta_{\alpha',\alpha}\delta_{\beta',\beta} \tag{8.18}$$

In a manner similar to equation (8.9), the Fourier-transformed alternative Coulomb Sturmians can be shown to obey

$$\int d\mathbf{p}\; \tilde{\chi}^{*t}_{n',\alpha',\beta'}(\mathbf{p}) \left(\frac{p^2+k^2}{2k^2}\right) \chi^t_{n,\alpha,\beta}(\mathbf{p}) = \delta_{n',n}\delta_{\alpha',\alpha}\delta_{\beta',\beta} \tag{8.19}$$

The momentum-space orthonormality relation for Coulomb Sturmians can be used to derive an expansion of a plane wave in terms of Coulomb Sturmians and their Fourier transforms [Avery and Avery, 2006]:

$$\begin{aligned} e^{i\mathbf{p}\cdot\mathbf{x}} &= (2\pi)^{3/2}\left(\frac{p^2+k^2}{2k^2}\right)\sum_{n,l,m}\chi^{*t}_{n,l,m}(\mathbf{p})\chi_{n,l,m}(\mathbf{x}) \\ &= (2\pi)^{3/2}\left(\frac{p^2+k^2}{2k^2}\right)\sum_{n,l,m}\chi_{n,l,m}(\mathbf{x})M(p)Y^*_{n-1,l,m}(\mathbf{u}_p) \\ &= \left(\frac{2\pi}{k}\right)^{3/2}\sum_{n,l,m}\chi_{n,l,m}(\mathbf{x})(1+u_{4(p)})Y^*_{n-1,l,m}(\mathbf{u}_p) \end{aligned} \tag{8.20}$$

This expansion is not absolutely convergent, but it is valid in the sense of distributions. The momentum-space orthonormality relations (34) can also be used to derive a similar relation for alternative Coulomb Sturmian basis sets:

$$\begin{aligned} e^{i\mathbf{p}\cdot\mathbf{x}} &= (2\pi)^{3/2}\left(\frac{p^2+k^2}{2k^2}\right)\sum_{n,\alpha,\beta}\tilde{\chi}^{*t}_{n,\alpha,\beta}(\mathbf{p})\tilde{\chi}_{n,\alpha,\beta}(\mathbf{x}) \\ &= (2\pi)^{3/2}\left(\frac{p^2+k^2}{2k^2}\right)\sum_{n,\alpha,\beta}\tilde{\chi}_{n,\alpha,\beta}(\mathbf{x})M(p)\mathcal{Y}^*_{n-1,\alpha,\beta}(\mathbf{u}_p) \\ &= \left(\frac{2\pi}{k}\right)^{3/2}\sum_{n,\alpha,\beta}\tilde{\chi}_{n,\alpha,\beta}(\mathbf{x})(1+u_{4(p)})\mathcal{Y}^*_{n-1,\alpha,\beta}(\mathbf{u}_p) \end{aligned} \tag{8.21}$$

In the form originally studied by Shibuya and Wulfman [Shibuya and Wulfman, 1965], the Shibuya-Wulfman integrals can be written as:

$$\mathfrak{S}_{n',l',m';n,l,m}(\mathbf{R}) \equiv \int d\Omega_4 \; e^{i\mathbf{p}\cdot\mathbf{R}} \; Y^*_{n'-1,l',m'}(\mathbf{u}) Y_{n-1,l,m}(\mathbf{u}) \tag{8.22}$$

where $\mathbf{R} \equiv \mathbf{X}_{a'} - \mathbf{X}_a$. Substituting our Sturmian expansion of a plane wave into this definition, we obtain:

$$\mathfrak{S}_{n',l',m';n,l,m}(\mathbf{R}) = \left(\frac{2\pi}{k}\right)^{3/2} \sum_{n'',l'',m''} \chi_{n'',l'',m''}(\mathbf{R}) \times \int d\Omega_4 \; (1+u_4) Y^*_{n''-1,l'',m''}(\mathbf{u}) Y^*_{n'-1,l',m'}(\mathbf{u}) Y_{n-1,l,m}(\mathbf{u}) \tag{8.23}$$

If we introduce the definition:

$$\tilde{\mathfrak{S}}_{n',\alpha',\beta';n,\alpha,\beta}(\mathbf{R}) \equiv \int d\Omega_4 \; e^{i\mathbf{p}\cdot\mathbf{R}} \mathcal{Y}^*_{n'-1,\alpha',\beta'}(\mathbf{u}) \mathcal{Y}_{n-1,\alpha,\beta}(\mathbf{u}) \tag{8.24}$$

then our plane-wave expansion in terms of the alternative hyperspherical harmonics yields

$$\tilde{\mathfrak{S}}_{n',\alpha',\beta';n,\alpha,\beta}(\mathbf{R}) = \left(\frac{2\pi}{k}\right)^{3/2} \sum_{n'',\alpha'',\beta''} \tilde{\chi}_{n'',\alpha'',\beta''}(\mathbf{R}) \times \int d\Omega_4 \; (1+u_4) \mathcal{Y}^*_{n''-1,\alpha'',\beta''}(\mathbf{u}) \mathcal{Y}^*_{n'-1,\alpha',\beta'}(\mathbf{u}) \mathcal{Y}_{n-1,\alpha,\beta}(\mathbf{u}) \tag{8.25}$$

8.2 Transformations between matrix representations

Since

$$\begin{aligned} \mathcal{Y}^*_{n'-1,\alpha',\beta'}(\mathbf{u}) &= \sum_{l',m'} U^{*(n'-1)}_{\alpha',\beta',l',m'} Y^*_{n'-1,l',m'}(\mathbf{u}) \\ \mathcal{Y}_{n-1,\alpha,\beta}(\mathbf{u}) &= \sum_{l,m} U^{(n-1)}_{\alpha,\beta,l,m} Y_{n-1,l,m}(\mathbf{u}) \end{aligned} \tag{8.26}$$

it follows that

$$\tilde{\mathfrak{S}}_{n',\alpha',\beta';n,\alpha,\beta}(\mathbf{S}) = \sum_{l',m'} \sum_{l,m} U^{*(n'-1)}_{\alpha',\beta',l',m'} \; \mathfrak{S}_{n',l',m';n,l,m}(\mathbf{S}) \; U^{(n-1)}_{\alpha,\beta,l,m} \tag{8.27}$$

and similarly the overlap matrix for alternative Coulomb Sturmians obeys the relationship

$$\tilde{m}_{n',\alpha',\beta';n,\alpha,\beta}(\mathbf{S}) = \sum_{l',m'} \sum_{l,m} U^{*(n'-1)}_{\alpha',\beta',l',m'} \; m_{n',l',m';n,l,m}(\mathbf{S}) \; U^{(n-1)}_{\alpha,\beta,l,m} \tag{8.28}$$

where $m_{n',l',m';n,l,m}(\mathbf{S})$ is the overlap matrix in terms of standard Coulomb Sturmians

8.3 An example: Alternative Coulomb Sturmians corresponding to real spherical harmonics

As is well known, the real spherical harmonics, $\mathcal{Y}_{l,\bar{m}}$, are related to the complex ones by

$$\mathcal{Y}_{l,\bar{m}} = \begin{cases} \frac{1}{\sqrt{2}}\left(Y_{l,\bar{m}} + (-1)^{\bar{m}} Y_{l,-\bar{m}}\right) & \bar{m} > 0 \\ Y_{l,\bar{m}} & \bar{m} = 0 \\ \frac{1}{i\sqrt{2}}\left(Y_{l,-\bar{m}} - (-1)^{\bar{m}} Y_{l,\bar{m}}\right) & \bar{m} < 0 \end{cases} \tag{8.29}$$

They obey the orthonormality relations

$$\int d\Omega_3 \; \mathcal{Y}^*_{l',\bar{m}'}(\mathbf{u}) \mathcal{Y}_{l,\bar{m}}(\mathbf{u}) = \delta_{l',l}\delta_{\bar{m}',\bar{m}} \tag{8.30}$$

and the sum rule

$$\sum_{\bar{m}=-l}^{l} \mathcal{Y}^*_{l,\bar{m}}(\mathbf{u}) \mathcal{Y}_{l,\bar{m}}(\mathbf{u}_p) = \frac{2l+1}{4\pi} P_l(\mathbf{u}\cdot\mathbf{u}_p) \tag{8.31}$$

Thus the expansion of a plane wave

$$e^{i\mathbf{p}\cdot\mathbf{x}} = \sum_{l=0}^{\infty} i^l (2l+1) j_l(pr) P_l(\mathbf{u}\cdot\mathbf{u}_p) \tag{8.32}$$

can be written in the form

$$e^{i\mathbf{p}\cdot\mathbf{x}} = 4\pi \sum_{l=0}^{\infty} \sum_{\bar{m}=-l}^{l} i^l j_l(pr) \mathcal{Y}^*_{l,\bar{m}}(\mathbf{u}) \mathcal{Y}_{l,\bar{m}}(\mathbf{u}_p) \tag{8.33}$$

We can see that many of the usual theorems for spherical harmonics hold for their real form.

Now consider a Coulomb Sturmian whose angular part is a complex spherical harmonic

$$\chi_{n,l,m}(\mathbf{x}) = R_{n,l}(r)Y_{l,m}(\mathbf{u}) \tag{8.34}$$

We know that its Fourier transform is related to the usual 4-dimensional hyperspherical harmonics by

$$\begin{aligned}\chi^t_{n,l,m}(\mathbf{p}) &= \frac{1}{(2\pi)^{3/2}} \int d\mathbf{x}\; e^{-i\mathbf{p}\cdot\mathbf{x}} R_{n,l}(r)Y_{l,m}(\mathbf{u}) \\ &= M(p)Y_{n-1,l,m}(\mathbf{u})\end{aligned} \tag{8.35}$$

It follows that if we let

$$\mathcal{Y}_{\lambda,l,\bar{m}} = \begin{cases} \frac{1}{\sqrt{2}}\left(Y_{\lambda,l,\bar{m}} + (-1)^{\bar{m}} Y_{\lambda,l,-\bar{m}}\right) & \bar{m} > 0 \\ Y_{\lambda,l,\bar{m}} & \bar{m} = 0 \\ \frac{1}{i\sqrt{2}}\left(Y_{\lambda,l,-\bar{m}} - (-1)^{\bar{m}} Y_{\lambda,l,\bar{m}}\right) & \bar{m} < 0 \end{cases} \tag{8.36}$$

then

$$\frac{1}{(2\pi)^{3/2}} \int d\mathbf{x}\; e^{-i\mathbf{p}\cdot\mathbf{x}} R_{n,l}(r)\mathcal{Y}_{l,\bar{m}}(\mathbf{u}) = M(p)\mathcal{Y}_{n-1,l,\bar{m}}(\mathbf{u}) \tag{8.37}$$

The set of 4-dimensional hyperspherical harmonics $\mathcal{Y}_{\lambda,l,\bar{m}}$ are sometimes imaginary, although they are related to the real 3-dimensional spherical harmonics. Their complex conjugates are given by

$$\mathcal{Y}^*_{\lambda,l,\bar{m}} = \begin{cases} \frac{1}{\sqrt{2}}\left(Y_{\lambda,l,\bar{m}} + (-1)^{\bar{m}} Y_{\lambda,l,-\bar{m}}\right) & \bar{m} > 0 \\ Y_{\lambda,l,m} & \bar{m} = 0 \\ \frac{-1}{i\sqrt{2}}\left(Y_{\lambda,l,-\bar{m}} - (-1)^{\bar{m}} Y_{\lambda,l,\bar{m}}\right) & \bar{m} < 0 \end{cases} \tag{8.38}$$

The 4-dimensional hyperspherical harmonics related to the standard ones by the transformation shown above obey the orthonormality relations

$$\int d\Omega_4\; \mathcal{Y}^*_{\lambda',l',\bar{m}'}(\mathbf{u})\mathcal{Y}_{\lambda,l,\bar{m}}(\mathbf{u}) = \delta_{\lambda',\lambda}\delta_{l',l}\delta_{\bar{m}',\bar{m}} \tag{8.39}$$

Since the transformation preserves orthonormality, it must be unitary. Thus all of the remarks of the preceding section hold for the set of Coulomb

Sturmians based on real spherical harmonics. These are defined by the relationship:

$$\begin{aligned}\tilde{\chi}_{n,l,\bar{m}}(\mathbf{x}) &= \frac{1}{(2\pi)^{3/2}}\int d\mathbf{p}\; e^{i\mathbf{p}\cdot\mathbf{x}}M(p)\mathcal{Y}_{n-1,l,\bar{m}}(\mathbf{u}) \\ &= R_{n,l}(r)\mathcal{Y}_{l,\bar{m}}(\mathbf{u})\end{aligned} \tag{8.40}$$

where $R_{n,l}(r)$ is the usual Coulomb Sturmian wave function, and where the angular functions $\mathcal{Y}_{l,\bar{m}}(\mathbf{u})$ are the real spherical harmonics. The unitary transformation matrix which takes us from the standard Coulomb Sturmians to the real Coulomb Sturmians is given by

$$U^{(\lambda)}_{l,m;l',\bar{m}} = \int d\Omega_4 Y^*_{\lambda,l,m}(\mathbf{u})\mathcal{Y}_{\lambda,l',\bar{m}}(\mathbf{u}) \tag{8.41}$$

Then, in terms of this unitary transformation matrix, we can write:

$$\tilde{\chi}_{n,l',\bar{m}}(\mathbf{x}) = \sum_{l,m}\chi_{n,l,m}(\mathbf{x})U^{(n-1)}_{l,m;l',\bar{m}} \tag{8.42}$$

When, for example, $n = 2$ and $\lambda = 1$, the unitary transformation matrix is:

$$U^{(1)}_{l,m;l',\bar{m}} = \begin{pmatrix} 1 & 0 & 0 & 0 \\ 0 & \frac{-i}{\sqrt{2}} & 0 & \frac{-1}{\sqrt{2}} \\ 0 & 0 & 1 & 0 \\ 0 & \frac{-i}{\sqrt{2}} & 0 & \frac{1}{\sqrt{2}} \end{pmatrix} \tag{8.43}$$

The perturbation method for generating alternative Sturmians

We can generate alternative Sturmian basis sets by diagonalizing matrices of perturbations in a basis consisting of standard Coulomb Sturmians corresponding to a particular value of n. For example, we might introduce the perturbation:

$$v'(\mathbf{x}) = z \tag{8.44}$$

Then if we diagonalize the Hermitian matrix

$$v^{(\lambda)}_{l',m';l,m} \equiv \int d\mathbf{x}\; \chi^*_{n,l',m'}(\mathbf{x})\frac{v'(\mathbf{x})}{r}\chi_{n,l,m}(\mathbf{x}) \tag{8.45}$$

the set of eigenvectors form a unitary transformation which takes us from the standard Coulomb Sturmians to an alternative set which are adapted to the symmetry of the potential (as has been discussed in our recent book, *Symmetry-Adapted Basis Sets: Automatic Generation for Problems in Physics and Chemistry*, World Scientific, (2012) [Avery, Rettrup and Avery, 2012]). In the case of the simple potential shown in equation (8.44), we are led to the parabolic Coulomb Sturmians discussed by Aquilanti, Calderini and Coletti. We could, however, introduce a potential of the form:

$$\begin{aligned} v'(\mathbf{x}) &= \frac{1}{|\mathbf{x}-\mathbf{X}|} \\ &= \sum_{l=0}^{\infty} \frac{r_<^l}{r_>^{l+1}} P_l(\mathbf{u}\cdot\mathbf{u}_X) \\ &= \frac{4\pi}{2l+1} \sum_{l=0}^{\infty} \frac{r_<^l}{r_>^{l+1}} \sum_{m=-l}^{l} Y_{l,m}^*(\mathbf{u}) Y_{l,m}(\mathbf{u}_X) \end{aligned} \tag{8.46}$$

In that case, the matrix to be diagonalized would have the form:

$$\begin{aligned} &\int d\mathbf{x}\ \chi_{n,l',m'}^*(\mathbf{x})\ \frac{v'(\mathbf{x})}{r}\ \chi_{n,l,m}(\mathbf{x}) \\ &= \sum_{l''=0}^{\infty} \frac{4\pi}{2l''+1} \int_0^\infty dr\ r\ \frac{r_<^{l''}}{r_>^{l''+1}} R_{n,l'}(r) R_{n,l}(r) \\ &\times \sum_{m=-l''}^{l''} Y_{l'',m''}(\mathbf{u}_X) \int d\Omega_3\ Y_{l'',m''}^*(\mathbf{u}) Y_{l',m'}^*(\mathbf{u}) Y_{l,m}(\mathbf{u}) \\ &\equiv 4\pi \sum_{l''=|l-l'|}^{l+l'} a_{l'',l',l}^{(n)}\ Y_{l'',m-m'}(\mathbf{u}_X) \int d\Omega_3\ Y_{l'',m-m'}^*(\mathbf{u}) Y_{l',m'}^*(\mathbf{u}) Y_{l,m}(\mathbf{u}) \end{aligned} \tag{8.47}$$

where

$$\begin{aligned} (2l''+1)\ a_{l'',l',l}^{(n)} &\equiv \int_0^\infty dr\ r\ \frac{r_<^{l''}}{r_>^{l''+1}} R_{n,l'}(r) R_{n,l}(r) \\ &\equiv \frac{1}{|\mathbf{X}|^{l''+1}} \int_0^{|\mathbf{X}|} dr\ r^{l''+1}\ R_{n,l'}(r) R_{n,l}(r) \\ &+ |\mathbf{X}|^{l''} \int_{|\mathbf{X}|}^\infty dr\ r^{-l''}\ R_{n,l'}(r) R_{n,l}(r) \end{aligned} \tag{8.48}$$

In the particular case where $\hat{\mathbf{X}} = (0,0,1)$, equation (49) reduces to

$$\int d\mathbf{x}\; \chi^*_{n,l',m'}(\mathbf{x})\; \frac{v'(\mathbf{x})}{r}\; \chi_{n,l,m}(\mathbf{x})$$
$$= 4\pi\delta_{m',m} \sum_{l''=|l-l'|}^{l+l'} a^{(n)}_{l'',l',l}\; Y_{l'',0}((0,0,1)) \int d\Omega_3\; Y^*_{l'',0}(\mathbf{u})Y^*_{l',m'}(\mathbf{u})Y_{l,m}(\mathbf{u}) \tag{8.49}$$

Interestingly, this perturbation leads to elliptic Coulomb Sturmians with one focus at the origin of the coordinate system and the other focus at the distance $|\mathbf{X}|$ along the z-axis. We can, if we like, introduce a more complicated perturbation:

$$v'(\mathbf{x}) = \sum_j \frac{1}{|\mathbf{x} - \mathbf{X}_j|}$$
$$= \frac{4\pi}{2l+1} \sum_{l=0}^{\infty} \frac{r_<^l}{r_>^{l+1}} \sum_{m=-l}^{l} Y^*_{l,m}(\mathbf{u}) \sum_j Y_{l,m}(\mathbf{u}_j) \tag{8.50}$$

Then, if the distance $|\mathbf{X}_j|$ is the same for all values of j, the matrix to be diagonalized will have the form

$$\int d\mathbf{x}\; \chi^*_{n,l',m'}(\mathbf{x})\; \frac{v'(\mathbf{x})}{r}\; \chi_{n,l,m}(\mathbf{x})$$
$$= 4\pi \sum_{l''=|l-l'|}^{l+l'} a^{(n)}_{l'',l',l} \sum_j Y_{l'',m-m'}(\mathbf{u}_j) \int d\Omega\; Y^*_{l'',m-m'}(\mathbf{u})Y^*_{l',m'}(\mathbf{u})Y_{l,m}(\mathbf{u}) \tag{8.51}$$

As an example, we can consider an atom surrounded by 8 equal point charges q at the positions

$$\begin{aligned}
\mathbf{u}_1 &= (+1,+1,+1)/\sqrt{3} & \mathbf{u}_5 &= (+1,-1,-1)/\sqrt{3}\\
\mathbf{u}_2 &= (-1,+1,+1)/\sqrt{3} & \mathbf{u}_6 &= (-1,+1,-1)/\sqrt{3}\\
\mathbf{u}_3 &= (+1,-1,+1)/\sqrt{3} & \mathbf{u}_7 &= (-1,-1,+1)/\sqrt{3}\\
\mathbf{u}_4 &= (+1,+1,-1)/\sqrt{3} & \mathbf{u}_8 &= (-1,-1,-1)/\sqrt{3}
\end{aligned} \tag{8.52}$$

i.e., on the corners of a cube. We can also discuss the effect of a square-planar array

$$\begin{aligned}
\mathbf{u}_1 &= (+1,0,0) & \mathbf{u}_3 &= (0,+1,0)\\
\mathbf{u}_2 &= (-1,0,0) & \mathbf{u}_4 &= (0,-1,0)
\end{aligned} \tag{8.53}$$

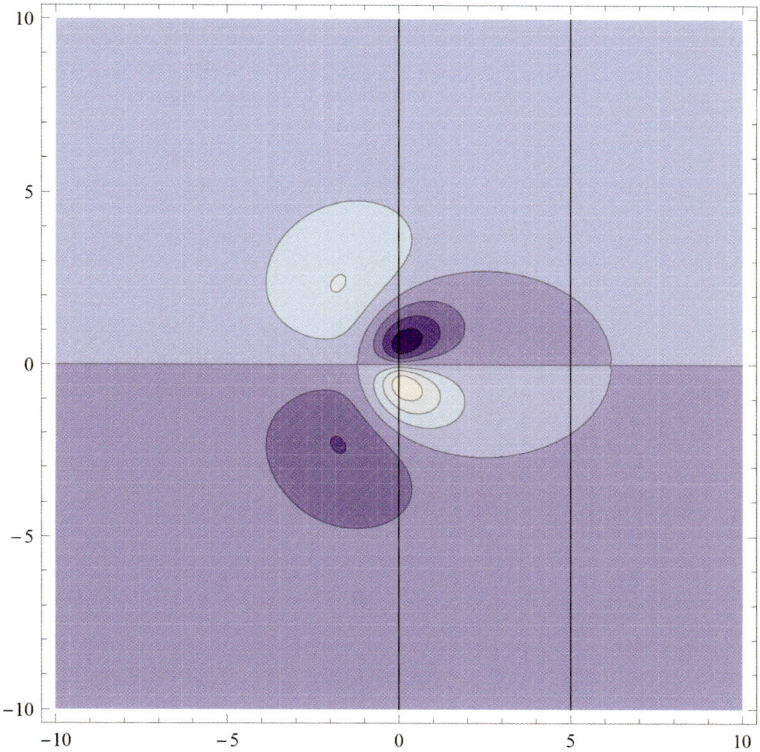

Fig. 8.1: An ellipsoidal Coulomb Sturmian, produced by the perturbing potential shown in equations (8.46)–(8.48) with $|\mathbf{X}| = 5$. *The focal points of the ellipse are the coordinate origin and the position of the perturbing point charge. The function shown here has the form:* $e^{-s} s_x [c_1 s_z + c_2(1 - s/2)]$. *The constants* c_1 *and* c_2*, which determine the position of the second focal point of the ellipse, are found by diagonalizing a matrix representation of the perturbation shown in equation (8.48), based on the real Coulomb Sturmians.*

a linear array:

$$\mathbf{u}_1 = (0, 0, +1) \qquad \mathbf{u}_2 = (0, 0, -1) \tag{8.54}$$

a trigonal array:

$$\begin{aligned} \mathbf{u}_1 &= (2\sqrt{2}/3, 0, -1/3) \\ \mathbf{u}_2 &= (-\sqrt{2}/3, \sqrt{6}/3, -1/3) \\ \mathbf{u}_3 &= (-\sqrt{2}/3, -\sqrt{6}/3, -1/3) \end{aligned} \tag{8.55}$$

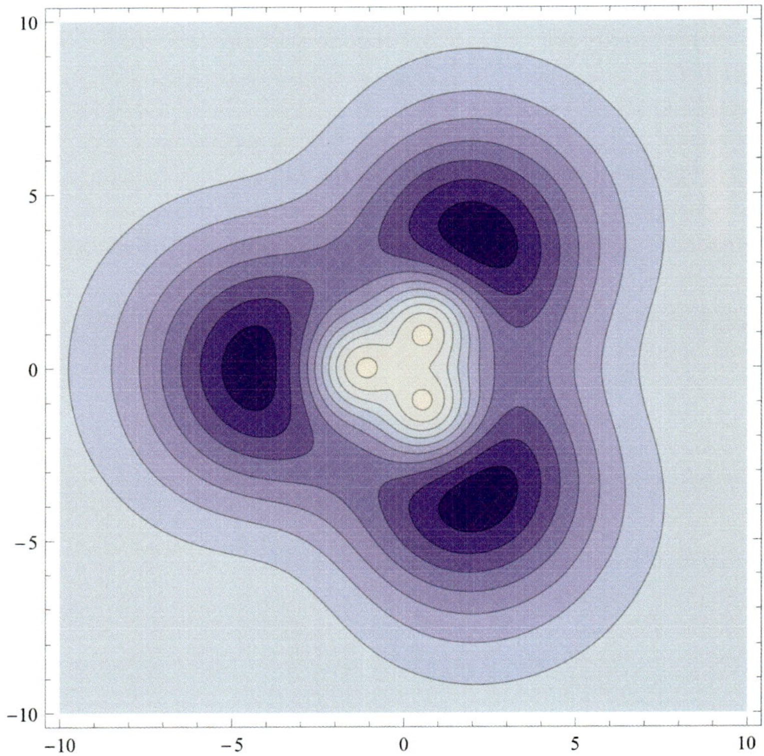

Fig. 8.2: An alternative Coulomb Sturmian produced by the perturbing potential due to a tetrahedral array of point charges, equation (8.56). The function is plotted in the $z = -1/3$ plane.

and a tetrahedral array

$$\begin{aligned}
\mathbf{u}_1 &= (0, 0, 1) \\
\mathbf{u}_2 &= (2\sqrt{2}/3, 0, -1/3) \\
\mathbf{u}_3 &= (-\sqrt{2}/3, \sqrt{2/3}, -1/3) \\
\mathbf{u}_4 &= (-\sqrt{2}/3, -\sqrt{2/3}, -1/3)
\end{aligned} \tag{8.56}$$

We have been considering matrix representations of perturbations based on the standard Coulomb Sturmians, but we could also consider matrix representations of $v'(\mathbf{x})$ based on the real Coulomb Sturmians discussed in the previous section, in which case we would be led to real alternative Coulomb Sturmians with the symmetry of the appropriate point-group.

8.4 The d-dimensional case

In a d-dimensional space, the Fourier transform of Coulomb Sturmian basis functions is given by

$$\psi^t_{n,\mu}(\mathbf{p}) = \frac{(2k)^{d/2+1}}{2^{1/2}(k^2+p^2)^{d/2+1/2}} Y_{n-1,\mu}(\mathbf{u}) \tag{8.57}$$

where

$$\begin{aligned}
u_1 &= \frac{2kp_1}{k^2+p^2} \equiv \sin\theta_1 \ \sin\theta_2 \cdots \ \sin\theta_{d-1} \ \cos\theta_d \\
u_2 &= \frac{2kp_2}{k^2+p^2} \equiv \sin\theta_1 \ \sin\theta_2 \cdots \ \sin\theta_{d-1} \ \sin\theta_d \\
u_3 &= \frac{2kp_3}{k^2+p^2} \equiv \sin\theta_1 \ \sin\theta_2 \cdots \ \cos\theta_{d-1} \\
&\vdots \qquad \vdots \qquad \vdots \\
u_d &= \frac{2kp_d}{k^2+p^2} \equiv \sin\theta_1 \ \cos\theta_2 \\
u_{d+1} &= \frac{k^2-p^2}{k^2+p^2} \equiv \cos\theta_1
\end{aligned} \tag{8.58}$$

The corresponding direct-space functions

$$\psi_{n,\mu}(\mathbf{x}) = \frac{1}{(2\pi)^{d/2}} \int d\mathbf{p} \ e^{i\mathbf{p}\cdot\mathbf{x}} \ \psi^t_{n,\mu}(\mathbf{p}) \tag{8.59}$$

will be solutions to the wave equation

$$\left[-\frac{1}{2}\Delta - \frac{Z}{r} + \frac{k^2}{2}\right] \psi_{n,\mu}(\mathbf{x}) = 0 \tag{8.60}$$

where Δ is the Laplace-Beltrami operator, r is the hyperradius and

$$k = \frac{2Z}{(d+2\lambda-1)} \tag{8.61}$$

These relationships will hold regardless of how we organize the indices μ. In other words, if we let

$$\tilde{\psi}_{n,\mu'}(\mathbf{x}) = \sum_{\mu} \psi_{n,\mu}(\mathbf{x}) \ U_{\mu,\mu'} \tag{8.62}$$

where $U_{\mu,\mu'}$ is any unitary matrix whatever, equation (8.60) will hold with $\psi_{n,\mu}(\mathbf{x})$ replaced by $\tilde{\psi}_{n,\mu'}(\mathbf{x})$. Since we have great freedom in the choice of $U_{\mu,\mu'}$, we can, if we like, choose it in such a way that the unitary transformation diagonalizes the matrix

$$V_{\mu',\mu} = \int d\mathbf{x}\ \psi^*_{n,\mu'}(\mathbf{x})V(\mathbf{x})\psi_{n,\mu}(\mathbf{x}) \tag{8.63}$$

If we solve the set of secular equations

$$\sum_{\mu} [V_{\mu',\mu} - \delta_{\mu',\mu}\mathcal{V}_\kappa]\, U_{\mu,\kappa} = 0 \tag{8.64}$$

to determine which unitary transformation to use, then this transformation will give us a set of d-dimensional Coulomb Sturmian functions $\tilde{\psi}_{n,\mu'}(\mathbf{x})$ which are especially adapted basis functions for solving the Schrödinger equation when $V(\mathbf{x})$ is the potential. Having generated the matrix $V_{\mu',\mu}$ in direct space and diagonalized it, we are in possession of the desired unitary transform matrix $U_{\mu,\kappa}$. Our next step is to generate a new set of $(d+1)$-dimensional hyperspherical harmonics by means of the transformation

$$\mathcal{Y}_{n-1,\kappa}(\mathbf{u}) = \sum_{\mu} Y_{n-1,\mu}(\mathbf{u})U_{\mu,\kappa} \tag{8.65}$$

The Fourier transforms of the symmetry-adapted basis functions will be given by

$$\begin{aligned}\tilde{\psi}^t_{n,\kappa}(\mathbf{p}) &= \frac{(2k)^{d/2+1}}{2^{1/2}(k^2+p^2)^{d/2+1/2}}\mathcal{Y}_{n-1,\kappa}(\mathbf{u})\\ &= \sum_{\mu} \psi^t_{n,\mu}(\mathbf{p})U_{\mu,\kappa}\end{aligned} \tag{8.66}$$

Finally, we can generate the symmetry-adapted basis set in direct space by means of a d-dimensional Fourier transformation:

$$\begin{aligned}\tilde{\psi}_{n,\kappa}(\mathbf{x}) &= \frac{1}{(2\pi)^{d/2}}\int d\mathbf{x}\ e^{i\mathbf{p}\cdot\mathbf{x}}\ \tilde{\psi}^t_{n,\kappa}(\mathbf{p})\\ &= \frac{1}{(2\pi)^{d/2}}\int d\mathbf{x}\ e^{i\mathbf{p}\cdot\mathbf{x}}\ \sum_{\mu}\psi^t_{n,\mu}(\mathbf{p})U_{\mu,\kappa}\\ &= \sum_{\mu}\psi_{n,\mu}(\mathbf{x})U_{\mu,\kappa}\end{aligned} \tag{8.67}$$

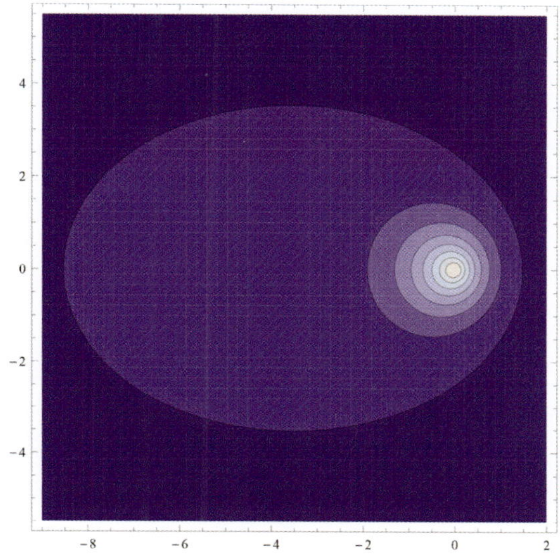

Fig. 8.3: This figure shows the first function in the array of equation (8.71). The function is shown in the x_1, x_3 plane, with $x_2 = x_4 = x_5 = x_6 = 0$.

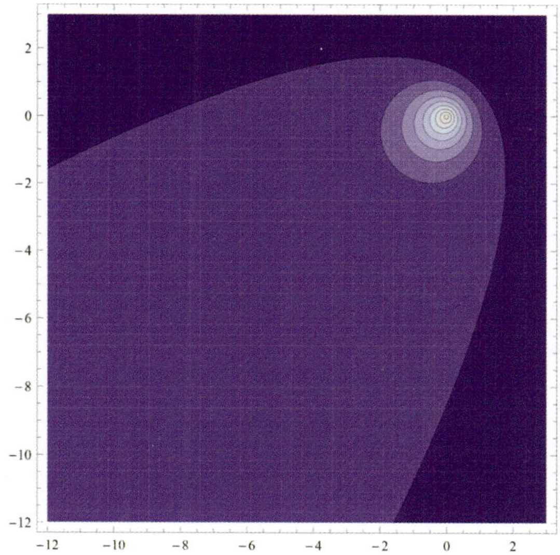

Fig. 8.4: The function of Figure 8.3 is shown here in the x_3, x_6 plane, with $x_1 = x_2 = x_4 = x_5 = 0$.

For example, when $d = 6$ and $n = 2$, the standard Coulomb Sturmian basis functions are

$$\begin{aligned}
\psi_{2;0,0,0,0,0}(\mathbf{x}) &= 2\sqrt{\frac{10}{21\pi^3}}k^3 e^{-kr}\left(1 - \frac{2kr}{5}\right) \\
\psi_{2;1,0,0,0,0}(\mathbf{x}) &= 4\sqrt{\frac{2}{105\pi^3}}k^4\ r\ e^{-kr}u_6 \\
\psi_{2;1,1,0,0,0}(\mathbf{x}) &= 4\sqrt{\frac{2}{105\pi^3}}k^4\ r\ e^{-kr}u_5 \\
\psi_{2;1,1,1,0,0}(\mathbf{x}) &= 4\sqrt{\frac{2}{105\pi^3}}k^4\ r\ e^{-kr}u_4 \\
\psi_{2;1,1,1,1,-1}(\mathbf{x}) &= 4\sqrt{\frac{1}{105\pi^3}}k^4\ r\ e^{-kr}(u_1 - iu_2) \\
\psi_{2;1,1,1,1,0}(\mathbf{x}) &= 4\sqrt{\frac{2}{105\pi^3}}k^4\ r\ e^{-kr}u_3 \\
\psi_{2;1,1,1,1,1}(\mathbf{x}) &= -4\sqrt{\frac{1}{105\pi^3}}k^4\ r\ e^{-kr}(u_1 + iu_2)
\end{aligned} \tag{8.68}$$

We now apply a perturbing potential, for example

$$V(\mathbf{x}) = r(u_3 + u_6) = x_3 + x_6 \tag{8.69}$$

In this simple example, the unitary transformation for the $n = 2$ Coulomb Sturmians is

$$U_{\mu,\kappa} = \begin{pmatrix}
\frac{1}{\sqrt{2}} & -\frac{1}{2} & 0 & 0 & 0 & -\frac{1}{2} & 0 \\
\frac{1}{\sqrt{2}} & \frac{1}{2} & 0 & 0 & 0 & \frac{1}{2} & 0 \\
0 & \frac{1}{\sqrt{2}} & 0 & 0 & 0 & -\frac{1}{\sqrt{2}} & 0 \\
0 & 0 & 0 & 1 & 0 & 0 & 0 \\
0 & 0 & 1 & 0 & 0 & 0 & 0 \\
0 & 0 & 0 & 0 & 1 & 0 & 0 \\
0 & 0 & 0 & 0 & 0 & 0 & 1
\end{pmatrix} \tag{8.70}$$

and the transformed (symmetry-adapted) 6-dimensional Coulomb Sturmians are

$$\tilde{\psi}_{2,\kappa}(\mathbf{x}) = \sum_{\mu=1}^{7} \psi_{2,\kappa}(\mathbf{x}) U_{\mu,\kappa}$$

$$= \frac{1}{\sqrt{105\pi^3}} \begin{pmatrix} -2e^{-kr}k^3\left(kr\left(\sqrt{2}u_3+\sqrt{2}u_6+2\right)-5\right) \\ 2e^{-kr}k^3\left(kr\left(\sqrt{2}u_3+\sqrt{2}u_6-2\right)+5\right) \\ -4e^{-kr}k^4r(u_3-u_6) \\ 4\sqrt{2}e^{-kr}k^4ru_4 \\ 4\sqrt{2}e^{-kr}k^4ru_5 \\ 4e^{-kr}k^4r(u_1-iu_2) \\ -4e^{-kr}k^4r(u_1+iu_2) \end{pmatrix} \tag{8.71}$$

8.5 The *d*-dimensional Schrödinger equation in momentum space

In order to discuss another type of orbital that might be appropriate in some calculations, let us review the arguments that led V. Fock to his famous reciprocal-space solution to the hydrogen-like wave equation (Chapter 5): If we let

$$e^{i\mathbf{p}\cdot\mathbf{x}} \equiv e^{i(p_1x_1+p_2x_2+\cdots+p_dx_d)} \tag{8.72}$$

represent a d-dimensional plane wave, then the wave function and its Fourier transform are related by

$$\begin{aligned} \psi(\mathbf{x}) &= \frac{1}{(2\pi)^{d/2}}\int d\mathbf{p}\; e^{i\mathbf{p}\cdot\mathbf{x}}\psi^t(\mathbf{p}) \\ \psi^t(\mathbf{p}) &= \frac{1}{(2\pi)^{d/2}}\int d\mathbf{x}\; e^{-i\mathbf{p}\cdot\mathbf{x}}\psi(\mathbf{x}) \end{aligned} \tag{8.73}$$

where

$$\begin{aligned} \psi(\mathbf{x}) &\equiv \psi(x_1, x_2, \cdots, x_d) \\ \psi^t(\mathbf{p}) &\equiv \psi(p_1, p_2, \cdots, p_d) \\ d\mathbf{x} &\equiv dx_1 dx_2 \cdots dx_d \\ d\mathbf{p} &\equiv dp_1 dp_2 \cdots dp_d \end{aligned} \tag{8.74}$$

In direct space, the d-dimensional Schrödinger equation can be written in the form

$$\left[-\Delta + p_0^2 + 2V(\mathbf{x})\right]\psi(\mathbf{x}) = 0 \tag{8.75}$$

where

$$p_0^2 \equiv -2E$$
$$\Delta \equiv \sum_{j=1}^{d} \frac{\partial^2}{\partial x_j^2} \tag{8.76}$$

If we substitute the expression for $\psi(\mathbf{x})$ in terms of its Fourier transform into equation (8.75), we have

$$\begin{aligned}&\left[-\Delta + p_0^2 + 2V(\mathbf{x})\right]\psi(\mathbf{x})\\ &= \left[-\Delta + p_0^2 + 2V(\mathbf{x})\right]\frac{1}{(2\pi)^{d/2}}\int d\mathbf{p}\ e^{i\mathbf{p}\cdot\mathbf{x}}\psi^t(\mathbf{p})\\ &= \frac{1}{(2\pi)^{d/2}}\int d\mathbf{p}\ \left[p^2 + p_0^2 + 2V(\mathbf{x})\right]e^{i\mathbf{p}\cdot\mathbf{x}}\psi^t(\mathbf{p}) = 0\end{aligned} \tag{8.77}$$

If we multiply the last line of (8.77) by $e^{-i\mathbf{p}'\cdot\mathbf{x}}/(2\pi)^{d/2}$ and integrate over $d\mathbf{x}$, we obtain

$$\frac{1}{(2\pi)^d}\int d\mathbf{x}\int d\mathbf{p}\ e^{i(\mathbf{p}-\mathbf{p}')\cdot\mathbf{x}}\left[p^2 + p_0^2 + 2V(\mathbf{x})\right]\psi^t(\mathbf{p}) = 0 \tag{8.78}$$

Remembering that

$$\frac{1}{(2\pi)^d}\int d\mathbf{x}\ e^{i(\mathbf{p}-\mathbf{p}')\cdot\mathbf{x}} = \delta(\mathbf{p}-\mathbf{p}') \tag{8.79}$$

we can rewrite (8.78) in the form

$$(p'^2 + p_0^2)\psi^t(\mathbf{p}') = -\frac{2}{(2\pi)^{d/2}}\int d\mathbf{p}\ V^t(\mathbf{p}'-\mathbf{p})\psi^t(\mathbf{p}) \tag{8.80}$$

where

$$V^t(\mathbf{p}'-\mathbf{p}) = \frac{1}{(2\pi)^{d/2}}\int d\mathbf{x}\ e^{-i(\mathbf{p}'-\mathbf{p})}V(\mathbf{x}) \tag{8.81}$$

Equation (8.80) is the momentum-space form of the d-dimensional Schrödinger equation.

8.6 Shibuya-Wulfman orbitals

In their famous 1965 paper [Shibuya and Wulfman, 1965], T.I. Shibuya and C.E. Wulfman were able to generalize Fock's result to molecules. They considered the case where $d = 3$ and

$$V(\mathbf{x}) = -\sum_a \frac{Z_a}{|\mathbf{x}-\mathbf{X}_a|} \tag{8.82}$$

Table 8.1: The table below shows around-state electronic energies of H_2^+ in Hartrees, calculated using the Shibuya-Wulfman method. The internuclear distance R is measured in Bohrs. The results obtained using 3 basis functions on each center are shown in column a. T. Koga and T. Matsuhashi have performed the same calculation with a large number of basis functions [Koga and Matsuhashi, 1988]. Their nearly exact results are shown in column b, compared with the best results obtained using ellipsoidal coordinates (column c).

R	a	b	c
0.1	−1.9782	−1.978242014	−1.9782421
0.2	−1.9285	−1.928620275	−1.9286202
0.4	−1.8001	−1.800754051	−1.8007539
0.6	−1.6703	−1.671484711	−1.6714846
0.8	−1.5531	−1.554480093	−1.5544801
1.0	−1.4503	−1.451786313	−1.451786313
2.0	−1.1018	−1.102634214	−1.102634214
3.0	− 0.9100	− 0.910896197	− 0.910896197
4.0	−0.7948	−0.796084884	−0.796084884
6.0	−0.6776	−0.678635715	−0.678635715
8.0	−0.6272	−0.627570389	−0.627570389

In that case, the momentum-space wave equation becomes

$$(p'^2 + k^2)\psi^t_\zeta(\mathbf{p}') = \frac{1}{\pi^2}\int d\mathbf{p}\ \frac{1}{|\mathbf{p}-\mathbf{p}'|^2}\sum_a Z_a e^{i(\mathbf{p}-\mathbf{p}')\cdot\mathbf{X}_a}\psi^t_\zeta(\mathbf{p}) \tag{8.83}$$

where k has taken the place of p_0. Following a path similar to that of Fock, Shibuya and Wulfman let

$$\begin{aligned} \psi^t_\zeta(\mathbf{p}) &= \frac{4k^{5/2}}{(k^2+p^2)^2} f_\zeta(\mathbf{u}) \\ u_1 &= \frac{2kp_1}{k^2+p^2} \\ u_2 &= \frac{2kp_2}{k^2+p^2} \\ u_3 &= \frac{2kp_3}{k^2+p^2} \\ u_4 &= \frac{k^2-p^2}{k^2+p^2} \end{aligned} \tag{8.84}$$

With this transformation, (8.83) becomes

$$kf_\zeta(\mathbf{u}') = \sum_\tau \eta_\tau(\mathbf{u}')\int d\Omega_4\ \eta^*_\tau(\mathbf{u}) f_\zeta(\mathbf{u}) \tag{8.85}$$

where

$$\tau \equiv (\lambda, l, m, a) \tag{8.86}$$

and

$$\eta_\tau(\mathbf{u}) \equiv \left(\frac{Z_a}{\lambda+1}\right)^{1/2} e^{-i\mathbf{p}\cdot\mathbf{X}_a} Y_{\lambda,l,m}(\mathbf{u}) \tag{8.87}$$

The hyperangular overlap integral between two such functions is

$$\begin{aligned} K_{\tau',\tau} &\equiv \int d\Omega_4\ \eta^*_{\tau'}(\mathbf{u})\eta_\tau(\mathbf{u}) \\ &= \left(\frac{Z_{a'}Z_a}{(\lambda'+1)(\lambda+1)}\right)^{1/2}\int d\Omega_4\ e^{i\mathbf{p}\cdot(\mathbf{X}_{a'}-\mathbf{X}_a)} Y^*_{\lambda',l',m'}(\mathbf{u}) Y_{\lambda,l,m}(\mathbf{u}) \end{aligned} \tag{8.88}$$

Shibuya and Wulfman then let

$$f_\zeta(\mathbf{u}') = \sum_\tau \eta_\tau(\mathbf{u}') C'_{\tau,\zeta} \tag{8.89}$$

With this substitution (8.85) becomes

$$k\sum_\tau \eta_\tau(\mathbf{u}')C'_{\tau,\zeta} = \sum_{\tau'} \eta_{\tau'}(\mathbf{u}') K_{\tau',\tau} C'_{\tau,\zeta} \tag{8.90}$$

The linear independence of the functions $\eta_\tau(\mathbf{u}')$ then implies that the solutions satisfy the secular equation

$$\sum_\tau \left[K_{\tau',\tau} - k\delta_{\tau',\tau}\right] C'_{\tau,\zeta} = 0 \tag{8.91}$$

where the energy of each of the solutions is given by

$$E_\zeta = -\frac{1}{2}k^2 \tag{8.92}$$

Of course in practice, the basis set needs to be truncated. The momentum-space molecular orbitals then become

$$\begin{aligned}
\psi^t_\zeta(\mathbf{p}) &= \frac{4k^{5/2}}{(k^2+p^2)^2} f_\zeta(\mathbf{u}) \\
&= \sum_\tau \sqrt{\frac{Z_a}{n}} \frac{4k^{5/2}}{(k^2+p^2)^2} e^{-i\mathbf{p}\cdot\mathbf{X}_a} Y_{\lambda,l,m}(\mathbf{u}) C'_{\tau,\zeta} \\
&= \sum_\tau \chi^t_\tau(\mathbf{p}) C_{\tau,\zeta}
\end{aligned} \tag{8.93}$$

with

$$C'_{\tau,\zeta} = \sqrt{\frac{Z_a}{n}} C_{\tau,\zeta} \tag{8.94}$$

In direct space this becomes

$$\begin{aligned}
\psi_\zeta(\mathbf{x}) &= \sum_\tau \chi_\tau(\mathbf{x}) C_{\tau,\zeta} \\
&\equiv \sum_{n,l,m,a} \chi_{n,l,m}(\mathbf{x} - \mathbf{X}_a) C_{\tau,\zeta}
\end{aligned} \tag{8.95}$$

The matrix $K_{\tau',\tau}$ can be written in the form

$$K_{\tau',\tau} = \sqrt{\frac{Z_{a'} Z_a}{n'n}} \,\mathfrak{S}_{\tau',\tau} \tag{8.96}$$

where

$$\mathfrak{S}_{\tau',\tau} \equiv \int d\Omega_4 \; e^{i\mathbf{p}\cdot\mathbf{R}} \; Y^*_{\mu'}(\mathbf{u}) Y_\mu(\mathbf{u}) \tag{8.97}$$

where $\mathbf{R} = \mathbf{X}_{a'} - \mathbf{X}_a$ and $Y_\mu(\mathbf{u}) \equiv Y_{n-1,l,m}(\mathbf{u})$. Shibuya and Wulfman [Shibuya and Wulfman, 1965] evaluated these integrals by expanding $e^{i\mathbf{p}\cdot\mathbf{R}}$ in terms of hyperspherical harmonics. Alternatively it is possible to show [Caligiana, 2003] that

$$\int d\Omega_4 \; e^{i\mathbf{p}\cdot\mathbf{R}} Y_\mu(\mathbf{u}) = (2\pi)^{3/2} f_{n,l}(S) Y_{l,m}(\hat{\mathbf{S}}) \tag{8.98}$$

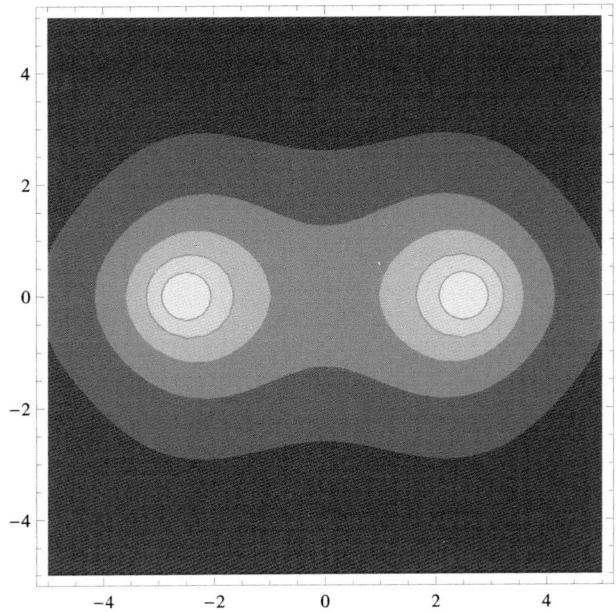

Fig. 8.5: This figure shows the Sturmian molecular orbital corresponding to the ground state of the H_2^+ ion, with S=6, k=1.16885 and R=5.13325 Bohrs.

where

$$\mathbf{S} = \{S_x, S_y, S_z\} \equiv k\mathbf{R} \qquad S = k|\mathbf{R}| \tag{8.99}$$

and

$$\mathbf{R} \equiv \mathbf{X}_{a'} - \mathbf{X}_a \tag{8.100}$$

and where

$$k^{3/2} f_{n,l} \equiv R_{n,l} - \frac{1}{2}\sqrt{\frac{(n-l)(n+l+1)}{n(n+1)}} R_{n+1,l} - \frac{1}{2}\sqrt{\frac{(n+l)(n-l-1)}{n(n-1)}} R_{n-1,l} \tag{8.101}$$

with

$$R_{n-1,l} \equiv 0 \qquad \text{if } l > n-1 \tag{8.102}$$

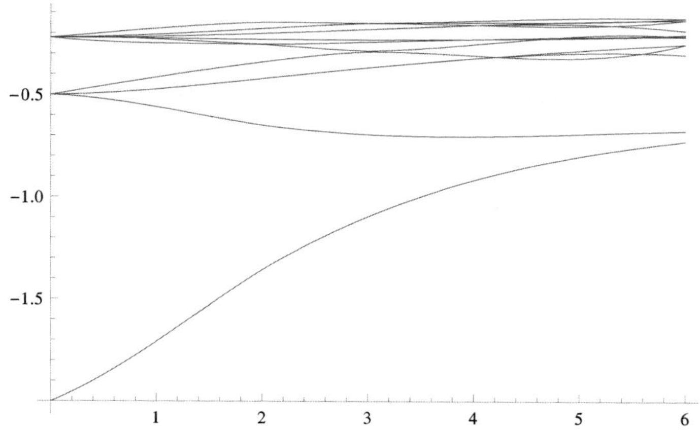

Fig. 8.6: This figure shows the first few energy levels of the H_2^+ ion as functions of $S = kR$, where R is the internuclear separation in atomic units. The energy levels were calculated using the Shibuya-Wulfman method. The energy levels were calculated at integral values of S, and the curves were found by interpolation.

Here the functions $R_{n,l}$ are the Coulomb Sturmian radial functions. If we let

$$c_{\mu'';\mu',\mu} = \int d\Omega_4 Y^*_{\mu''}(\mathbf{u})Y^*_{\mu'}(\mathbf{u})Y_\mu(\mathbf{u}) \tag{8.103}$$

then

$$Y^*_{\mu'}(\mathbf{u})Y_\mu(\mathbf{u}) = \sum_{\mu''} Y_{\mu''}(\mathbf{u})c_{\mu'';\mu',\mu} \tag{8.104}$$

and

$$\mathfrak{S}_{\tau',\tau} = (2\pi)^{3/2} \sum_{\mu''} Y_{l'',m''}(\hat{\mathbf{S}}) f_{n'',l''}(S) c_{\mu'';\mu',\mu} \tag{8.105}$$

The matrix $c_{\mu'';\mu',\mu}$ is a large but very sparse matrix that can be precalculated and stored. Thus equation ([Shibuya and Wulfman, 1965]) gives us a rapid and convenient way of evaluating the Shibuya-Wulfman integrals. Similarly, the Wulfman integrals $\mathfrak{W}_{\tau',\tau}$ can be evaluated by means of Fock's projection, as can the molecular Sturmian overlap integrals

$$m_{\tau',\tau} \equiv \int d\mathbf{x}\ \chi^*_{\tau'}(\mathbf{x})\chi_\tau(\mathbf{x}) \tag{8.106}$$

These are given by

$$m_{\tau',\tau} = (2\pi)^{3/2} \sum_{\mu''} Y_{l'',m''}(\hat{\mathbf{S}}) g_{n'',l''}(S) c_{\mu'';\mu',\mu} \tag{8.107}$$

where the radial functions $g_{n,l}$ are given by

$$g_{n,l} \equiv f_{n,l} - \frac{1}{2}\sqrt{\frac{(n-l)(n+l+1)}{n(n+1)}} f_{n+1,l} - \frac{1}{2}\sqrt{\frac{(n+l)(n-l-1)}{n(n-1)}} f_{n-1,l} \tag{8.108}$$

with

$$f_{n-1,l} \equiv 0 \qquad \text{if } l > n-1 \tag{8.109}$$

8.7 Exercises

(1) Calculate the unitary matrix analogous to equation (8.43) for the case where $\lambda = 2$.

(2) Type `?swI`. This command generates the Shibuya-Wulfman integrals defined by equation (8.97). Before using this command, we need a basis consisting of the quantum numbers of 4-dimensional hyperspherical harmonics. Generate a basis table listing all the possible quantum numbers up to $\lambda = 2$. How are these related to the quantum numbers of all the atomic orbitals up to $n = 3$?

(3) Use the basis generated in the previous exercise to generate a table of Shibuya-Wulfman integrals linking the basis functions. Show that when $S = 0$, this matrix reduces to the identity matrix.

(4) The Koga matrix is related to the Shibuya-Wulfman matrix by equation (8.96). In this equation, τ stands of the set of indices $\tau \equiv (\lambda, l, m, a)$, where a labels the atoms in the system. For diatomic atoms, this means that the Koga matrix has twice the dimension of the Shibuya-Wulfman matrix. Use equation (8.96) to construct the Koga matrix for a diatomic molecule from the Shibuya-Wulfman matrix of the previous exercise.

(5) A spectrum of k values can be obtained by diagonalizing the Koga matrix, as is shown in equation (8.96), and from these, the corresponding energies can be obtained by means of the relationship shown in equation (8.92). Using these relationships, make a plot of the electronic energy levels of the H_2^+ molecular ion as functions of $S \equiv kR$.

(6) Repeat the previous exercise for the HHe^{2+} molecular ion.

Chapter 9

MOLECULAR INTEGRALS FROM HYPERSPHERICAL HARMONICS

9.1 Molecular integrals using exponential-type orbitals

Calculations in quantum chemistry usually take place within the framework of the Born-Oppenheimer approximation, where electronic wave functions are calculated for a variety of fixed nuclear positions, and the electronic energies thus obtained are used as an effective potential in calculating the wave function and energies of the nuclei.

Present methods of mainstream quantum chemistry are built on "Gaussian technology", i.e. on the use of Cartesian Gaussian basis functions of the form $Nx^n y^{n'} z^{n''} e^{-\alpha r^2}$. Gaussian basis functions were introduced by S.F. Boys in 1950. He used the fact that the product of two Cartesian Gaussians centered at two different points is a small linear combination of Cartesian Gaussians centered at an intermediate point. This property allows many-center interelectron repulsion integrals to be evaluated with great ease and speed. Because of this, an entire technology has been formed around Gaussian orbitals, on which mainstream quantum chemistry almost exclusively relies.

Gaussian basis functions, however, have serious drawbacks, since very many of them are needed to approximate the molecular orbitals, and since the cusp at the nucleus is never adequately represented. Furthermore, Gaussian basis functions cannot accurately represent the exponential decay of the orbitals at large distances from the nuclei. Thus, while the mainstream effort of quantum chemistry today follows the path of Gaussian technology, there exists a group of researchers who struggle with the difficult mathematical problems involved in using exponential-type orbitals (ETO's) as basis functions. We hope that that the present chapter will contribute to this effort to make ETO calculations not only more accurate than those based on Gaussians, but also competitive in speed.

9.2 Many center Sturmians

In C.C.J. Roothaan's method for the calculation of molecular orbitals (for example the Hartree-Fock-Roothaan method followed by a configuration interaction calculation), molecular orbitals $\phi_\zeta(\mathbf{x})$ are built up from linear combinations of basis functions, $\chi_\tau(\mathbf{x})$. In the method that we will now discuss, we build up molecular orbitals from Coulomb Sturmians centered on the various atoms of a molecule: Let

$$\tau \equiv (a, l, m, n) \tag{9.1}$$

and

$$\chi_\tau(\mathbf{x}) \equiv \chi_{n,l,m}(\mathbf{x} - \mathbf{X}_a) \tag{9.2}$$

where $\mathbf{X}_a$ is a nuclear position, and $\chi_{n,l,m}(\mathbf{x})$ is a Coulomb Sturmian. We express a molecular orbital $\varphi_\zeta(\mathbf{x})$ as a linear combination of Coulomb Sturmians $\chi_\tau(\mathbf{x})$ centered on the atoms of the molecule:

$$\varphi_\zeta(\mathbf{x}) = \sum_{a,n,l,m} \chi_{n,l,m}(\mathbf{x} - \mathbf{X}_a) C_{(a,n,l,m),\zeta} \equiv \sum_\tau \chi_\tau(\mathbf{x}) C_{\tau,\zeta} \tag{9.3}$$

In the Hartree-Fock-Roothaan Self-Consistent Field method, one finds progressively improved solutions to the set of equations

$$\sum_{\tau=1}^{M} \left[F_{\tau',\tau} - m_{\tau',\tau}\epsilon_\zeta\right] C_{\tau,\zeta} = 0 \tag{9.4}$$

Here M is the number of basis functions, and

$$m_{\tau',\tau} = \int d\mathbf{x}\ \chi^\dagger_{\tau'}(\mathbf{x})\chi_\tau(\mathbf{x}) \tag{9.5}$$

is an overlap integral. $F_{\tau',\tau}$ is the *Fock matrix*,

$$F_{\tau',\tau} = H^c_{\tau',\tau} + \sum_{\tau''',\tau''=1}^{M} P_{\tau''',\tau''} \Gamma_{(\tau''',\tau''),\tau',\tau} \tag{9.6}$$

Here $H^c_{\tau',\tau}$ is the *core Hamiltonian:*

$$H^c_{\tau',\tau} = \int d\mathbf{x}\ \chi^\dagger_{\tau'}(\mathbf{x}) \left[-\frac{1}{2}\nabla^2 + v(\mathbf{x})\right] \chi_\tau(\mathbf{x}) \tag{9.7}$$

which contains a kinetic energy term and a potential energy term $v(\mathbf{x})$ due to the (sometimes screened) attraction of the nuclei in the molecule. The symbol $P_{\tau''',\tau''}$ represents the *density matrix*, defined by

$$P_{\tau''',\tau''} = \sum_{\zeta=1}^{M} \nu_\zeta\ C_{\tau''',\zeta}\ C_{\tau'',\zeta} \tag{9.8}$$

where ν_ζ is the *occupation number* of the spin-orbital $\phi_\zeta(\mathbf{x})$.

$$\nu_\zeta = \begin{cases} 1 \text{ for a filled spin orbital} \\ \\ 0 \text{ for an empty spin orbital} \end{cases} \tag{9.9}$$

The matrix $\Gamma_{(\tau''',\tau''),\tau',\tau}$ is called the *interelectron repulsion matrix*, and it is given by

$$\Gamma_{(\tau''',\tau''),\tau',\tau}$$
$$= \int d\mathbf{x}_1 \int d\mathbf{x}_2\ \chi^\dagger_{\tau'}(\mathbf{x}_1)\chi^\dagger_{\tau'''}(\mathbf{x}_2)\frac{1}{|\mathbf{x}_1 - \mathbf{x}_2|}\left[1 - \mathcal{P}_{12}\right]\chi_{\tau''}(\mathbf{x}_2)\chi_\tau(\mathbf{x}_1) \tag{9.10}$$

Here $\mathcal{P}_{12}$ is an operator which permutes the identities of electrons 1 and 2.

In the Hartree-Fock-Roothaan SCF method, a guessed density matrix is initially chosen. Then the secular equations are solved and an improved density matrix is calculated. One continues in this way until the change resulting from successive iterations is as small as desired. Finally, a configuration interaction calculation is performed. In this calculation, a matrix representation of the Hamiltonian of the system based on both occupied and virtual spin-orbitals is constructed and diagonalized. For the configuration interaction calculation, matrices such as the interelectron repulsion matrix Γ, can be taken from storage and used again.

9.3 Overlap integrals involving Coulomb Sturmians

As a simple example of how the relationship discovered by Fock can be used to evaluate molecular integrals, let us consider the overlap integral between two Coulomb Sturmians centered on different atoms: Let

$$\begin{aligned} m_{\tau',\tau} &\equiv \int d\mathbf{x}\ \chi^*_{\tau'}(\mathbf{x})\chi_\tau(\mathbf{x}) \\ &\equiv \int d\mathbf{x}\ \chi^*_{n',l',m'}(\mathbf{x} - \mathbf{X}_{a'})\chi_{n,l,m}(\mathbf{x} - \mathbf{X}_a) \end{aligned} \tag{9.11}$$

Using the fact that a scalar product is invariant under Fourier transforms (equation (4.8)), we obtain

$$m_{\tau',\tau} = \int d\mathbf{p}\ \chi^{*t}_{\tau'}(\mathbf{p})\chi^t_\tau(\mathbf{p}) \tag{9.12}$$

Then, with the abbreviation $(n, l, m) \equiv \mu$, we have

$$\begin{aligned}\chi_\tau^t(\mathbf{p}) &= \frac{1}{(2\pi)^{3/2}} \int d\mathbf{x}\ e^{-i\mathbf{p}\cdot\mathbf{x}} \chi_\mu(\mathbf{x} - \mathbf{X}_a) \\ &= \frac{e^{-i\mathbf{p}\cdot\mathbf{X}_a}}{(2\pi)^{3/2}} \int d\mathbf{x}\ e^{-i\mathbf{p}\cdot(\mathbf{x}-\mathbf{X}_a)} \chi_\mu(\mathbf{x} - \mathbf{X}_a) \\ &= e^{-i\mathbf{p}\cdot\mathbf{X}_a} \chi_\mu^t(\mathbf{p})\end{aligned} \tag{9.13}$$

and similarly

$$\chi_{\tau'}^{t*}(\mathbf{p}) = e^{i\mathbf{p}\cdot\mathbf{X}_{a'}} \chi_{\mu'}^{t*}(\mathbf{p}) \tag{9.14}$$

Combining these relations with (5.30), we obtain:

$$\begin{aligned}m_{\tau',\tau} &= \int d\mathbf{p}\ e^{-i\mathbf{p}\cdot(\mathbf{X}_a - \mathbf{X}_{a'})} \chi_{\mu'}^{*t}(\mathbf{p}) \chi_\mu^t(\mathbf{p}) \\ &= \int d\mathbf{p}\ e^{-i\mathbf{p}\cdot\mathbf{R}} M(p)^2 Y_{n'-1,l',m'}^*(\mathbf{u}) Y_{n-1,l,m}(\mathbf{u}) \\ &\equiv \int d\mathbf{p}\ e^{-i\mathbf{p}\cdot\mathbf{R}} M(p)^2 Y_{\mu'}^*(\mathbf{u}) Y_\mu(\mathbf{u})\end{aligned} \tag{9.15}$$

where $\mathbf{R} \equiv \mathbf{X}_a - \mathbf{X}_{a'}$ and $Y_\mu(\mathbf{u}) \equiv Y_{n-1,l,m}(\mathbf{u})$. The 4-dimensional hyperspherical harmonics, $Y_\mu(\mathbf{u})$, obey the orthonormality relationship

$$\int d\Omega_4\ Y_{\mu'}^*(\mathbf{u}) Y_\mu(\mathbf{u}) = \delta_{\mu',\mu} \tag{9.16}$$

where the solid angle element is related to $d\mathbf{p}$ by

$$d\Omega_4 = d\mathbf{p} \left(\frac{2k}{k^2 + p^2} \right)^3 \tag{9.17}$$

(equation (5.17)). Just as the product of two of the familiar 3-dimensional spherical harmonics can be expressed as a sum of spherical harmonics, so we can also write an analogous relationship for 4-dimensional hyperspherical harmonics:

$$Y_{\mu'}^*(\mathbf{u}) Y_\mu(\mathbf{u}) = \sum_{\mu''} Y_{\mu''}(\mathbf{u}) c_{\mu',\mu}^{\mu''} \tag{9.18}$$

From the orthonormality relation (9.16), it follows that the the expansion coefficient in (9.18) is given by

$$c_{\mu',\mu}^{\mu''} = \int d\Omega_4\ Y_{\mu''}^*(\mathbf{u}) Y_{\mu'}^*(\mathbf{u}) Y_\mu(\mathbf{u}) \tag{9.19}$$

(see equations (6.54) and (6.55)). Hyperangular integrals of this type are very easy to evaluate using the theorem discussed in Chapter 2. With the help of equation (9.18), the overlap integral can be rewritten as

$$m_{\tau',\tau} = \sum_{\mu''} c_{\mu',\mu}^{\mu''} \int d\mathbf{p}\ e^{-i\mathbf{p}\cdot\mathbf{R}} M(p)^2 Y_{\mu''}(\mathbf{u}) \tag{9.20}$$

Were it not for the extra factor of $M(p)$, we could immediately evaluate the integral in equation (9.20) by means of Fock's relationships. But we can get rid of this unwanted extra factor by noting that from equations (5.10) and (5.30) we can derive the relationship

$$M(p) = \frac{4k^{5/2}}{(k^2 + p^2)^2} = k^{-3/2}(1 + u_4)^2 \tag{9.21}$$

Then

$$M(p)Y_{\mu''}(\mathbf{u}) = k^{-3/2}(1 + u_4)^2 Y_{\mu''}(\mathbf{u}) \tag{9.22}$$

The right-hand side of equation (9.22) can be expressed as a linear combination of hyperspherical harmonics:

$$k^{-3/2}(1 + u_4)^2 Y_{\mu''}(\mathbf{u}) = \sum_{\mu'''} Y_{\mu'''}(\mathbf{u})\ \alpha_{\mu''',\mu''} \tag{9.23}$$

where

$$\alpha_{\mu''',\mu''} = k^{-3/2} \int d\Omega_4\ Y^*_{\mu'''}(\mathbf{u})(1 + u_4)^2 Y_{\mu''}(\mathbf{u}) \tag{9.24}$$

The hyperangular integral in equation (9.24) can be evaluated easily and rapidly by means of the general theorem of Section 2.5. Alternatively, it can be evaluated analytically by noting that $-2u_4 = \sqrt{2}\pi Y_{1,0,0}(\mathbf{u})$ (see Table 2.1) and making use of the coupling coefficients for hyperspherical harmonics, as we shall see below. Inserting equations (9.22)–(9.24) into equation (9.20) and using (5.30), we obtain:

$$\begin{aligned} m_{\tau',\tau} &= \sum_{\mu''} c^{\mu''}_{\mu',\mu} \sum_{\mu'''} \alpha_{\mu''',\mu''} \int d\mathbf{p}\ e^{-i\mathbf{p}\cdot\mathbf{R}} M(p) Y_{\mu'''}(\mathbf{u}) \\ &= (2\pi)^{3/2} \sum_{\mu''} c^{\mu''}_{\mu',\mu} \sum_{\mu'''} \chi_{\mu'''}(-\mathbf{R})\ \alpha_{\mu''',\mu''} \\ &\equiv (2\pi)^{3/2} \sum_{\mu''} g_{\mu''}(-\mathbf{R})\ c^{\mu''}_{\mu',\mu} \end{aligned} \tag{9.25}$$

As we shall see below, closed-form expressions for the functions $g_{\mu''}(\mathbf{R})$ can be derived from the coupling coefficients for 4-dimensional hyperspherical harmonics. The coefficients $c_{\mu'';\mu',\mu}$ form a large but sparse matrix, which can be pre-evaluated and stored.

9.4 Shibuya-Wulfman integrals

When Coulomb Sturmians are used as a basis for molecular calculations, it is important to be able to evaluate the Shibuya-Wulfman integrals, which are defined by

$$\mathfrak{S}_{\tau',\tau} \equiv \int d\mathbf{x}\chi^*_{\tau'}(\mathbf{x})\left(\frac{-\nabla^2+k^2}{2k^2}\right)\chi_\tau(\mathbf{x}) \tag{9.26}$$

These integrals were first studied in a different form by T. Shibuya and C. Wulfman, who pioneered the generalization to molecules of Fock's momentum-space solutions to the 1-electron hydrogenlike wave equation. We can convert the Shibuya-Wulfman integrals to a momentum-space form:

$$\begin{aligned}
\mathfrak{S}_{\tau',\tau} &= \frac{1}{(2\pi)^{3/2}}\int d\mathbf{x}\chi^*_{\tau'}(\mathbf{x})\left(\frac{-\nabla^2+k^2}{2k^2}\right)\int d\mathbf{p}\ e^{i\mathbf{p}\cdot\mathbf{x}}\chi^t_\tau(\mathbf{p}) \\
&= \int d\mathbf{p}\ \frac{1}{(2\pi)^{3/2}}\int d\mathbf{x}\ e^{i\mathbf{p}\cdot\mathbf{x}}\chi^*_{\tau'}(\mathbf{x})\left(\frac{p^2+k^2}{2k^2}\right)\chi^t_\tau(\mathbf{p}) \\
&= \int d\mathbf{p}\ \chi^{t*}_{\tau'}(\mathbf{p})\left(\frac{p^2+k^2}{2k^2}\right)\chi^t_\tau(\mathbf{p})
\end{aligned} \tag{9.27}$$

Using (9.13), (9.14) and (9.18), and from Fock's relationship, (5.30), we then obtain:

$$\begin{aligned}
\mathfrak{S}_{\tau',\tau} &= \int d\mathbf{p}\ e^{i\mathbf{p}\cdot\mathbf{R}}M(p)^2\left(\frac{p^2+k^2}{2k^2}\right)Y^*_{\mu'}(\mathbf{u})Y_\mu(\mathbf{u}) \\
&= \sum_{\mu''} c^{\mu''}_{\mu',\mu}\int d\mathbf{p}\ e^{-i\mathbf{p}\cdot\mathbf{R}}M(p)^2\left(\frac{p^2+k^2}{2k^2}\right)Y_{\mu''}(\mathbf{u})
\end{aligned} \tag{9.28}$$

If we let

$$\begin{aligned}
\beta_{\mu''',\mu''} &\equiv k^{-3/2}\int d\Omega_4\ Y^*_{\mu'''}(\mathbf{u})(1+u_4)Y_{\mu''}(\mathbf{u}) \\
&= \int d\Omega_4\ Y^*_{\mu'''}(\mathbf{u})M(p)\left(\frac{p^2+k^2}{2k^2}\right)Y_{\mu''}(\mathbf{u})
\end{aligned} \tag{9.29}$$

then we can express the Shibuya-Wulfman integrals in the form:

$$\begin{aligned}
\mathfrak{S}_{\tau',\tau} &= \sum_{\mu''} c^{\mu''}_{\mu',\mu}\sum_{\mu'''}\beta_{\mu''',\mu''}\int d\mathbf{p}\ e^{-i\mathbf{p}\cdot\mathbf{R}}M(p)Y_{\mu'''}(\mathbf{u}) \\
&= (2\pi)^{3/2}\sum_{\mu''} c^{\mu''}_{\mu',\mu}\sum_{\mu'''}\chi_{\mu'''}(\mathbf{R})\ \beta_{\mu''',\mu''} \\
&\equiv (2\pi)^{3/2}\sum_{\mu''} f_{\mu''}(\mathbf{R})\ c^{\mu''}_{\mu',\mu}
\end{aligned} \tag{9.30}$$

where

$$\begin{aligned}
f_{\mu''}(\mathbf{R}) &\equiv \sum_{\mu'''} \chi_{\mu'''}(\mathbf{R})\ \beta_{\mu''',\mu''} \\
&\equiv \sum_{\mu'''} k^{-3/2}\chi_{\mu'''}(\mathbf{R}) \int d\Omega_4\ Y^*_{\mu'''}(\mathbf{u})(1+u_4)Y_{\mu''}(\mathbf{u}) \\
&\equiv \sum_{\mu'''} \tilde{R}_{n''',l'''}(S)Y_{l''',m'''}(\mathbf{u}_S) \int d\Omega_4\ Y^*_{\mu'''}(\mathbf{u})(1+u_4)Y_{\mu''}(\mathbf{u}) \\
&= f_{n'',l''}(S)Y_{l'',m''}(\mathbf{u}_S) \qquad (9.31)
\end{aligned}$$

and where $\mathbf{S} \equiv k\mathbf{R}$. Similarly, we can write the function $g_{\mu''}(\mathbf{R})$, which was used in equation (9.25) for the calculation of overlap integrals, as

$$g_{\mu''}(\mathbf{R}) = g_{n'',l''}(S)Y_{l'',m''}(\mathbf{u}_S) \qquad (9.32)$$

It is possible to use the coupling coefficients for 4-dimensional hyperspherical harmonics to derive closed-form expressions for the functions $f_{n'',l''}(S)$ and $g_{n'',l''}(S)$. One finds that

$$\begin{aligned}
f_{n,l}(S) \equiv \tilde{R}_{n,l}(S) &- \frac{1}{2}\sqrt{\frac{(n-l)(n+l+1)}{n(n+1)}}\tilde{R}_{n+1,l}(S) \\
&-\frac{1}{2}\sqrt{\frac{(n+l)(n-l-1)}{n(n-1)}}\tilde{R}_{n-1,l}(S) \qquad (9.33)
\end{aligned}$$

where $k^{3/2}\tilde{R}_{n,l}$ is the radial function of the Coulomb Sturmians (Table 1), and where

$$\tilde{R}_{n-1,l}(S) \equiv 0 \qquad \text{if } l > n-1 \qquad (9.34)$$

Similarly, one can show that

$$\int d\mathbf{p}\ e^{-i\mathbf{p}\cdot\mathbf{R}}M(p)^2Y_\mu(\mathbf{u}) = (2\pi)^{3/2}g_{n,l}(S)Y_{l,m}(\hat{\mathbf{S}}) \qquad (9.35)$$

where

$$\begin{aligned}
g_{n,l}(S) \equiv f_{n,l}(S) &- \frac{1}{2}\sqrt{\frac{(n-l)(n+l+1)}{n(n+1)}}f_{n+1,l}(S) \\
&-\frac{1}{2}\sqrt{\frac{(n+l)(n-l-1)}{n(n-1)}}f_{n-1,l}(S) \qquad (9.36)
\end{aligned}$$

where we define

$$f_{n-1,l}(S) \equiv 0 \qquad \text{if } l > n-1 \qquad (9.37)$$

The first few values of $f_{n,l}(S)$ and $g_{n,l}(S)$ are shown in the following table.

Table 9.1: $g_{n,l}(S)$ and $f_{n,l}(S)$, where $S \equiv k|\mathbf{X}_a - \mathbf{X}_{a'}|$. The functions $g_{n,l}(S)$ and $f_{n,l}(S)$ appear respectively in the two-center overlap integrals and the Shibuya-Wulfman integrals.

n	l	$g_{n,l}(S)$	$f_{n,l}(S)$
1	0	$\frac{e^{-S}\left(3+3S+S^2\right)}{3}$	$e^{-S}\left(1+S\right)$
2	0	$-\frac{1}{6}e^{-S}\left(3+3S+2S^2+S^3\right)$	$-\frac{2}{3}e^{-S}S^2$
2	1	$\frac{e^{-S}S\left(3+3S+S^2\right)}{6\sqrt{3}}$	$\frac{2e^{-S}S(1+S)}{3\sqrt{3}}$
3	0	$\frac{1}{15}e^{-S}S^4$	$\frac{1}{3}e^{-S}S^2(-2+S)$
3	1	$-\frac{e^{-S}S^3(1+S)}{15\sqrt{2}}$	$\frac{e^{-S}S(1+S-S^2)}{3\sqrt{2}}$
3	2	$\frac{e^{-S}S^2(3+3S+S^2)}{15\sqrt{10}}$	$\frac{e^{-S}S^2(1+S)}{3\sqrt{10}}$

9.5 Matrices representing kinetic energy and nuclear attraction

We have seen above that

$$\begin{aligned} m_{\tau',\tau} &\equiv \int d\mathbf{x}\ \chi^*_{\tau'}(\mathbf{x})\chi_\tau(\mathbf{x}) \\ &= (2\pi)^{3/2}\sum_{\mu''} g_{n'',l''}(S)Y_{l'',m''}(\mathbf{u}_S)\ c^{\mu''}_{\mu',\mu} \end{aligned} \tag{9.38}$$

and that

$$\begin{aligned} \mathfrak{S}_{\tau',\tau} &\equiv \int d\mathbf{x}\ \chi^*_{\tau'}(\mathbf{x})\left(\frac{-\nabla^2+k^2}{2k^2}\right)\chi_\tau(\mathbf{x}) \\ &= (2\pi)^{3/2}\sum_{\mu''} f_{n'',l''}(S)Y_{l'',m''}(\mathbf{u}_S)\ c^{\mu''}_{\mu',\mu} \end{aligned} \tag{9.39}$$

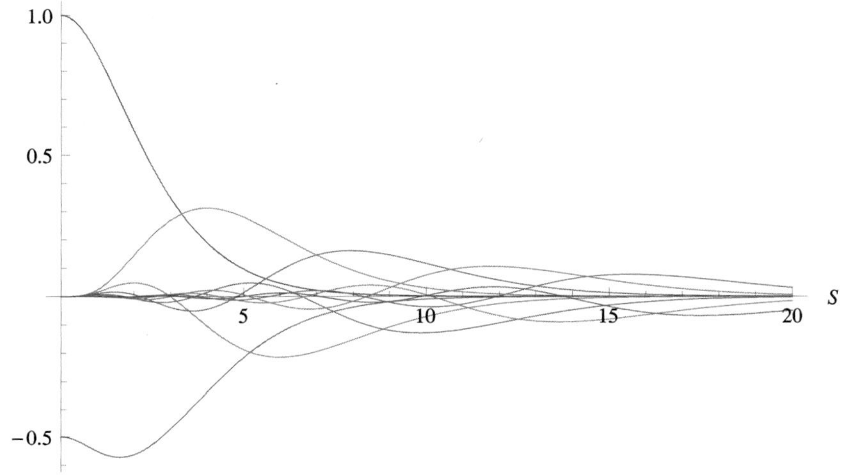

Fig. 9.1: Overlap integrals between the 1s Coulomb Sturmian and ns Coulomb Sturmians. The integrals are shown as functions of $S \equiv k|\mathbf{X}_2 - \mathbf{X}_1|$*, and they are exact. The overlap integral* $m_{1s,1s} = e^{-S}(1+S+S^2/3)$ *is 1 when* $S = 0$*. When* $n = 2$*, we have* $m_{1s,2s} = -e^{-S}(3+3S+2S^2+S^3)/6$ *which is -.5 when* $S = 0$*. The remaining 1s-ns overlap integrals vanish when* $S = 0$*.*

where $c^{\mu''}_{\mu',\mu}$ are the coefficients defined by equation (9.19) and where $f_{n'',l''}(S)$ and $g_{n'',l''}(S)$ are related to the Coulomb Sturmian radial functions by equations (9.33) and (9.36). We can combine equations (9.38) and (9.39) to obtain an expression for the matrix representing 1-electron kinetic energy:

$$\begin{aligned} T_{\tau',\tau} &\equiv \int d\mathbf{x}\, \chi^*_{\tau'}(\mathbf{x}) \left(-\frac{1}{2}\nabla^2\right) \chi_\tau(\mathbf{x}) \\ &= k^2 \left(\mathfrak{S}_{\tau',\tau} - \frac{1}{2} m_{\tau',\tau}\right) \\ &= k^2 (2\pi)^{3/2} \sum_{\mu''} \left(f_{n'',l''}(S) - \frac{1}{2} g_{n'',l''}(S)\right) Y_{l'',m''}(\mathbf{u}_S)\, c^{\mu''}_{\mu',\mu} \qquad (9.40) \end{aligned}$$

The Shibuya-Wulfman integrals also yield the matrices representing a certain class of nuclear attraction integrals: From equation (6.36), we have

$$\left(-\frac{1}{2}\nabla^2 - \frac{kn}{r} + \frac{k^2}{2}\right) \chi_{n,l,m}(\mathbf{x}) = 0 \qquad (9.41)$$

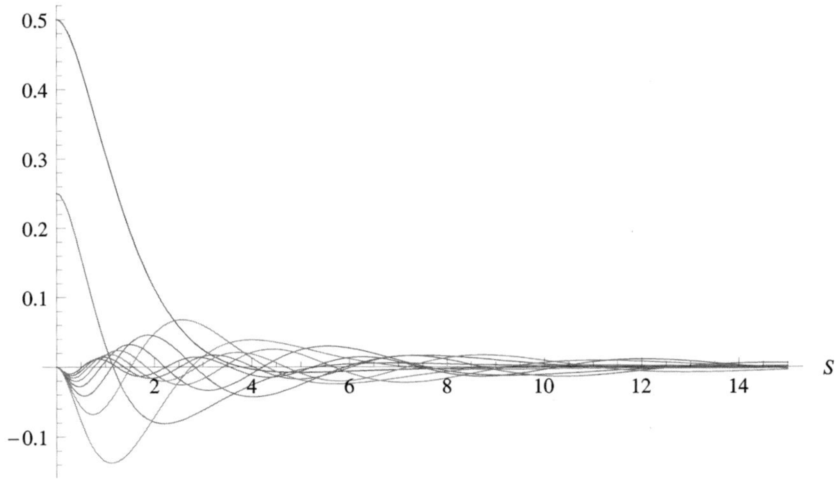

Fig. 9.2: Kinetic energy integrals for 1s-ns Coulomb Sturmians divided by k^2. As in the previous figure, the integrals are shown as functions of $S \equiv k|\mathbf{X}_2 - \mathbf{X}_1|$, and they are exact. The kinetic energy integral $T_{1s,1s}/k^2 = e^{-S}(3+3S-S^2)/6$ is .5 when $S = 0$, while $T_{1s,2s}/k^2 = e^{-S}(3+3S-6S^2+S^3)/12 = .25$ when $S = 0$. The remaining 1s-ns kinetic energy integrals vanish when $S = 0$.

and therefore

$$\left(\frac{-\nabla^2 + k^2}{2k^2}\right)\chi_\tau(\mathbf{x}) \equiv \left(\frac{-\nabla^2 + k^2}{2k^2}\right)\chi_\mu(\mathbf{x} - \mathbf{X}_a)$$
$$= \frac{n}{k|\mathbf{x} - \mathbf{X}_a|}\chi_\mu(\mathbf{x} - \mathbf{X}_a) \tag{9.42}$$

so that

$$\mathfrak{S}_{\tau',\tau} \equiv \int d\mathbf{x}\ \chi^*_{\tau'}(\mathbf{x})\left(\frac{-\nabla^2 + k^2}{2k^2}\right)\chi_\tau(\mathbf{x})$$
$$= \frac{n}{k}\int d\mathbf{x}\ \chi^*_{\mu'}(\mathbf{x} - \mathbf{X}_{a'})\frac{1}{|\mathbf{x} - \mathbf{X}_a|}\chi_\mu(\mathbf{x} - \mathbf{X}_a) \tag{9.43}$$

Using the potential-weighted orthonormality relation, combined with equation (9.43), we can see that when $\mathbf{X}_{a'} = \mathbf{X}_a$, i.e. when $S = 0$, we have

$$\mathfrak{S}_{\tau',\tau}(S = 0) = \frac{n}{k}\int d\mathbf{x}\ \chi^*_{\mu'}(\mathbf{x} - \mathbf{X}_a)\frac{1}{|\mathbf{x} - \mathbf{X}_a|}\chi_\mu(\mathbf{x} - \mathbf{X}_a)$$
$$= \delta_{\mu',\mu} \tag{9.44}$$

Table 9.2: This table shows the first few Coulomb Sturmian overlap integrals, $\int d\mathbf{x}\chi^*_{\mu'}(\mathbf{x}-\mathbf{X}_1)\chi_\mu(\mathbf{x}-\mathbf{X}_2)$ as functions of $\mathbf{S}\equiv k(\mathbf{X}_2-\mathbf{X}_1)$. The integrals are exact.

$1s$	$1s$	$\frac{1}{3}\, e^{-S}(3+3S+S^2)$
$1s$	$2s$	$-\frac{1}{6}\, e^{-S}(3+3S+2S^2+S^3)$
$1s$	$2p_z$	$-\frac{1}{6}\, e^{-S}(3+3S+S^2)S_z$
$2s$	$2s$	$\frac{1}{15}\, e^{-S}(15+15S+5S^2+S^4)$
$2s$	$2p_z$	$\frac{1}{15}\, e^{-S}S^2(1+S)S_z$

Equations (9.44) and (9.27) can be combined to give the weighted orthonormality relation obeyed by the Fourier transforms of Coulomb Sturmians:

$$\mathfrak{S}_{\tau',\tau}(S=0)=\int d\mathbf{p}\;\chi^{t*}_{\mu'}(\mathbf{p})\left(\frac{p^2+k^2}{2k^2}\right)\chi^t_\mu(\mathbf{p})=\delta_{\mu',\mu} \tag{9.45}$$

In Figure 9.1 and Figure 9.2, we can notice that many of the curves go to zero at $S=0$. We can understand why this must be the case by looking at equations (9.33). None of the terms in (9.33) have a constant term if $l>0$. If $l=0$, the constant terms from each of the three terms cancel unless $n=1$. Thus, the only $f_{n,l}(S=0)$ not equal to zero is $f_{1,0}(S=0)$. As a consequence, if we look at (9.36), the only $g_{n,l}(S=0)$ not equal to zero are $g_{2,0}(S=0)$ and $g_{1,0}(S=0)$.

9.6 1-center densities in terms of 2k Sturmians

In order to use Fourier transform methods to evaluate molecular integrals involving

$$\rho_{\tau_1,\tau_2}(\mathbf{x})\equiv\chi^*_{\tau_1}(\mathbf{x})\chi_{\tau_2}(\mathbf{x})\equiv\chi^*_{\mu_1}(\mathbf{x}-\mathbf{X}_{a_1})\chi_{\mu_2}(\mathbf{x}-\mathbf{X}_{a_2})\qquad \mu\equiv(n,l,m) \tag{9.46}$$

Table 9.3: A similar table for the first few Coulomb Sturmian kinetic energy integrals, $\int d\mathbf{x}\chi^*_{\mu'}(\mathbf{x}-\mathbf{X}_1)\left[-\nabla^2/(2k^2)\right]\chi_\mu(\mathbf{x}-\mathbf{X}_2)$.

1s	1s	$\frac{1}{6}\, e^{-S}(3+3S-S^2)$
1s	2s	$\frac{1}{12}\, e^{-S}(3+3S-6S^2+S^3)$
1s	$2p_z$	$\frac{1}{12}\, e^{-S}(-5-5S+S^2)S_z$
2s	2s	$\frac{1}{30}\, e^{-S}(15+15S-25S^2+10S^3-S^4)$
2s	$2p_z$	$\frac{1}{30}\, e^{-S}(-10-10S+9S^2-S^3)S_z$

where $\mathbf{X}_{a_1}$ and $\mathbf{X}_{a_2}$ are nuclear positions, we need to evaluate integrals of the form

$$\rho^t_{\tau_1,\tau_2}(\mathbf{p}) \equiv \frac{1}{(2\pi)^{3/2}}\int d\mathbf{x}\, e^{-i\mathbf{p}\cdot\mathbf{x}}\rho_{\tau_1,\tau_2}(\mathbf{x}) \tag{9.47}$$

In the special case where $\mathbf{X}_{a_2}=\mathbf{X}_{a_1}$, this Fourier transform is very easy to evaluate: We expand the product of Coulomb Sturmians in terms of single Sturmians with double the k-value:

$$\begin{aligned}\rho_{\mu_1,\mu_2}(\mathbf{x}) &= \chi^*_{\mu_1}(\mathbf{x})\chi_{\mu_2}(\mathbf{x})\\ &= \sum_\mu \chi_\mu(2k,\mathbf{x})\, \mathcal{C}^\mu_{\mu_1,\mu_2}\end{aligned} \tag{9.48}$$

where, using the potential-weighted orthonormality relation, equation (6.39), we find that

$$\begin{aligned}\mathcal{C}^\mu_{\mu_1,\mu_2} &= \frac{n}{2k}\int_0^\infty dr\; r\; R_{n,l}(2k,r)R_{n_1,l_1}(r)R_{n_2,l_2}(r)\\ &\quad\times\int d\Omega_3\; Y^*_{l,m}(\mathbf{u})Y^*_{l_1,m_1}(\mathbf{u})Y_{l_2,m_2}(\mathbf{u})\end{aligned} \tag{9.49}$$

Table 9.4: This table shows the first few Coulomb Sturmian Shibuya-Wulfman integrals $\int d\mathbf{x}\chi^*_{\mu'}(\mathbf{x}-\mathbf{X}_1)\left[-\nabla^2/(2k^2)+1/2\right]\chi_\mu(\mathbf{x}-\mathbf{X}_2)$ *as functions of* $\mathbf{S} \equiv k(\mathbf{X}_2-\mathbf{X}_1)$. *The integrals are exact, and the z-axis is taken in the direction of* $\mathbf{S}$.

$1s$	$1s$	$e^{-S}(1+S)$
$1s$	$2s$	$-\frac{2}{3}\,e^{-S}S^2$
$1s$	$2p_z$	$-\frac{2}{3}\,e^{-S}(1+S)S_z$
$1s$	$3s$	$\frac{1}{3}\,e^{-S}(-2+S)S^2$
$2s$	$2s$	$\frac{1}{3}\,e^{-S}(3+3S-2S^2+S^3)$
$2s$	$2p_z$	$\frac{1}{3}\,e^{-S}(-1-S+S^2)S_z$
$2s$	$3s$	$\frac{2}{15}\,e^{-S}S^2(-10+5S-S^2)$
$2p_z$	$2p_z$	$\frac{1}{3}\,e^{-S}(-3-3S-S^2+S_z^2+SS_z^2)$
$2p_z$	$3s$	$\frac{2}{15}\,e^{-S}(4-S)S^2S_z$
$3s$	$3s$	$\frac{1}{45}\,e^{-S}(45+45S-60S^2+60S^3-18S^4+2S^5)$

Mathematica is able to evaluate the radial integral exactly for particular values of the indices. The coefficients $\mathcal{C}_{\mu;\mu_1,\mu_2}$ form a large but very sparse matrix that can be pre-calculated and stored. The relationship shown in equation (9.48) is not an approximation. It is exact. We next make use of the form of Fock's relationship that is appropriate for the $2k$ Sturmian basis:

$$\rho^t_\mu(\mathbf{p}) = M(2k,p)Y_{n-1,l,m}(\hat{\mathbf{w}}) \equiv M(2k,p)Y_\mu(\hat{\mathbf{w}})$$
$$\hat{\mathbf{w}} = \left(\frac{4kp_1}{4k^2+p^2}, \frac{4kp_2}{4k^2+p^2}, \frac{4kp_3}{4k^2+p^2}, \frac{4k^2-p^2}{4k^2+p^2}\right)$$
$$M(2k,p) \equiv \frac{4(2k)^{5/2}}{(4k^2+p^2)^2} \tag{9.50}$$

Here $Y_\mu(\hat{\mathbf{w}})$ is a 4-dimensional hyperspherical harmonic. Thus we obtain the desired Fourier transform in terms of hyperspherical harmonics:

$$\begin{aligned}\rho^t_{\tau_1,\tau_2}(\mathbf{p}) &= \frac{1}{(2\pi)^{3/2}}\int d\mathbf{x}\; e^{-i\mathbf{p}\cdot\mathbf{x}}\rho_{\tau_1,\tau_2}(\mathbf{x}-\mathbf{X}_{a_1}) \\ &= \frac{e^{-i\mathbf{p}\cdot\mathbf{X}_{a_1}}}{(2\pi)^{3/2}}\int d\mathbf{x}\; e^{-i\mathbf{p}\cdot(\mathbf{x}-\mathbf{X}_{a_1})}\rho_{\tau_1,\tau_2}(\mathbf{x}-\mathbf{X}_{a_1}) \\ &= e^{-i\mathbf{p}\cdot\mathbf{X}_{a_1}}\sum_\mu M(2k,p)Y_\mu(\hat{\mathbf{w}})\mathcal{C}^\mu_{\mu_1,\mu_2}\end{aligned} \tag{9.51}$$

9.7 Interelectron repulsion integrals between two 1-center densities

We can use the Fourier transform discussed in the previous section to evaluate the interelectron repulsion integral between two 1-center Coulomb Sturmian densities located respectively on the centers $\mathbf{X}_{a_1}$ and $\mathbf{X}_{a_2}$. Let

$$J_{\tau_1,\tau_2;\tau_3,\tau_4} \equiv \int d\mathbf{x}\int d\mathbf{x}'\; \rho_{\mu_1,\mu_2}(\mathbf{x}-\mathbf{X}_{a_1})\frac{1}{|\mathbf{x}-\mathbf{x}'|}\rho_{\mu_3,\mu_4}(\mathbf{x}'-\mathbf{X}_{a_2}) \tag{9.52}$$

If we introduce the Fourier transform representation of the Coulomb interaction operator:

$$\frac{1}{|\mathbf{x}-\mathbf{x}'|} = \frac{1}{2\pi^2}\int d\mathbf{p}\;\frac{1}{p^2}e^{i\mathbf{p}\cdot(\mathbf{x}-\mathbf{x}')} \tag{9.53}$$

equation (4.33), then equation (9.52) can be rewritten in the form

$$\begin{aligned}&J_{\tau_1,\tau_2;\tau_3,\tau_4}(\mathbf{R}) \\ &= \frac{1}{2\pi^2}\int d\mathbf{p}\;\frac{1}{p^2}\int d\mathbf{x}\int d\mathbf{x}'\;\rho_{\tau_1,\tau_2}(\mathbf{x})\rho_{\tau_3,\tau_4}(\mathbf{x}')e^{i\mathbf{p}\cdot(\mathbf{x}-\mathbf{x}')} \\ &= 4\pi\int d\mathbf{p}\;\frac{1}{p^2}\;\rho^t_{\tau_1,\tau_2}(\mathbf{p})\;\rho^t_{\tau_3,\tau_4}(-\mathbf{p})\end{aligned} \tag{9.54}$$

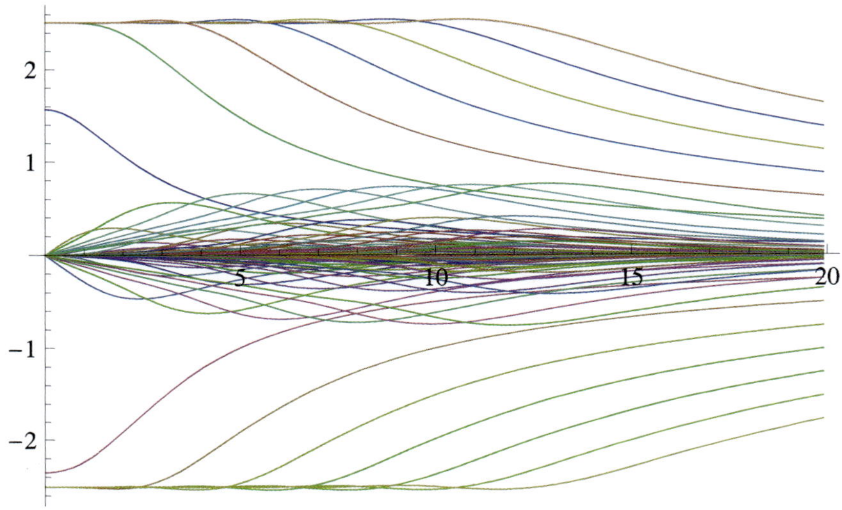

Fig. 9.3: The integrals $k^2 \int_0^\infty dp\ M(2k,p)R^t_{n,l}(2k,p)j_l(pR)$ of Equation (9.59) are shown here plotted as functions of $S \equiv kR$. There are 105 functions, corresponding $n = 1, 2, \ldots, 14$ and $l = 0, 1, \ldots, n-1$.

Using (9.51) and the relation $Y_{l,m}(-\mathbf{w}) = (-1)^l Y_{l,m}(-\mathbf{w})$, we obtain the interelectron repulsion integral between two 1-center densities, located respectively at positions $\mathbf{X}_{a_1}$ and $\mathbf{X}_{a_2}$:

$$J_{\tau_1,\tau_2,\tau_3,\tau_4}(\mathbf{R}) = \sum_{\mu'} \sum_{\mu} J_{\mu',\mu}(\mathbf{R})\ \mathcal{C}^{\mu'}_{\mu_1,\mu_2} \mathcal{C}^{\mu}_{\mu_3,\mu_4} \tag{9.55}$$

where

$$\begin{aligned} J_{\mu',\mu}(\mathbf{R}) &= 4\pi \int_0^\infty dp\ M(2k,p)^2 \int d\Omega_{3(p)}\ e^{i\mathbf{p}\cdot\mathbf{R}}(-1)^l Y_{\mu'}(\hat{\mathbf{w}}) Y_\mu(\hat{\mathbf{w}}) \\ M(2k,p)^2 &\equiv \left(\frac{4(2k)^{5/2}}{(4k^2+p^2)^2}\right)^2 \\ \hat{\mathbf{w}} &\equiv \left(\frac{4kp_1}{4k^2+p^2}, \frac{4kp_2}{4k^2+p^2}, \frac{4kp_3}{4k^2+p^2}, \frac{4k^2-p^2}{4k^2+p^2}\right) \end{aligned} \tag{9.56}$$

and where $\mathbf{R} \equiv \mathbf{X}_{a_2} - \mathbf{X}_{a_1}$. The integral $J_{\mu',\mu}$ can be further simplified.

Expanding the product $Y_{\mu'}(\hat{\mathbf{w}})Y_\mu(\hat{\mathbf{w}})$ in terms of single 4-dimensional hyperspherical harmonics $Y_{\mu''}(\hat{\mathbf{w}})$, and making use of equation (5.30), we

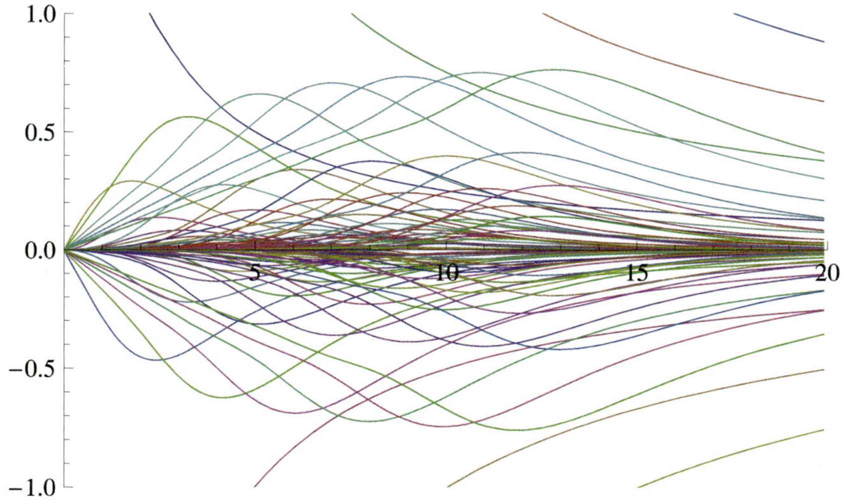

Fig. 9.4: The integrals $k^2 \int_0^\infty dp\ M(2k,p)R^t_{n,l}(2k,p)j_l(pR)$ shown in more detail. For small values of S the integrals are proportional to S^l, while for large values they are proportional to $1/S^{l+1}$.

have

$$\begin{aligned}
J_{\mu',\mu}(\mathbf{R}) &= (-1)^l \sum_{\mu''} J_{\mu''}(\mathbf{R}) \int d\Omega_4 Y^*_{\mu''}(\hat{\mathbf{w}}) Y_{\mu'}(\hat{\mathbf{w}}) Y_{\mu}(\hat{\mathbf{w}}) \\
J_{\mu''}(\mathbf{R}) &= 4\pi \int_0^\infty dp\ M(2k,p) \int d\Omega_{3(p)}\ e^{i\mathbf{p}\cdot\mathbf{R}} M(2k,p) Y_{\mu''}(\hat{\mathbf{w}}) \\
&= 4\pi i^{-l''} \int_0^\infty dp\ M(2k,p) \int d\Omega_{3(p)}\ e^{i\mathbf{p}\cdot\mathbf{R}} R^t_{n'',l''}(2k,p) Y_{l'',m''}(\mathbf{u}_p)
\end{aligned} \tag{9.57}$$

where $\int d\Omega_4 Y^*_{\mu''}(\hat{\mathbf{w}}) Y_{\mu'}(\hat{\mathbf{w}}) Y_{\mu}(\hat{\mathbf{w}})$ is a large but very sparse matrix that can be pre-calculated and stored. The next step is to expand $e^{i\mathbf{p}\cdot\mathbf{R}}$ in the series of equation (4.26):

$$e^{i\mathbf{p}\cdot\mathbf{R}} = 4\pi \sum_{l=0}^{\infty} i^l j_l(pR) \sum_{m=-l}^{l} Y_{l,m}(\mathbf{u}_R) Y^*_{l,m}(\mathbf{u}_p) \tag{9.58}$$

Performing the angular integration over $d\Omega_{3(p)}$ then yields

$$J_{\mu''}(\mathbf{R}) = (4\pi)^2 Y_{l'',m''}(\mathbf{u}_R) \int_0^\infty dp\ M(2k,p) R^t_{n'',l''}(2k,p) j_{l''}(pR) \tag{9.59}$$

Table 9.5: This table shows the radial Fourier transform of equation (9.60) for the first few values of n and l. Notice that k has been replaced by 2k.

n	l	$\sqrt{\frac{2}{\pi}}\int_0^\infty dr\ r^2 j_l(pr) R_{n,l}(2k,r)$
1	0	$\frac{32k^{5/2}}{(4k^2+p^2)^2\sqrt{\pi}}$
2	0	$-\frac{64k^{5/2}(4k^2-p^2)}{(4k^2+p^2)^3\sqrt{\pi}}$
2	1	$\frac{256k^{7/2}p}{(4k^2+p^2)^3\sqrt{3\pi}}$
3	0	$\frac{32k^{5/2}(48k^4-40k^2p^2+3p^4)}{(4k^2+p^2)^4\sqrt{\pi}}$
3	1	$-\frac{256k^{7/2}(4k^2p-p^3)\sqrt{2}}{(4k^2+p^2)^4\sqrt{\pi}}$
3	2	$\frac{1024k^{9/2}p^2\sqrt{2}}{(4k^2+p^2)^4\sqrt{5\pi}}$

where, using equation (4.44),

$$R^t_{n'',l''}(2k,p) \equiv \sqrt{\frac{2}{\pi}}\int_0^\infty dr\ r^2 j_{l''}(pr) R_{n'',l''}(2k,r) \tag{9.60}$$

as is illustrated in Table 8.5. Mathematica is able to evaluate the dp integration in equation (9.59) exactly. Some results are shown in Figures 9.3 and 9.4. These integrals can be pre-evaluated and stored. This gives us an exact and very rapid method for calculating interelectron repulsion integrals involving 1-center densities.

9.8 2-center densities in terms of 2k Sturmians

2-center densities are more difficult, but they too can be expanded in terms of a $2k$ Sturmian basis. We would like to represent a density $\rho_{\tau_1,\tau_2}(\mathbf{x}) =$

Table 9.6: Here we see, for the first few values of n and l, the integral which appears in equation (9.59), and which is illustrated in Figures 9.3 and 9.4.

n	l	$k^2 \int_0^\infty dp\ M(2k,p) R^t_{n,l}(p) j_l(pR) \qquad S \equiv kR$
1	0	$\frac{e^{-2S}}{12S}\sqrt{\frac{\pi}{2}}\left(24\ e^{2S} - 24 - 33S - 18S^2 - 4S^3\right)$
2	0	$\frac{e^{-2S}}{24S}\sqrt{\frac{\pi}{2}}\left(-96\ e^{2S} + 96 + 147S + 102S^2 + 40S^3 + 8S^4\right)$
2	1	$\frac{e^{-2S}}{12S^2}\sqrt{\frac{\pi}{6}}\left(48\ e^{2S} - 48 - 96S - 96S^2 - 59S^3 - 22S^4 - 4S^5\right)$
3	0	$\frac{e^{-2S}\sqrt{2\pi}}{15S}\left(45\ e^{2S} - 45 - 75S - 60S^2 - 30S^3 - 10S^4 - 2S^5\right)$
3	1	$\frac{e^{-2S}\sqrt{\pi}}{30S^2}\left(-120\ e^{2S} + 120 + 240S + 240S^2 + 155S^3 + 70S^4 + 22S^5 + 4S^6\right)$

$\chi^*_{\mu_1}(\mathbf{x}-\mathbf{X}_1)\chi_{\mu_2}(\mathbf{x}-\mathbf{X}_2)$ as a series of the form

$$\begin{aligned}&\chi^*_{\mu_1}(\mathbf{x}-\mathbf{X}_1)\chi_{\mu_2}(\mathbf{x}-\mathbf{X}_2)\\ &= k^{3/2}\sum_\mu \left[\chi_\mu(2k,\mathbf{x}-\mathbf{X}_1)c^{1,\mu}_{\mu_1;\mu_2}(\mathbf{S}) + \chi_\mu(2k,\mathbf{x}-\mathbf{X}_2)c^{2,\mu}_{\mu_1;\mu_2}(\mathbf{S})\right]\end{aligned} \tag{9.61}$$

where $\mathbf{X}_1$ and $\mathbf{X}_2$ are the positions of the two centers, and the expansion coefficients $c^{a,\mu}_{\mu_1;\mu_2}(\mathbf{S})$ are pure functions of $\mathbf{S} = k\mathbf{R}$. Multiplying equation (9.60) from the left by a conjugate $2k$-basis function and integrating, we have

$$\begin{aligned}&k^{3/2}\int d\mathbf{x}\ \chi^*_{\mu'}(2k,\mathbf{x}-\mathbf{X}_a)\sum_\mu\\ &\times\left[\chi_\mu(2k,\mathbf{x}-\mathbf{X}_1)c^{1,\mu}_{\mu_1;\mu_2}(\mathbf{S}) + \chi_\mu(2k,\mathbf{x}-\mathbf{X}_2)c^{2,\mu}_{\mu_1;\mu_2}(\mathbf{S})\right]\\ &= \int d\mathbf{x}\ \chi^*_{\mu'}(2k,\mathbf{x}-\mathbf{X}_a)\chi^*_{\mu_1}(\mathbf{x}-\mathbf{X}_1)\chi_{\mu_2}(\mathbf{x}-\mathbf{X}_2)\\ &\equiv k^{3/2}\ I^{a,\mu'}_{\mu_1,\mu_2}(\mathbf{S})\end{aligned} \tag{9.62}$$

We next use the definition of the overlap integral to rewrite the right hand side:

$$
\begin{aligned}
k^{3/2}\, I^{a,\mu'}_{\mu_1,\mu_2}(\mathbf{S}) &= k^{3/2}\sum_{\mu}\int d\mathbf{x}\ \chi^*_{\mu'}(2k,\mathbf{x}-\mathbf{X}_a)\\
&\times\left[\chi_\mu(2k,\mathbf{x}-\mathbf{X}_1)c^{1,\mu}_{\mu_1;\mu_2}(\mathbf{S})+\chi_\mu(2k,\mathbf{x}-\mathbf{X}_2)c^{2,\mu}_{\mu_1;\mu_2}(\mathbf{S})\right]\\
&= k^{3/2}\sum_{\mu}\sum_{j=1}^{2} m_{\mu',\mu}(2k,\mathbf{X}_j-\mathbf{X}_a)c^{j,\mu}_{\mu_1;\mu_2}(\mathbf{S})
\end{aligned}
\tag{9.63}
$$

where $m_{\mu',\mu}(2\mathbf{S})$ is the overlap integral between the $2k$ Coulomb Sturmians. Then, inserting $a=1$ and $a=2$ into equation (9.63), we obtain the linear matrix equations

$$
\begin{aligned}
I^{1,\mu'}_{\mu_1,\mu_2}(\mathbf{S}) &= \sum_{\mu}\left[m_{\mu'\mu}(\mathbf{0})\,c^{1,\mu}_{\mu_1;\mu_2}(\mathbf{S})+m_{\mu'\mu}(2\mathbf{S})c^{2,\mu}_{\mu_1;\mu_2}(\mathbf{S})\right]\\
I^{2,\mu'}_{\mu_1,\mu_2}(\mathbf{S}) &= \sum_{\mu}\left[m_{\mu'\mu}(-2\mathbf{S})\,c^{1,\mu}_{\mu_1;\mu_2}(\mathbf{S})+m_{\mu'\mu}(\mathbf{0})c^{2,\mu}_{\mu_1;\mu_2}(\mathbf{S})\right]
\end{aligned}
\tag{9.64}
$$

When S is large, it may be useful to multiply both sides of these equations by the reciprocal of the Coulomb Sturmian overlap matrix. If we do this, we obtain:

$$
\begin{aligned}
&\sum_{\mu'} m^{-1}_{\mu'',\mu'}(\mathbf{0})I^{1,\mu'}_{\mu_1,\mu_2}(\mathbf{S})\\
&= \sum_{\mu'} m^{-1}_{\mu'',\mu'}(\mathbf{0})\sum_{\mu}\left[m_{\mu'\mu}(\mathbf{0})\,c^{1,\mu}_{\mu_1;\mu_2}(\mathbf{S})+m_{\mu'\mu}(2\mathbf{S})c^{2,\mu}_{\mu_1;\mu_2}(\mathbf{S})\right]\\
&= c^{1,\mu''}_{\mu_1;\mu_2}(\mathbf{S})+\sum_{\mu}\left(\sum_{\mu'} m^{-1}_{\mu'',\mu'}(\mathbf{0})m_{\mu'\mu}(2\mathbf{S})\right)c^{2,\mu}_{\mu_1;\mu_2}(\mathbf{S})
\end{aligned}
\tag{9.65}
$$

and

$$
\begin{aligned}
&\sum_{\mu'} m^{-1}_{\mu'',\mu'}(\mathbf{0})I^{2,\mu'}_{\mu_1,\mu_2}(\mathbf{S})\\
&= \sum_{\mu'} m^{-1}_{\mu'',\mu'}(\mathbf{0})\sum_{\mu}\left[m_{\mu'\mu}(-2\mathbf{S})\,c^{1,\mu}_{\mu_1;\mu_2}(\mathbf{S})+m_{\mu'\mu}(\mathbf{0})c^{2,\mu}_{\mu_1;\mu_2}(\mathbf{S})\right]\\
&= c^{2,\mu''}_{\mu_1;\mu_2}(\mathbf{S})+\sum_{\mu}\left(\sum_{\mu'} m^{-1}_{\mu'',\mu'}(\mathbf{0})m_{\mu'\mu}(-2\mathbf{S})\right)c^{1,\mu}_{\mu_1;\mu_2}(\mathbf{S})
\end{aligned}
\tag{9.66}
$$

Rearranging the terms we have

$$c^{1,\mu''}_{\mu_1;\mu_2}(\mathbf{S}) = \sum_{\mu'} m^{-1}_{\mu'',\mu'}(\mathbf{0}) I^{1,\mu'}_{\mu_1,\mu_2}(\mathbf{S})$$
$$- \sum_{\mu} \left(\sum_{\mu'} m^{-1}_{\mu'',\mu'}(\mathbf{0}) m_{\mu'\mu}(2\mathbf{S}) \right) c^{2,\mu}_{\mu_1;\mu_2}(\mathbf{S})$$
$$c^{2,\mu''}_{\mu_1;\mu_2}(\mathbf{S}) = \sum_{\mu'} m^{-1}_{\mu'',\mu'}(\mathbf{0}) I^{2,\mu'}_{\mu_1,\mu_2}(\mathbf{S})$$
$$- \sum_{\mu} \left(\sum_{\mu'} m^{-1}_{\mu'',\mu'}(\mathbf{0}) m_{\mu'\mu}(-2\mathbf{S}) \right) c^{1,\mu}_{\mu_1;\mu_2}(\mathbf{S}) \tag{9.67}$$

If we let

$$W^{1}_{\mu'',\mu}(\mathbf{S}) \equiv \sum_{\mu'} m^{-1}_{\mu'',\mu'}(\mathbf{0}) m_{\mu'\mu}(2\mathbf{S})$$
$$W^{2}_{\mu'',\mu}(\mathbf{S}) \equiv \sum_{\mu'} m^{-1}_{\mu'',\mu'}(\mathbf{0}) m_{\mu'\mu}(-2\mathbf{S}) \tag{9.68}$$

then this becomes:

$$c^{1,\mu''}_{\mu_1;\mu_2}(\mathbf{S}) = \sum_{\mu'} m^{-1}_{\mu'',\mu'}(\mathbf{0}) I^{1,\mu'}_{\mu_1,\mu_2}(\mathbf{S}) - \sum_{\mu} W^{1}_{\mu'',\mu}(\mathbf{S}) c^{2,\mu}_{\mu_1;\mu_2}(\mathbf{S})$$
$$c^{2,\mu''}_{\mu_1;\mu_2}(\mathbf{S}) = \sum_{\mu'} m^{-1}_{\mu'',\mu'}(\mathbf{0}) I^{2,\mu'}_{\mu_1,\mu_2}(\mathbf{S}) - \sum_{\mu'} W^{2}_{\mu'',\mu'}(\mathbf{S}) c^{1,\mu'}_{\mu_1;\mu_2}(\mathbf{S}) \tag{9.69}$$

or, in a more condensed notation,

$$\mathbb{C}^1(\mathbf{S}) = \mathbb{M}^{-1}(\mathbf{0})\mathbb{I}^1(\mathbf{S}) - \mathbb{W}^1(\mathbf{S})\mathbb{C}^2(\mathbf{S})$$
$$\mathbb{C}^2(\mathbf{S}) = \mathbb{M}^{-1}(\mathbf{0})\mathbb{I}^2(\mathbf{S}) - \mathbb{W}^2(\mathbf{S})\mathbb{C}^1(\mathbf{S}) \tag{9.70}$$

We can also join the two equations of (9.70) into a single equation:

$$\begin{pmatrix} \mathbb{C}^1(\mathbf{S}) \\ \mathbb{C}^2(\mathbf{S}) \end{pmatrix} = \begin{pmatrix} \mathbb{M}^{-1}(\mathbf{0})\mathbb{I}^1(\mathbf{S}) \\ \mathbb{M}^{-1}(\mathbf{0})\mathbb{I}^2(\mathbf{S}) \end{pmatrix} - \begin{pmatrix} 0 & \mathbb{W}^1(\mathbf{S}) \\ \mathbb{W}^2(\mathbf{S}) & 0 \end{pmatrix} \begin{pmatrix} \mathbb{C}^1(\mathbf{S}) \\ \mathbb{C}^2(\mathbf{S}) \end{pmatrix} \tag{9.71}$$

or, in a still more abbreviated notation,

$$\mathbb{C} = \mathbb{M}^{-1}\mathbb{I} - \mathbb{W}\mathbb{C} \tag{9.72}$$

Looking at the form of equations (9.72), we can see that for moderately large values of S, it can be iterated:

$$\begin{aligned}
[\mathbb{C}]_0 &= \mathbb{M}^{-1}\mathbb{I} \\
[\mathbb{C}]_1 &= \mathbb{M}^{-1}\mathbb{I} - \mathbb{W}[\mathbb{C}]_0 = (1 - \mathbb{W})\mathbb{M}^{-1}\mathbb{I} \\
[\mathbb{C}]_2 &= \mathbb{M}^{-1}\mathbb{I} - \mathbb{W}[\mathbb{C}]_1 = (1 - \mathbb{W} + \mathbb{W}^2)\mathbb{M}^{-1}\mathbb{I} \\
[\mathbb{C}]_3 &= \mathbb{M}^{-1}\mathbb{I} - \mathbb{W}[\mathbb{C}]_2 = (1 - \mathbb{W} + \mathbb{W}^2 - \mathbb{W}^3)\mathbb{M}^{-1}\mathbb{I} \\
\vdots \quad & \quad \vdots \qquad \quad \vdots
\end{aligned} \tag{9.73}$$

where the subscript on the square brackets indicates the order of iteration. Alternatively, equation (9.72) can be written in the form

$$(1+\mathbb{W})\mathbb{C} = \mathbb{M}^{-1}\mathbb{I} \tag{9.74}$$

A formal solution for $\mathbb{C}$ is then given by

$$\mathbb{C} = \frac{1}{(1+\mathbb{W})}\,\mathbb{M}^{-1}\mathbb{I} = (1-\mathbb{W}+\mathbb{W}^2-\mathbb{W}^3+\cdots)\,\mathbb{M}^{-1}\mathbb{I} \tag{9.75}$$

where we have expanded $(1+\mathbb{W})^{-1}$ as a Taylor series in powers of $\mathbb{W}$. The series can also be generated by iteration, as is shown in equation (9.73). The inverse matrix, $(1+\mathbb{W})^{-1}$, can be obtained without any problems using an ordinary matrix inversion program when S is moderately large. However, as S becomes smaller, $(S<4)$, the matrix $(1+\mathbb{W})$ becomes more and more ill-conditioned because of overcompleteness of the basis set, and its inverse can only be obtained by following a special procedure in which $(1+\mathbb{W})$ is first transformed to its diagonal representation, and the smallest eigenvalues and the corresponding eigenvectors are gradually phased out. This special procedure is described in [Avery, 2013]. The iterative solutions of equation (9.73) can be obtained for all values of S except $S=0$. Convergence of the series $1-\mathbb{W}+\mathbb{W}^2-\mathbb{W}^3+,,,$ is very rapid when $S>2$, but becomes slower for smaller values of S. Finally, the convergence fails entirely when $S=0$, but the solution here is not needed, since in this special case, the density is just represented by the product of two Coulomb Sturmians located at the same point. All of the solutions can be checked by comparing them with the 2-center density that we wish to represent.

Before we are able to calculate the coefficients $c^{a,\mu}_{\mu_1;\mu_2}(\mathbf{S})$, we must perform integrals of the form

$$\begin{aligned}
I^{1,\mu}_{\mu_1;\mu_2}(\mathbf{S}) &= k^{-3/2}\int d\mathbf{x}\;\chi^*_\mu(2k,\mathbf{x}-\mathbf{X}_1)\chi^*_{\mu_1}(\mathbf{x}-\mathbf{X}_1)\chi_{\mu_2}(\mathbf{x}-\mathbf{X}_2)\\
I^{2,\mu}_{\mu_1;\mu_2}(\mathbf{S}) &= k^{-3/2}\int d\mathbf{x}\;\chi^*_\mu(2k,\mathbf{x}-\mathbf{X}_2)\chi_{\mu_2}(\mathbf{x}-\mathbf{X}_2)\chi^*_{\mu_1}(\mathbf{x}-\mathbf{X}_1)
\end{aligned} \tag{9.76}$$

For calculating $I^{1,\mu}_{\mu_1;\mu_2}(\mathbf{S})$ and $I^{2,\mu}_{\mu_1;\mu_2}(\mathbf{S})$, we use the exact expansion

$$\begin{aligned}
\chi^*_\mu(2k,\mathbf{x})\chi^*_{\mu_1}(\mathbf{x}) &= \sum_{\mu'}\chi^*_{\mu'}(3k,\mathbf{x})\;b^{1,\mu'}_{\mu,\mu_1}\\
\chi^*_\mu(2k,\mathbf{x})\chi_{\mu_2}(\mathbf{x}) &= \sum_{\mu'}\chi_{\mu'}(3k,\mathbf{x})\;b^{2,\mu'}_{\mu,\mu_2}
\end{aligned} \tag{9.77}$$

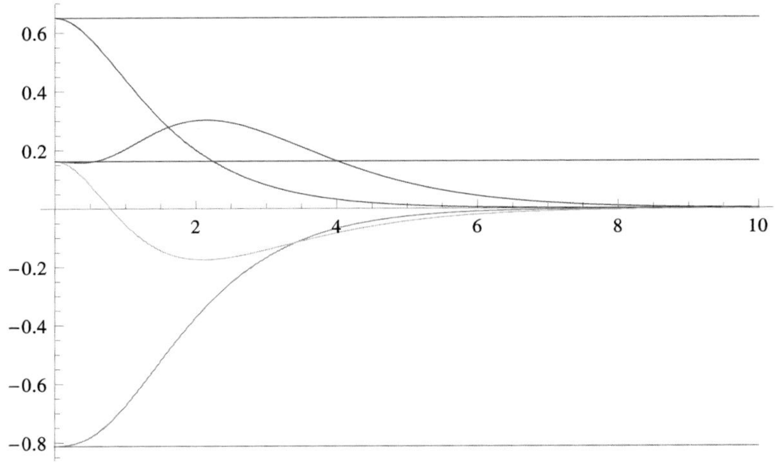

Fig. 9.5: This figure shows the overlap integrals $m^{3k,k}_{\mu',\mu_2}(S)$ of equation (9.79) for a few values of μ' and μ. The curves shown correspond to the 1s-1s, 2s-1s, 1s-2s, and 2s-2s overlaps. The evaluation of the overlap integrals when $S=0$ can be evaluated exactly as a check. By a coincidence, both the 2s-2s and the 1s-2s overlaps have the same value in this limit.

where, using the potential-weighted orthonormality of the Coulomb Sturmians and the orthonormality of the spherical harmonics, we find that

$$\begin{aligned}
b^{1,\mu'}_{\mu,\mu_1} &= \frac{n'}{3k}\int_0^\infty dr\; r\; R_{n',l'}(3k,r)R_{n,l}(2k,r)R_{n_1,l_1}(r) \\
&\times \int d\Omega_3\; Y_{l',m+m_1}(\mathbf{u})Y^*_{l,m}(\mathbf{u})Y^*_{l_1,m_1}(\mathbf{u}) \\
b^{2,\mu'}_{\mu,\mu_2} &= \frac{n'}{3k}\int_0^\infty dr\; r\; R_{n',l'}(3k,r)R_{n,l}(2k,r)R_{n_2,l_2}(r) \\
&\times \int d\Omega_3\; Y^*_{l',m_2-m}(\mathbf{u})Y^*_{l,m}(\mathbf{u})Y_{l_2,m_2}(\mathbf{u})
\end{aligned} \tag{9.78}$$

Then, from equation (4.61) we have

$$\begin{aligned}
k^{3/2}I^{1,\mu}_{\mu_1,\mu_2}(\mathbf{S}) &= \sum_{\mu'} b^{1,\mu'}_{\mu,\mu_1}\int d\mathbf{x}\; \chi^*_{\mu'}(3k,\mathbf{x}-\mathbf{X}_1)\chi_{\mu_2}(\mathbf{x}-\mathbf{X}_2) \\
&= \sum_{\mu'} b^{1,\mu'}_{\mu,\mu_1}\int d\mathbf{p}\; \chi^{*t}_{\mu'}(3k,\mathbf{p})\chi^t_{\mu_2}(\mathbf{p})\; e^{-i\mathbf{p}\cdot\mathbf{R}} \\
&\equiv \sum_{\mu'} b^{1,\mu'}_{\mu,\mu_1}\; m^{3k,k}_{\mu',\mu_2}(\mathbf{S})
\end{aligned} \tag{9.79}$$

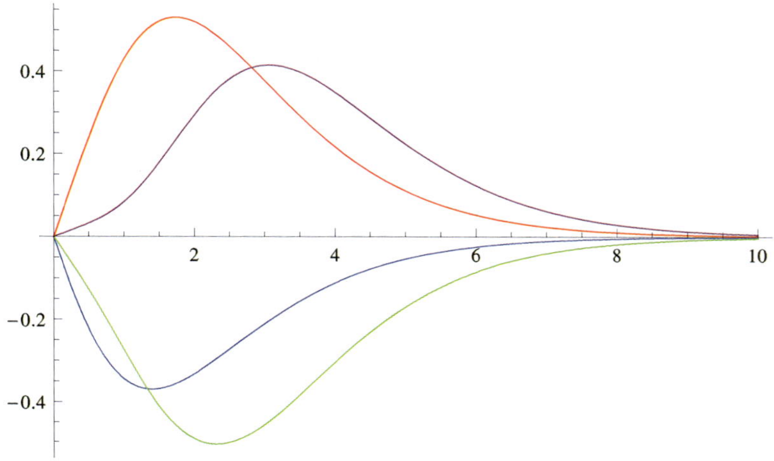

Fig. 9.6: This figure shows the overlap integrals $m^{3k,k}_{\mu',\mu_2}(S)$ *of equation (9.79) for* $\mu_2 = (2,1,0)$ *with* $\mu' = (n,0,0)$. *The values of* n *shown in the figure are 1 (blue), 2 (red), 3 (green) and 4 (purple).*

The equation for $I^{2,\mu}_{\mu_1,\mu_2}(\mathbf{S})$ is similar, but not identical:

$$\begin{aligned} k^{3/2} I^{2,\mu}_{\mu_1,\mu_2}(\mathbf{S}) &= \sum_{\mu'} b^{2,\mu'}_{\mu,\mu_2} \int d\mathbf{x}\; \chi_{\mu'}(3k, \mathbf{x} - \mathbf{X}_2) \chi^*_{\mu_1}(\mathbf{x} - \mathbf{X}_1) \\ &= \sum_{\mu'} b^{2,\mu'}_{\mu,\mu_2} \int d\mathbf{p}\; \chi^t_{\mu'}(3k, \mathbf{p}) \chi^{*t}_{\mu_1}(\mathbf{p})\; e^{-i\mathbf{p}\cdot\mathbf{R}} \\ &\equiv \sum_{\mu'} b^{2,\mu'}_{\mu,\mu_2}\; m^{*3k,k}_{\mu',\mu_1}(-\mathbf{S}) \end{aligned} \tag{9.80}$$

Using the expansion of a plane wave in terms of spherical harmonics and spherical Bessel functions,

$$e^{-i\mathbf{p}\cdot\mathbf{R}} = 4\pi \sum_{l=0}^{\infty} (-i)^l j_l(pR) \sum_{m=-l}^{l} Y_{l,m}(\mathbf{u}_R) Y^*_{l,m}(\mathbf{u}_p) \tag{9.81}$$

we obtain

$$\begin{aligned} \chi^t_{n_2,l_2,m_2}(\mathbf{p}) &= (-i)^{l_2} Y_{l_2,m_2}(\mathbf{u}_p) \sqrt{\frac{2}{\pi}} \int_0^{\infty} dr\; r^2\; j_{l_2}(pr)\; R_{n_2,l_2}(r) \\ &\equiv (-i)^{l_2} Y_{l_2,m_2}(\mathbf{u}_p) R^t_{n_2,l_2}(p) \\ \chi^{*t}_{n',l',m'}(3k, \mathbf{p}) &= (-i)^{l'} Y^*_{l',m'}(\mathbf{u}_p) \sqrt{\frac{2}{\pi}} \int_0^{\infty} dr\; r^2\; j_{l'}(pr)\; R_{n',l'}(3k, r) \\ &\equiv (-i)^{l'} Y^*_{l',m'}(\mathbf{u}_p) R^t_{n',l'}(3k, p) \end{aligned} \tag{9.82}$$

With the help of these Fourier transforms, we can write $m^{3k,k}_{\mu',\mu_2}(\mathbf{S})$ as

$$\begin{aligned}
m^{3k,k}_{\mu',\mu_2}(\mathbf{S}) &\equiv \int d\mathbf{x}\ \chi^*_{\mu'}(3k,\mathbf{x}-\mathbf{X}_1)\chi_{\mu_2}(\mathbf{x}-\mathbf{X}_2) \\
&= \int d\mathbf{p}\ \chi^{t*}_{\mu'}(3k,\mathbf{p})\,\chi^t_{\mu_2}(-\mathbf{p})\,e^{-i\mathbf{p}\cdot\mathbf{R}} \\
&= 4\pi\sum_{l=0}^{\infty}(-i)^l\sum_{m=-l}^{l} Y_{l,m}(\mathbf{u}_S) \\
&\times \int_0^\infty dp\ p^2 j_l(pR)\int d\Omega_{3(p)}\chi^{*t}_{\mu'}(3k,\mathbf{p})\chi^t_{\mu_2}(-\mathbf{p})\ Y^*_{l,m}(\mathbf{u}_p) \\
&= 4\pi\ (-1)^{l_2}\sum_{l=|l'-l_2|}^{l'+l_2}(-i)^{l+l'+l_2}Y_{l,m_2-m'}(\mathbf{u}_S) \\
&\times \int d\Omega_{3(p)}Y^*_{l,m_2-m'}(\mathbf{u}_p)Y^*_{l',m'}(\mathbf{u}_p)Y_{l_2,m_2}(\mathbf{u}_p) \\
&\times \int_0^\infty dp\ p^2 j_l(pR)\ R^t_{n',l'}(3k,p)\ R^t_{n_2,l_2}(p)
\end{aligned} \tag{9.83}$$

Alternatively, $m^{3k,k}_{\mu',\mu_2}(\mathbf{S})$ can be evacuated by expanding $\chi^*_{\mu'}(3k,\mathbf{x})$ in terms of $\chi^*_{\mu'}(1k,\mathbf{x})$:

$$\begin{aligned}
m^{3k,k}_{\mu',\mu_2}(\mathbf{S}) &\equiv \int d\mathbf{x}\ \chi^*_{\mu'}(3k,\mathbf{x}-\mathbf{X}_1)\chi_{\mu_2}(\mathbf{x}-\mathbf{X}_2) \\
&= \sum_{\mu} a^{\mu}_{\mu'}\int d\mathbf{x}\ \chi^*_{\mu}(\mathbf{x}-\mathbf{X}_1)\chi_{\mu_2}(\mathbf{x}-\mathbf{X}_2) \\
&= \sum_{\mu} a^{\mu}_{\mu'}\ m_{\mu,\mu_2}(\mathbf{S})
\end{aligned} \tag{9.84}$$

Having found the coefficients $c^{a,\mu}_{\mu_1,\mu_2}$ in the expression

$$\begin{aligned}
&\chi^*_{\mu_1}(\mathbf{x}-\mathbf{X}_1)\chi_{\mu_2}(\mathbf{x}-\mathbf{X}_2) \\
&\approx \sum_{\mu}\left[\chi_\mu(2k,\mathbf{x}-\mathbf{X}_1)c^{1,\mu}_{\mu_1,\mu_2}+\chi_\mu(2k,\mathbf{x}-\mathbf{X}_2)c^{2,\mu}_{\mu_1,\nu_2}\right]
\end{aligned} \tag{9.85}$$

we can store them and use them to obtain interelectron repulsion integrals involving 2-center densities in a manner similar to our previous calculation for 1-center densities: However, the result is an approximation rather than an exact equality. Furthermore, the matrices $c^{a,\mu}_{\mu_1,\mu_2}$ are not sparse.

9.9 Generalized scattering factors

In recent years, very high quality X-ray diffraction data has become available from experiments where the crystal being examined is held at liquid helium temperatures, and from experiments using synchrotron radiation. In order to obtain the maximum amount of information from such high quality data, it is desirable to use generalized scattering factors in the refinement of the structures. In such a refinement, the charge density in the crystal is represented by means of the first-order density matrix $P_{\tau',\tau}$:

$$\begin{aligned}\rho(\mathbf{x}) &= \sum_{\tau'}\sum_{\tau} P_{\tau',\tau}\ \chi^*_{\tau'}(\mathbf{x})\chi_\tau(\mathbf{x}) \\ &\equiv \sum_{\mu',a'}\sum_{\mu,a} P_{\mu,a';\mu,a}\ \chi^*_{\mu'}(\mathbf{x}-\mathbf{X}_{a'})\chi_\mu(\mathbf{x}-\mathbf{X}_a)\end{aligned} \tag{9.86}$$

where

$$\tau \equiv (\mu, a) \equiv (n, l, m, a) \tag{9.87}$$

and where

$$\chi_\mu(\mathbf{x}-\mathbf{X}_a) \equiv \chi_{n,l,m}(\mathbf{x}-\mathbf{X}_a) \tag{9.88}$$

is an atomic orbital centered on $\mathbf{X}_a$, the position of the ath atom. Then the Fourier transform of the density, which is the quantity measured in X-ray diffraction experiments, is given by

$$\int d\mathbf{x}\ e^{i\mathbf{q}\cdot\mathbf{x}}\rho(\mathbf{x}) = \sum_{\mu',a'}\sum_{\mu,a} P_{\mu,a';\mu,a}\int d\mathbf{x}\ e^{i\mathbf{q}\cdot\mathbf{x}}\ \chi^*_{\mu'}(\mathbf{x}-\mathbf{X}_{a'})\chi_\mu(\mathbf{x}-\mathbf{X}_a) \tag{9.89}$$

The quantity

$$\int d\mathbf{x}\ e^{i\mathbf{q}\cdot\mathbf{x}}\ \chi^*_{\mu'}(\mathbf{x}-\mathbf{X}_{a'})\chi_\mu(\mathbf{x}-\mathbf{X}_a) \tag{9.90}$$

is called a *generalized scattering factor*. Here $\mathbf{q} = \mathbf{k}' - \mathbf{k}$ is the difference between the incoming and scattered wave vectors. If one is able to evaluate an appropriate set of generalized scattering factors, then the first-order density matrix may be obtained experimentally by refining the high-quality diffraction data so as to obtain the best possible fit. When this is possible, it is an important achievement, since the first-order density matrix is a starting point for density functional calculations of the properties of materials, or alternatively for Hartree-Fock calculations. The calculation of generalized scattering factors for Coulomb Sturmian basis sets is already programmed and working, since the Fourier transforms of 2-center partial densities are needed for the calculation of interelectron repulsion integrals.

The density of heavy atoms is produced by a core, which is almost undisturbed when the atom is incorporated into a crystal, and a valence shell, where considerable charge flow occurs into the bonding regions, or even from one atom to another. Therefore in analyzing X-ray diffraction data, it is a good approximation to assume that only the valence shell is affected, while the core is "frozen", and only the generalized scattering factors for valence-shell orbitals need be calculated. The part of the density that differs from the density of a superposition atoms is called the "deformation density", and it is almost entirely due to orbitals in valence shells.

9.10 Exercises

(1) Type "`?mIc`" to obtain the definition of the function which generates displaced Coulomb Sturmian overlap integrals using Cartesian coordinates. Make a table of (λ, l, m) values up to $\lambda = 2$. Use this table, together with the function `mIc` to generate a 14×14 table of displaced Coulomb Sturmian overlap integrals. Compare these results with Table 9.2.

(2) Type `?swIc` to obtain the definition of the function which generates Shibuya-Wulfman integrals using Cartesian coordinates. Make a 14x14 table of these integrals. Show that when $S = 0$ the table becomes the identity matrix.

(3) Use equation (9.40) to make a table of kinetic energy integrals for many center Coulomb Sturmian basis sets. Compare with Table 9.3.

(4) Type `?fC36` and `?g56` to obtain definitions of the functions that generate $f_{nl}(S)$ and $g_{n,l}(S)$ (equations (9.33) and (9.36)). Make tables of these functions and compare them with Table 9.1.

(5) Use equation (9.31) to generate the functions $f_{nl}(S)$ using 4-dimensional hyperspherical harmonics.

Chapter 10

LAGRANGIANS FOR PARTICLES AND FIELDS

10.1 Introduction

In our discussion of hyperspherical harmonics in previous chapters, the generalized Laplacian operator played an important role. We asserted that when expressed in terms of the hyperradius and hyperangular coordinates, the generalized Laplacian operator can be written in the form shown in equation (2.1). In the present chapter, we will supply the mathematical background for (2.1), and we will show how the generalized Laplacian (or Laplace-Beltrami) operator can be expressed in any curvilinear coordinate system whatever. To do this, we will need to know something about the calculus of variations.

In a typical problem of the calculus of variations, one considers an integral of the form

$$\delta S \equiv \delta \int L\left(x^1, x^2, \cdots, x^d, \frac{dx^1}{dt}, \frac{dx^2}{dt}, \cdots, \frac{dx^d}{dt},\right) dt = 0 \qquad (10.1)$$

L is some function of the coordinates, $x^1, \cdots, x^d$ and their t-derivatives. The problem is to find the coordinates as functions of t which will give a minimum or maximum value to the integral S. For example, the principle of Pierre Fermat (1601–1665) states that *in geometrical optics, the actual path of a ray of light is the one that takes the least time.* The infinitesimal time dt required for the light signal to move an infinitesimal distance dl along its path is

$$dt = \frac{n(\mathbf{x})}{c} dl \qquad (10.2)$$

where c is the velocity of light in a vacuum and $n(\mathbf{x})$ is the index of refraction. From the Pythagorean Theorem we have

$$
\begin{aligned}
dl &= \sqrt{dx^2 + dy + dz^2} = \sqrt{d\mathbf{x} \cdot d\mathbf{x}} \\
&= \sqrt{\frac{d\mathbf{x}}{dl} \cdot \frac{d\mathbf{x}}{dl}}\, dl = \frac{d\mathbf{x}}{dl} \cdot \frac{d\mathbf{x}}{dl}\, dl
\end{aligned}
\tag{10.3}
$$

Thus we can write Fermat's principle in the form:

$$
S = \int L\left(x, y, z, \frac{dx}{dl}, \frac{dy}{dl}, \frac{dz}{dl}\right) dl = \text{minimum} \tag{10.4}
$$

where

$$
L = n(\mathbf{x}) \frac{d\mathbf{x}}{dl} \cdot \frac{d\mathbf{x}}{dl} \tag{10.5}
$$

A similar principle was discovered by the great Irish mathematician Sir William Rowan Hamilton (1805–1865). In 1835, he showed that for a system of particles whose state in Newtonian mechanics is specified at a given time by the coordinates $x^1, x^2, \cdots, x^d$, and the velocities $dx^1/dt, dx^2/dt, \cdots, dx^d/dt$. the integral

$$
S = \int L\ dt = \int (T - V)\ dt \tag{10.6}
$$

is an extremum, where T is the kinetic energy

$$
T = \frac{1}{2} \sum_{i=1}^{d} \sum_{j=1}^{d} m_{i,j} \frac{dx^i}{dt} \frac{dx^j}{dt} \equiv \frac{1}{2} \sum_{i=1}^{d} \sum_{j=1}^{d} m_{i,j} \dot{x}^i \dot{x}^j \tag{10.7}
$$

and where $V(x^1, x^2, \cdots, x^d)$ is the potential energy. Leonhard Euler (1707–1783) and Joseph-Louis Lagrange (1736–1813), who developed the calculus of variations, had shown that if the coordinates and their time derivatives obey the differential equations

$$
\frac{d}{dt} \frac{\partial L}{\partial \dot{x}^i} - \frac{\partial L}{\partial x^i} = 0 \qquad i = 1, 2, \cdots, d \tag{10.8}
$$

Then the integral $S = \int L\ dt$ will be an extremum, and vice versa. The way that they showed this was as follows: Suppose that we have found the true path, $x^i(t)$, for which $S = \int L\ dt$ is an extremum. Now consider what happens to S when we wander slightly away from the true path. The situation is analogous to calculating the change of a function as we move very slightly away from one of its maxima or minima. If we are at the top of a mountain, or at the bottom of a valley, then taking a very slight step in any direction will not change our altitude, since at that point the ground is level. In the same way, if we alter the path by an amount δx^i, the resulting alteration in $\int L\,dt$ will be zero:

$$\delta \int L\,dt = \int \delta L\,dt = 0 \tag{10.9}$$

The variation of the Lagrangian function L resulting directly from the variation of the coordinates, or indirectly through the consequent variation of the velocities is

$$\delta L = \sum_{i=1}^{d} \left[\frac{\partial L}{\partial x^i} \delta x^i + \frac{\partial L}{\partial \dot{x}^i} \frac{d}{dt} (\delta x^i) \right] \tag{10.10}$$

We next integrate by parts using the relationship

$$\int_a^b u\,dv = [uv]_a^b - \int_a^b v\,du \tag{10.11}$$

This give us the relationship

$$\int_a^b \sum_{i=1}^{d} \frac{\partial L}{\partial \dot{x}^i} \frac{d}{dt} (\delta x^i)\,dt = \left[\sum_{i=1}^{d} \frac{\partial L}{\partial \dot{x}^i} \delta x^i \right]_a^b - \int_a^b \sum_{i=1}^{d} \frac{d}{dt} \frac{\partial L}{\partial \dot{x}^i} \delta x^i\,dt \tag{10.12}$$

Since the slightly altered path must still reach the end points a and b, the variation from the true path must vanish at those points, and therefore

$$\left[\sum_{i=1}^{d} \frac{\partial L}{\partial \dot{x}^i} \delta x^i \right]_a^b = 0 \tag{10.13}$$

Finally, combining equations (10.10), (10.12) and (10.13). we obtain

$$\int_a^b \delta L\,dt = \int_a^b \sum_{i=1}^{d} \left[-\frac{d}{dt} \frac{\partial L}{\partial \dot{x}^i} + \frac{\partial L}{\partial x^i} \right] \delta x^i\,dt = 0 \tag{10.14}$$

To ensure that the integral in (10.14) will vanish for an arbitrary slight variation of path δx^i, it is necessary that

$$\frac{d}{dt} \frac{\partial L}{\partial \dot{x}^i} - \frac{\partial L}{\partial x^i} = 0 \qquad i = 1, 2, \cdots, d \tag{10.15}$$

Therefore the Euler-Lagrange equations (10.8) are a consequence of action principle (10.1).

10.2 Cyclic coordinates

The Lagrangian formalism allows us to obtain conservation laws with great ease. As an example, we can think of a single particle moving in a central potential, $V(r)$. This is a case where it is convenient to express the particle's Lagrangian in terms of spherical polar coordinates. Let

$$\begin{aligned} x &= r\sin\theta\cos\phi \\ y &= r\sin\theta\sin\phi \\ z &= r\cos\theta \end{aligned} \tag{10.16}$$

In Cartesian coordinates, the element of length is given by

$$dl^2 = dx^2 + dy^2 + dz^2 \tag{10.17}$$

Combining (10.16) and (10.17) we find that in spherical polar coordinates, the element of length is

$$dl^2 = dr^2 + r^2 d\theta^2 + r^2\sin^2\theta\ d\phi^2 \tag{10.18}$$

We can now write down the Lagrangian of the particle in terms of r, θ and ϕ:

$$L = \frac{1}{2}m\left[\left(\frac{dr}{dt}\right)^2 + r^2\left(\frac{d\theta}{dt}\right)^2 + r^2\sin^2\theta\left(\frac{d\phi}{dt}\right)^2\right] - V(r) \tag{10.19}$$

The Euler-Lagrange equations of the particle then become

$$\begin{aligned} \frac{d}{dt}\frac{\partial L}{\partial \dot{r}} = \frac{\partial L}{\partial r} \quad &\rightarrow \quad m\frac{d^2 r}{dt^2} = mr\dot{\theta}^2 + mr\sin^2\theta\dot{\phi}^2 - \frac{\partial V}{\partial r} \\ \frac{d}{dt}\frac{\partial L}{\partial \dot{\theta}} = \frac{\partial L}{\partial \theta} \quad &\rightarrow \quad \frac{d}{dt}\left(mr^2\frac{d\theta}{dt}\right) = mr^2\sin\theta\cos\theta\dot{\phi}^2 \\ \frac{d}{dt}\frac{\partial L}{\partial \dot{\phi}} = \frac{\partial L}{\partial \phi} \quad &\rightarrow \quad \frac{d}{dt}\left(mr^2\sin^2\theta\frac{d\phi}{dt}\right) = 0 \end{aligned} \tag{10.20}$$

The second and third of the equations in this array are conservation laws. In fact, if the coordinate system is chosen in such a way that $\dot{\phi} = 0$, the second equation is Kepler's second law. When a coordinate does not appear in the Lagrangian, but only its time derivative, that coordinate is said to be *cyclic*. For each cyclic coordinate, there is a conservation law. The momentum conjugate to a coordinate is defined to be the partial derivative of the Lagrangian with respect to the time derivative of that coordinate.

In the example which we are considering here, the momenta conjugate to the coordinates r, θ and ϕ are

$$\begin{aligned} p_r &\equiv \frac{\partial L}{\partial \dot{r}} = m\frac{dr}{dt} \\ p_\theta &\equiv \frac{\partial L}{\partial \dot{\theta}} = mr^2\frac{d\theta}{dt} \\ p_\phi &\equiv \frac{\partial L}{\partial \dot{\phi}} = mr^2 \sin^2\theta\frac{d\phi}{dt} \end{aligned} \tag{10.21}$$

We can see from this example that the momenta which are conjugate to cyclic coordinates are conserved. The Euler-Lagrange equations ensure that this is true in general. We can also see that transformation to coordinates, as many as possible of which are cyclic, is a big step towards solving the equations of motion of a system.

As a second example of a transformation to coordinates, some of which are cyclic, we can think of a two particles interacting through a potential which depends only on the distance between them. In that case, the Lagrangian, expressed in Cartesian coordinates, is given by

$$L = \frac{1}{2}m_1\frac{d\mathbf{x}_1}{dt}\cdot\frac{d\mathbf{x}_1}{dt} + \frac{1}{2}m_2\frac{d\mathbf{x}_2}{dt}\cdot\frac{d\mathbf{x}_2}{dt} - V(|\mathbf{x}_1 - \mathbf{x}_2|) \tag{10.22}$$

The Lagrangian formulation allows us to introduce a new set of coordinates which are much more convenient. Let

$$\begin{aligned} \mathbf{X}_{c.m.} &\equiv \frac{m_1\mathbf{x}_1 + m_2\mathbf{x}_2}{m_1 + m_2} \\ \mathbf{X}_{12} &\equiv \mathbf{x}_1 - \mathbf{x}_2 \end{aligned} \tag{10.23}$$

In terms of the center of mass coordinates $\mathbf{X}_{c.m.}$ and the relative position coordinates $\mathbf{X}_{12}$, the Lagrangian of the system becomes:

$$\begin{aligned} L = {} & \frac{1}{2}(m_1 + m_2)\frac{d\mathbf{X}_{c.m.}}{dt}\cdot\frac{d\mathbf{X}_{c.m.}}{dt} + \frac{1}{2}\left(\frac{m_1 m_2}{m_1 + m_2}\right)\frac{d\mathbf{X}_{12}}{dt}\cdot\frac{d\mathbf{X}_{12}}{dt} \\ & - V(|\mathbf{X}_{12}|) \end{aligned} \tag{10.24}$$

Since the Lagrangian does not depend on $\mathbf{X}_{c.m}$, the center of mass coordinates are cyclic, and the momenta conjugate to them are conserved:

$$\frac{d}{dt}(\mathbf{p}_{c.m.}) = \frac{d}{dt}\left(\frac{\partial L}{\partial \dot{\mathbf{X}}_{c.m.}}\right) = \frac{d}{dt}\left((m_1 + m_2)\frac{d\mathbf{X}_{c.m.}}{dt}\right) = 0 \tag{10.25}$$

Since the potential energy does not depend on the orientation of the vector $\mathbf{X}_{12}$, but only on its magnitude, we could complete our transformation to cyclic coordinates by expressing $\mathbf{X}_{12}$ in terms of spherical polar coordinates. Then the only non-cyclic coordinate would be $r_{12} \equiv |\mathbf{X}_{12}|$. It is a general rule that if the Lagrangian is independent of some generalized coordinate X^μ, i.e if

$$\frac{\partial L}{\partial X^\mu} = 0 \tag{10.26}$$

then the momentum conjugate to it is conserved:

$$p_\mu \equiv \frac{\partial L}{\partial \dot{X}^\mu} = \text{constant} \tag{10.27}$$

10.3 Hamilton's unified formulation

The work of Sir William Rowan Hamilton (1805–1865) contains some remarkably modern insights, foreshadowing quantum mechanics and relativity theory. His treatment of mechanics and optics unified the two disciplines in a manner that foreshadows wave mechanics. In his first paper on systems of rays in geometrical optics, he considered rays coming from a point source which flashes on at a certain instant of time. If the light is propagating in a uniform medium, the rays will form system of straight lines, pointing outward from the point source of the light. Perpendicular to these lines, will be a set of concentric spherical surfaces, which represent the maximum distance that can be reached at any given time. In a non-uniform medium, the system of rays will not be straight lines, and surfaces will not be spheres, but nevertheless, the lines representing rays will always be perpendicular to the surfaces representing wave fronts. Hamilton introduced the integral

$$S(\mathbf{x}) = \int dt = \frac{1}{c}\int n(\mathbf{x})\, dl \tag{10.28}$$

This integral, taken along the path of a ray, gives the time needed for the wave front of a flash to reach a particular point $\mathbf{x}$. Hamilton called $S(\mathbf{x})$ the *eikonal function*, taking the name from the Greek word for "image", and he showed that it satisfies the differential equation

$$\frac{1}{[2n(\mathbf{x})]^2}\left[\left(\frac{\partial S}{\partial x}\right)^2 + \left(\frac{\partial S}{\partial y}\right)^2 + \left(\frac{\partial S}{\partial z}\right)^2\right] = 1 \tag{10.29}$$

Equation (10.29) follows from Fermat's principle, which states that the actual path of a ray of light is the one that takes the least time:

$$S(\mathbf{x}) = \int dt = \frac{1}{c}\int n(\mathbf{x})\, dl = \frac{1}{c}\int n(\mathbf{x})\frac{d\mathbf{x}}{dl}\cdot\frac{d\mathbf{x}}{dl}\, dl = \text{minimum} \tag{10.30}$$

The Euler-Lagrange equations corresponding to (10.30) are

$$\frac{d}{dl}\left[\frac{\partial L}{\partial\left(\frac{dx^i}{dl}\right)}\right] - \frac{\partial L}{\partial x^i} = 0 \tag{10.31}$$

so that

$$\frac{\partial S}{\partial x^i} = \int \frac{\partial L}{\partial x^i}\, dt = \int \frac{d}{dl}\left[\frac{\partial L}{\partial\left(\frac{dx^i}{dl}\right)}\right] dl = \frac{\partial L}{\partial\left(\frac{dx^i}{dl}\right)}$$

$$L = n(\mathbf{x})\frac{d\mathbf{x}}{dl}\cdot\frac{d\mathbf{x}}{dl} \tag{10.32}$$

Thus

$$\frac{\partial S}{\partial x^i} = 2n(\mathbf{x})\frac{dx^i}{dl} \qquad\qquad i = 1, 2, 3 \tag{10.33}$$

By combining (10.33) with the relation

$$\frac{d\mathbf{x}}{dl}\cdot\frac{d\mathbf{x}}{dl} = 1 \tag{10.34}$$

we obtain Hamilton's eikonal equation, (10.29). With remarkable intuition, Hamilton saw the analogy between the rays of geometrical optics and the trajectories of point masses in classical mechanics. His next step was to put mechanics on the same footing as optics by defining what he called the *characteristic function* for a system of trajectories. We can obtain an understanding of Hamilton's characteristic function by thinking of the fragments of an exploding skyrocket. If all of the fragments leave the point of the explosion with equal velocity, then they will form the sort of system which Hamilton studied. The upward-moving fragments are decelerated by gravity, while the downward-moving ones are accelerated. The positions of the fragments at successive instants of time are on spheres drawn around the falling center of mass of the system. Hamilton defined the characteristic function $S(\mathbf{x})$ by the relationship

$$S(\mathbf{x}) = \int_{\mathbf{x}_0}^{\mathbf{x}} L\, dt \tag{10.35}$$

taken along the system of trajectories. From the Euler-Lagrange equations, it follows that

$$\frac{\partial S}{\partial x^i} = \int \frac{\partial L}{\partial x^i}\, dt = \int \frac{d}{dt}\frac{\partial L}{\partial \dot{x}^i}\, dt = \frac{\partial L}{\partial \dot{x}^i} = p_i \tag{10.36}$$

Hamilton used this relationship to show that his characteristic function satisfies a differential equation similar to his eikonal equation (10.29). He

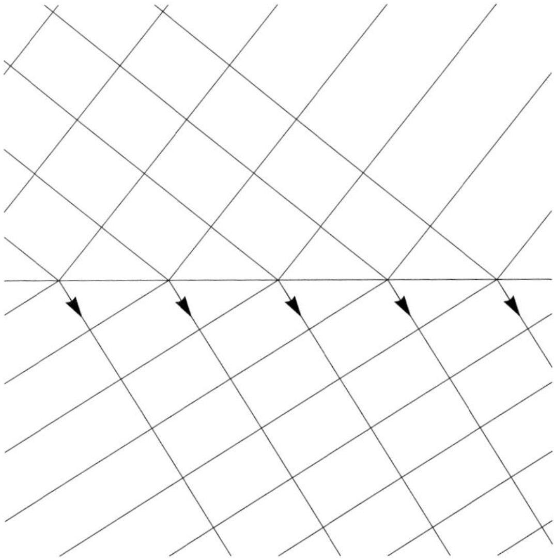

Fig. 10.1: This figure shows a system of parallel light rays entering a medium with a different index of refraction. The rays of light are perpendicular to the wave fronts at all points in space. The wave fronts correspond to surfaces with constant values of Hamilton's eikonal function.

first defined the total energy function (we call it the Hamiltonian) of a mechanical system as

$$H = \sum_i p_i \dot{x}^i - L \tag{10.37}$$

It follows from equation (10.37) that

$$\frac{dp_i}{dt} \equiv \frac{d}{dt}\left(\frac{\partial L}{\partial \dot{x}^i}\right) = \frac{\partial L}{\partial x^i} = -\frac{\partial H}{\partial x^i} \tag{10.38}$$

From (10.37) it also follows that

$$\frac{\partial H}{\partial p_i} = \dot{x}^i \tag{10.39}$$

Equations (10.38) and (10.39) are called Hamilton's equations of motion. From these equations, it follows that for systems where the potential energy is independent of time and where there are no velocity-dependent forces, the Hamiltonian function (10.37) is a constant of the motion. For such *conservative systems*, H is a constant of the motion.

$$\begin{aligned}\frac{dH}{dt} &= \frac{\partial H}{\partial t} + \sum_i \left[\frac{\partial H}{\partial p_i}\frac{dp_i}{dt} + \frac{\partial H}{\partial x^i}\frac{dx^i}{dt}\right] \\ &= \sum_i \left[-\frac{\partial H}{\partial p_i}\frac{\partial H}{\partial x^i} + \frac{\partial H}{\partial x^i}\frac{\partial H}{\partial p_i}\right] = 0\end{aligned} \tag{10.40}$$

Thus

$$H(x^i, p_i) = E \tag{10.41}$$

where E is a constant. Hamilton then substituted $\partial S/\partial x^i$ for p_i. In this way he obtained an equation which has become known as the *Hamiltonian-Jacobi equation*:

$$H\left(x^i, \frac{\partial S}{\partial x^i}\right) = E \tag{10.42}$$

For example, in the case where the mechanical system is a single point mass moving in the potential $V(\mathbf{x})$, the Hamiltonian of the system is

$$H = \frac{m}{2}\frac{d\mathbf{x}}{dt} \cdot \frac{d\mathbf{x}}{dt} + V(\mathbf{x}) = \frac{1}{2m}\mathbf{p} \cdot \mathbf{p} + V(\mathbf{x}) \tag{10.43}$$

and the Hamilton-Jacobi equation is

$$\frac{1}{2m}\left[\left(\frac{\partial S}{\partial x}\right)^2 + \left(\frac{\partial S}{\partial y}\right)^2 + \left(\frac{\partial S}{\partial z}\right)^2\right] + V(\mathbf{x}) = E \tag{10.44}$$

which is analogous to Hamilton's eikonal equation, (10.29).

10.4 Normal modes

We next consider the small vibrations of a classical system of particles about the equilibrium positions. Suppose that the kinetic energy of the system is given by

$$T = \frac{1}{2}\sum_{i=1}^{d}\sum_{j=1}^{d} m_i \delta_{i,j} \frac{dx^i}{dt}\frac{dx^j}{dt} \tag{10.45}$$

while the leading term in a Taylor series expansion of the potential energy has the form

$$V = \frac{1}{2}\sum_{i=1}^{d}\sum_{j=1}^{d} V_{i,j} x^i x^j \tag{10.46}$$

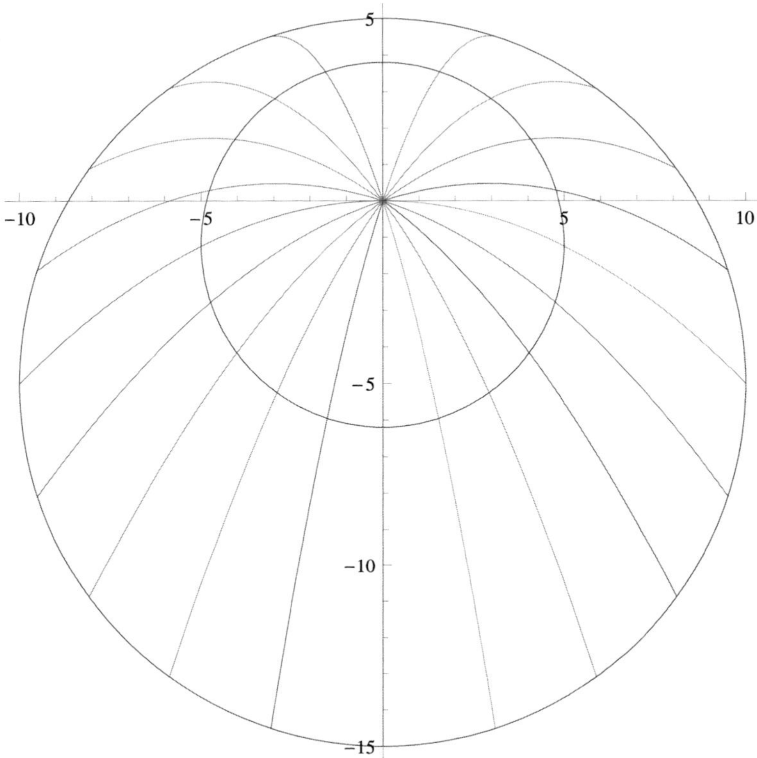

Fig. 10.2: This figure shows a system of particle trajectories of the kind visualized by Hamilton. Here the system might be produced by the fragments of an exploding sky-rocket, assuming that they are all of equal mass and are thrown out with equal velocities. At various times after the explosion, the fragments will reach points given by spheres drawn around the falling center of mass.

The coordinates $x^1, x^2 \cdots, x^d$, which represent small displacements from the equilibrium positions of the particles, are by no means the most convenient ones for solving the equations of motion of the system. We can bring the kinetic energy into a more convenient form by going over to the mass-weighted coordinates defined by

$$X^i \equiv \sqrt{m_i}\, x^i \qquad\qquad i = 1, 2, \cdots, d \tag{10.47}$$

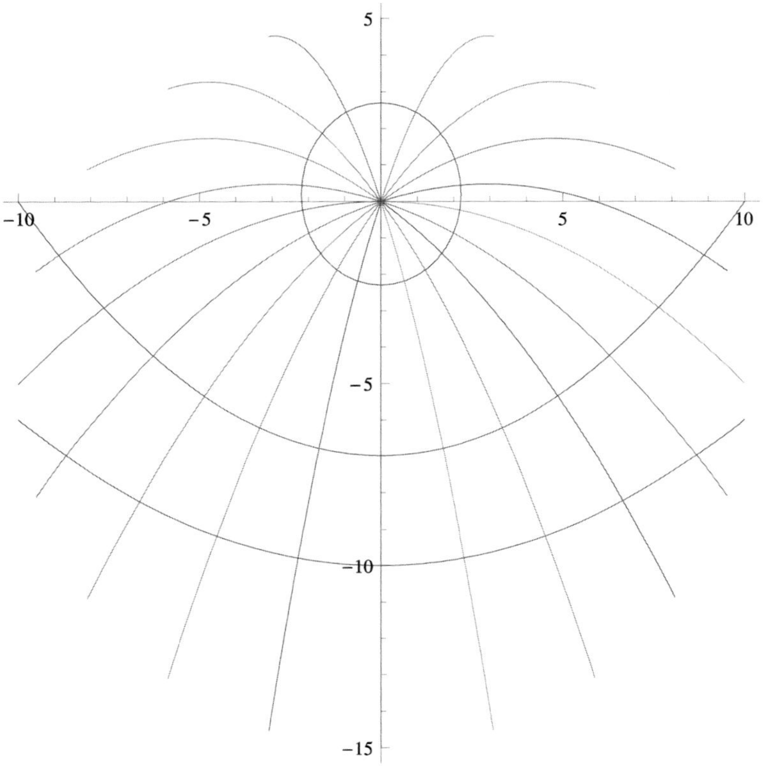

Fig. 10.3: This figure shows surfaces corresponding to constant values of Hamilton's characteristic function S. These surfaces are everywhere perpendicular to the trajectories discussed in the previous figure.

In terms of these coordinates, the kinetic energy has the form

$$T = \frac{1}{2}\sum_{i=1}^{d}\sum_{j=1}^{d}\delta_{i,j}\frac{dX^i}{dt}\frac{dX^j}{dt} \tag{10.48}$$

while the potential energy becomes

$$V = \frac{1}{2}\sum_{i=1}^{d}\sum_{j=1}^{d}\frac{V_{i,j}}{\sqrt{m_i m_j}}X^i X^j \tag{10.49}$$

The mass-weighted coordinates are still not the most convenient ones that we can find, since the potential energy matrix $V_{i,j}$ may contain off-diagonal terms. and we would like to get rid of these. We can find a unitary transformation which diagonalizes $V_{i,j}/\sqrt{m_i m_j}$ by solving the secular equations

$$\sum_{j=1}^{d}\left(\frac{V_{i,j}}{\sqrt{m_i m_j}} - \mathcal{V}(k)\delta_{i,j}\right)U_{j,k} = 0 \tag{10.50}$$

Having performed the diagonalization, we can express the potential energy and the kinetic energy of the system in terms of the *normal coordinates* defined by

$$q^k = \sum_{i=1}^{d} X^i U_{i,k} = \sum_{i=1}^{d} \sqrt{m_i}\, x^i U_{i,k} \tag{10.51}$$

When we do this, the kinetic energy retains its diagonal form because of the unitarity of $U_{j,k}$:

$$T = \frac{1}{2}\sum_{k=1}^{d}\left(\frac{dq^k}{dt}\right)^2 \tag{10.52}$$

but the off-diagonal terms in the potential energy disappear:

$$V = \frac{1}{2}\sum_{k=1}^{d}\mathcal{V}(k)(q^k)^2 \tag{10.53}$$

From (10.52) and (10.53) we can see that the Lagrangian of the system can be written in the form

$$L = T - V = \sum_{k=1}^{d} L_k \tag{10.54}$$

where

$$L_k = \frac{1}{2}\left[\left(\frac{dq^k}{dt}\right)^2 - \mathcal{V}(k)(q^k)^2\right] \tag{10.55}$$

The canonically conjugate momentum paired with the coordinate q^k is defined in mechanics to be

$$p_k = \frac{\partial L}{\partial \dot{q}^k} = \frac{dq^k}{dt} \tag{10.56}$$

he Hamiltonian of the system can be written in the form

$$H = T + V = \sum_{k=1}^{d} H_k \tag{10.57}$$

where

$$H_k = \frac{1}{2}\left(p_k^2 + \omega_k^2 q_k^2\right) \tag{10.58}$$

and

$$\omega_k = \sqrt{\mathcal{V}(k)} \tag{10.59}$$

In other words, when the Hamiltonian which represents small vibrations of a classical system is expressed in terms of the normal coordinates (or normal modes), it reduces to a sum of simple harmonic oscillator Hamiltonians. The normal coordinates are found by diagonalizing the mass-weighted potential energy matrix. The harmonic oscillator frequency of each is found by taking the square root of the corresponding eigenvalue of the mass-weighted potential energy matrix.

To illustrate this procedure, we can think of a system, whose Lagrangian is given by

$$L = \frac{1}{2}\sum_{i=1}^{d}\sum_{j=1}^{d}\left(m\ \delta_{i,j}\frac{dx^i}{dt}\frac{dx^j}{dt} - V_{i,j}x^i x^j\right) \tag{10.60}$$

where

$$V_{i,j} = \begin{cases} 2\kappa & i = j \\ -\kappa & i = j \pm 1 \end{cases} \tag{10.61}$$

This Lagrangian corresponds to a linear system of point masses, each joined elastically to the next. Then the secular equations (10.50) have the form

$$-\kappa U_{k-1,k} + [2\kappa - \mathcal{V}(k)]\, U_{k,k} - \kappa U_{k+1,k} = 0 \qquad\qquad k = 2, \ldots, d-1 \tag{10.62}$$

The trial solution

$$U_{j,k} = \sqrt{\frac{2}{d+1}}\ \sin(jka) \tag{10.63}$$

makes all of the secular equations redundant, All of them redundantly require that

$$\mathcal{V}(k) = \kappa\,[1 - \cos(ka)] \tag{10.64}$$

Imposing homogeneous boundary conditions (i.e. clamping the two ends of the line) restricts the allowed values of k, and we must have

$$k = \frac{\pi}{(d+1)a}, \frac{2\pi}{(d+1)a}, \ldots, \frac{\pi d}{(d+1)a} \tag{10.65}$$

where $(d+1)a$ is the length of the chain. The frequency spectrum of the normal modes is given by

$$\omega_k = \sqrt{\frac{\mathcal{V}(k)}{m}} = \sqrt{\frac{2\kappa\,[1 - \cos(2ka)]}{m}} \tag{10.66}$$

In terms of the normal mode coordinates and their time derivatives, the Lagrangian of the system becomes

$$L = \frac{1}{2}\sum_k \left[\left(\frac{dq^k}{dt} \right)^2 - (\omega_k q^k)^2 \right] \tag{10.67}$$

which can be recognized as a sum of harmonic oscillator Lagrangians.

10.5 Molecular vibrations and rotations

In the simplest possible approximation, we can regard a molecule (or a cluster in a non-melted state) as a collection of point masses held together by springlike bonds. When we calculate the normal modes of such a system, we always find that there are six zero-frequency modes. Three of these correspond to the degrees of freedom associated with translation of the whole system, and three with rotation. Let us use the symbols $\mathbf{R}_s$ to represent the equilibrium position of the atom s, and $\mathbf{x}_s$ to represent the displacement of the atom from its equilibrium position. Then in our simple model, the classical potential energy of the molecule can be written in the form

$$V = \frac{1}{2}\sum_{t>s}^{N}\sum_{s=1}^{N} k_{st} \left(|\mathbf{x}_s + \mathbf{R}_s - \mathbf{x}_t - \mathbf{R}_t| - |\mathbf{R}_s - \mathbf{R}_t| \right)^2 \tag{10.68}$$

Here k_{st} represents the force constant of the "spring" which connects atom s with atom t. Let us also introduce the notation

$$\mathbf{R}_{st} \equiv \mathbf{R}_s - \mathbf{R}_t$$

$$\mathbf{x}_{st} \equiv \mathbf{x}_s - \mathbf{x}_t \tag{10.69}$$

Then, if we assume that $|\mathbf{x}_{st}| << |\mathbf{R}_{st}|$ and expand V in a Taylor series, we obtain the leading term

$$\begin{aligned} V &\approx \sum_{t>s}^{N}\sum_{s=1}^{N} \frac{k_{st}}{|\mathbf{R}_{st}|^2} (\mathbf{R}_{st} \cdot \mathbf{x}_{st})^2 \\ &\equiv \frac{1}{2}\sum_{t>s}^{N}\sum_{s=1}^{N}\sum_{\mu=1}^{3}\sum_{\nu=1}^{3} V_{s,\mu;t,\nu}\, x_{s,\mu} x_{t,\nu} \end{aligned} \tag{10.70}$$

where

$$\mathbf{x}_s \equiv (x_{s1}, x_{s2}, x_{s3}) \tag{10.71}$$

By diagonalizing the mass-weighted potential energy matrix

$$\frac{V_{s,\mu;t,\nu}}{\sqrt{m_s m_t}} \tag{10.72}$$

we can find the normal modes of the system, and as mentioned, six of them will be zero-frequency modes corresponding to translations and rotations of the entire system.

10.6 Lagrangian densities for fields

Let us now turn our attention to the problem of how to treat continuous systems, or fields, using the Lagrangian formalism. We shall begin by considering a 3-dimensional space, with coordinates x, y and z, and later we will generalize to spaces of higher dimension and also to spaces where the metric is non-Euclidean. We have just been discussing normal modes for small vibrations of a classical system of particles about the equilibrium positions. We might also wish to discuss the normal modes of a violin string or a drumhead, or the harmonics of sound waves inside a closed container. In these examples, the medium can be regarded as continuous as long as we are considering low frequencies, but at high frequencies, where the wavelength is comparable to the distance between the particles of which the system is composed, this picture breaks down. But at low frequencies, the displacement from equilibrium can be treated as a field.

There are other fields that one might wish to consider, such as electromagnetic fields. What we now wish to discuss is the problem of how to start with a wave equation in classical mechanics, and how to pass over to a quantum mechanical treatment of the system. In order to do this, it is useful to introduce the concept of a Lagrangian density, $\mathcal{L}$, This is sometimes defined as the kinetic energy per unit volume of space, minus the density of potential energy. However, in some cases, for example when the Lagrangian must be invariant under a Lorentz transformation, this definition cannot be used, and we must instead define the Lagrangian density as that which gives the correct wave equations. The usual Lagrangian of the system is the Lagrangian density integrated over the whole volume of the system.

$$L = \int\int\int \mathcal{L}\ dx\ dy\ dz \tag{10.73}$$

The variational principle from which the equations of motion can be derived can be written in the form

$$\int \delta L\ dt = \int\int\int\int \delta\mathcal{L}\ dx\ dy\ dz\ dt = 0 \tag{10.74}$$

Suppose that the continuous system is described by a number of fields, $\phi_1, \phi_2, \cdots, \phi_n$, which are continuous functions of the space and time coordinates x, y, z and t. For example, $\phi_1, \phi_2, \cdots$ might represent the components of the electromagnetic field. Suppose also that the Lagrangian density can be expressed in terms of the fields and their first derivatives with respect to the space and time coordinates, so that

$$\mathcal{L} = \mathcal{L}\left(\phi_j, \frac{\partial \phi_j}{\partial x^\mu}\right) \qquad \begin{cases} j = 1, 2, \cdots, n \\ \mu = 1, 2, 3, 4 \\ x^4 = ict \end{cases} \tag{10.75}$$

If we change one of the fields by a small amount, the resulting change in the Lagrangian density is given by

$$\delta\mathcal{L} = \frac{\partial \mathcal{L}}{\partial \phi_i}\delta\phi_i + \sum_{\mu=1}^{4} \frac{\partial \mathcal{L}}{\partial(\partial \phi_i/\partial x^\mu)} \frac{\partial(\delta\phi_i)}{\partial x^\mu} \tag{10.76}$$

The variational principle then becomes

$$\int \delta L \; dt = \int\int\int\int \left[\frac{\partial \mathcal{L}}{\partial \phi_i} + \sum_{\mu=1}^{4} \frac{\partial \mathcal{L}}{\partial(\partial \phi_i/\partial x^\mu)} \frac{\partial(\delta\phi_i)}{\partial x^\mu}\right] dx \; dy \; dz \; dt \tag{10.77}$$

The second term on the right-hand side of (10.77) can be integrated by parts in a manner analogous to equations (10.11)–(10.14). Thus, for example, we can write

$$\begin{aligned} &\int\int\int\int \frac{\partial \mathcal{L}}{\partial(\partial \phi_i/\partial x)} \frac{\partial(\delta\phi_i)}{\partial x} dx \; dy \; dz \; dt \\ &= \left[\int\int\int \frac{\partial \mathcal{L}}{\partial(\partial \phi_i/\partial x)}\delta\phi_i\right]_{x_a}^{x_b} dy \; dz \; dt \\ &- \int\int\int\int \frac{\partial}{\partial x} \frac{\partial \mathcal{L}}{\partial(\partial \phi_i/\partial x)} \; \delta\phi_i \; dx \; dy \; dz \; dt \end{aligned} \tag{10.78}$$

The first term to the right of the = sign is required to vanish because the variation must be zero at the end points. In this way, we can rewrite the variational principle (10.74) in the form:

$$\int \delta L \; dt = \int\int\int\int \left[\frac{\partial \mathcal{L}}{\partial \phi_i} - \sum_{\mu=1}^{4} \frac{\partial}{\partial x^\mu} \frac{\partial \mathcal{L}}{\partial(\partial \phi_i/\partial x^\mu)}\right] \delta\phi_i \; dx \; dy \; dz \; dt = 0 \tag{10.79}$$

Since the variation $\delta\phi_i$ is arbitrary, (10.79) can be fulfilled Only if

$$\frac{\partial \mathcal{L}}{\partial \phi_i} - \sum_{\mu=1}^{4} \frac{\partial}{\partial x^\mu} \frac{\partial \mathcal{L}}{\partial(\partial \phi_i/\partial x^\mu)} = 0 \tag{10.80}$$

Thus if we are able to write down the Lagrangian density of a continuous system, the the variational principle leads us to the equations of motion

(10.80). Conversely, knowledge of the equations of motion of a continuous system can be used to infer the Lagrangian density from which they are derived. The canonical conjugate of the field ϕ^i is called the *momentum density*, and is defined as

$$\pi_i \equiv \frac{\partial \mathcal{L}}{\partial(\partial\phi_i/\partial t)} \tag{10.81}$$

and the Hamiltonian density $\mathcal{H}$, corresponding to the Lagrangian density $\mathcal{L}$, is given by

$$\mathcal{H} = \sum_{i=1}^{n} \pi_i \frac{\partial \phi^i}{\partial t} - \mathcal{L} \tag{10.82}$$

Notice that equation (10.80) is invariant under a Lorenz transformation. but equation (10.82) is not, since the time coordinate is singled out for special treatment. This corresponds to the fact that energy, by itself, is not a Lorenz-invariant quantity, but only the 4th component of a 4-vector. It is possible to perform approximately relativistic calculations using the Hamiltonian formulation, but the Lagrangian formalism should be used rather than the Hamiltonian in cases where Lorentz invariance is important. Having found the equations of motion (10.80), we can next solve them, subject to the boundary conditions, and find solutions corresponding to pure harmonic oscillations. The fields are then expressed as sums of these harmonic modes. When the sums are substituted into the Lagrangian $L = \int\int\int \mathcal{L}\, dx\, dy\, dz$, then L can be recognized as a sum of simple harmonic oscillator Lagrangians. The quantization of these can proceed in the standard way.

10.7 Electromagnetic potentials

Let us now attempt to apply these methods to the electromagnetic potential 4-vector, A_λ. This is a case where the definition of the Lagrangian density as the density of kinetic energy minus the density of potential energy cannot be used. We must instead use the concept of the Lagrangian density as a quantity which leads, through the variational principle, to the correct wave equations. In the case of the electromagnetic potentials, we know that they obey the wave equation

$$\Box^2 A_\lambda = -\frac{4\pi}{c} j_\lambda \qquad j_\lambda = 1, 2, 3, 4 \tag{10.83}$$

where

$$\Box^2 \equiv \sum_{\mu=1}^{4} \frac{\partial^2}{\partial x^{\mu 2}} = \nabla^2 - \frac{1}{c^2}\frac{\partial^2}{\partial t^2} \qquad x^4 = ict \tag{10.84}$$

We must invent a Lagrangian density that leads to the wave equation (10.83). The way to do this is to try all possibilities, starting with the simplest ones, always remembering that the forms must be Lorentz invariant. The Lagrangian density which actually leads to equation (10.83) is

$$\mathcal{L} = -\frac{1}{2}\sum_{\lambda=1}^{4}\sum_{\mu=1}^{4}\left(\frac{\partial A_\lambda}{\partial x^\mu}\right)^2 + \frac{4\pi}{c}\sum_{\lambda=1}^{4} A_\lambda j_\lambda \tag{10.85}$$

From (10.80) we have:

$$\frac{\partial \mathcal{L}}{\partial A_\lambda} - \sum_{\mu=1}^{4}\frac{\partial}{\partial x^\mu}\frac{\partial \mathcal{L}}{\partial(\partial A_\lambda/\partial x^\mu)} = 0 \tag{10.86}$$

With the trial Lagrangian density of equation (10.85), we have

$$\frac{\partial \mathcal{L}}{\partial A_\lambda} = \frac{4\pi}{c} j_\lambda$$

$$-\sum_{\mu=1}^{4}\frac{\partial}{\partial x^\mu}\frac{\partial \mathcal{L}}{\partial(\partial A_\lambda/\partial x^\mu)} = \sum_{\mu=1}^{4}\frac{\partial}{\partial x^\mu}\frac{\partial A_\lambda}{\partial x^\mu} = \Box^2 A_\lambda \tag{10.87}$$

Substituting these results into the Euler-Lagrange equations (10.86) gives

$$\frac{4\pi}{c} j_\lambda + \Box^2 A_\lambda = 0 \tag{10.88}$$

which is just the wave equation that we wanted to reproduce. The next step is to solve the wave equation and find the harmonic normal modes of the system. However, we can see that if we include the term involving the current density, the solutions will not necessarily be harmonic. Therefore we omit the interaction term, hoping to include it later by means of perturbation theory. This leaves us with the homogeneous wave equation:

$$\Box^2 A_\lambda = 0 \tag{10.89}$$

which has harmonic solutions of the form

$$A_{\mathbf{k},\lambda} = \mathcal{N}\left\{a_{\mathbf{k},\lambda}(t)e^{i\mathbf{k}\cdot\mathbf{x}} + a^*_{\mathbf{k},\lambda}(t)e^{-i\mathbf{k}\cdot\mathbf{x}}\right\} \tag{10.90}$$

Here the values of $\mathbf{k}$ are restricted by periodic boundary conditions imposed at the faces of a box of normalization of volume V. $\mathcal{N}$ is a constant which we will choose later in a convenient way. $a_{\mathbf{k},\lambda}$ and its complex conjugate $a^*_{\mathbf{k},\lambda}$ are Fourier coefficients which are harmonic functions of time. Substituting (10.90) into (10.89), we can see that $A_{\mathbf{k},\lambda}$ will be a solution, provided that

$$\begin{aligned}\frac{\partial^2 a_{\mathbf{k},\lambda}}{\partial t} + \omega_k^2 a_{\mathbf{k},\lambda} &= 0\\ \frac{\partial^2 a^*_{\mathbf{k},\lambda}}{\partial t} + \omega_k^2 a^*_{\mathbf{k},\lambda} &= 0\end{aligned} \tag{10.91}$$

where $\omega_k = |\mathbf{k}|/c$. Having found harmonic solutions to the wave equation, we must expand the potentials in terms of these, and substitute them back into the Lagrangian. We then hope to be able to recognize the Lagrangian as a sum of simple harmonic oscillator Lagrangians. Thus, expanding the potentials in terms of their normal modes, we write

$$A_\lambda(\mathbf{x}, t) = \mathcal{N} \sum_{\mathbf{k}}{}' \left(a_{\mathbf{k},\lambda} e^{i\mathbf{k}\cdot\mathbf{x}} + a^*_{\mathbf{k},\lambda} e^{-i\mathbf{k}\cdot\mathbf{x}} \right) \tag{10.92}$$

In (10.92), the sum runs over only half of k-space for the following reason: By writing the harmonic solutions in the form (10.90), (i.e. as a function plus its complex conjugate), we ensure that they shall be real. However, we do not want the solution corresponding to positive $\mathbf{k}$ to repeat that corresponding to $-\mathbf{k}$. Therefore we restrict the sum to half of k-space, and we denote this with a prime on the summation. From (10.92), we have

$$\frac{\partial A_\lambda}{\partial x^\mu} = \mathcal{N} \sum_{\mathbf{k}}{}' \left(ik_\mu a_{\mathbf{k},\lambda} e^{i\mathbf{k}\cdot\mathbf{x}} - ik_\mu a^*_{\mathbf{k},\lambda} e^{-i\mathbf{k}\cdot\mathbf{x}} \right) \qquad \mu = 1, 2, 3 \tag{10.93}$$

and

$$\frac{\partial A_\lambda}{\partial x^4} = \frac{\mathcal{N}}{ic} \sum_{\mathbf{k}}{}' \left(\frac{\partial a_{\mathbf{k},\lambda}}{\partial t} e^{i\mathbf{k}\cdot\mathbf{x}} + \frac{\partial a^*_{\mathbf{k},\lambda}}{\partial t} e^{-i\mathbf{k}\cdot\mathbf{x}} \right) \tag{10.94}$$

Let us now calculate the Hamiltonian density $\mathcal{H}$ associated with the electromagnetic potentials. According to our earlier discussion, the Hamiltonian density is given by

$$\begin{aligned} \mathcal{H} &= \sum_{\lambda=1}^{4} \frac{\partial A_\lambda}{\partial t} \left\{ \frac{\partial \mathcal{L}}{\partial(\partial A_\lambda/\partial t)} \right\} - \mathcal{L} \\ &= \frac{1}{2} \sum_{\lambda=1}^{4} \left[\sum_{\mu=1}^{3} \left(\frac{\partial A_\lambda}{\partial x^\mu} \right)^2 - \left(\frac{\partial A_\lambda}{\partial x^4} \right)^2 \right]' - \frac{4\pi}{c} \sum_{\lambda=1}^{4} j_\lambda A_\lambda \\ &= \mathcal{H}_{\text{radiation}} + \mathcal{H}_{\text{interaction}} \equiv \mathcal{H}_{\text{r}} + \mathcal{H}_{\text{i}} \end{aligned} \tag{10.95}$$

Now we know that the Fourier coefficients $a_{\mathbf{k},\lambda}$ depend harmonically on time, so that

$$\begin{aligned} \frac{\partial a_{\mathbf{k},\lambda}}{\partial t} &= \pm ikca_{\mathbf{k},\lambda} \\ \frac{\partial a^*_{\mathbf{k},\lambda}}{\partial t} &= \mp ikca^*_{\mathbf{k},\lambda} \end{aligned} \tag{10.96}$$

Also we know that the box-normalized functions $e^{i\mathbf{k}\cdot\mathbf{x}}/\sqrt{V}$ obey the orthonormality relations

$$\frac{1}{V}\int d\mathbf{x}\; e^{i(\mathbf{k}-\mathbf{k}')\cdot\mathbf{x}} = \delta_{\mathbf{k},\mathbf{k}'} \tag{10.97}$$

Substituting (10.97) and (10.96) into (10.95), we obtain the Hamiltonian for the free electromagnetic field in the form

$$H_{\rm r} = \int d\mathbf{x}\; \mathcal{H}_{\rm r} = V \sum_{\mathbf{k}}{}' \sum_{\lambda=1}^{4} \mathcal{N}^2 \left(a_{\mathbf{k},\lambda} a^*_{\mathbf{k},\lambda} + a^*_{\mathbf{k},\lambda} a_{\mathbf{k},\lambda}\right) \tag{10.98}$$

Notice that the terms proportional to $a_{\mathbf{k},\lambda}a_{\mathbf{k},\lambda}$ and $a^*_{\mathbf{k},\lambda}a^*_{\mathbf{k},\lambda}$ do not occur in $H_{\rm r}$ because the sum over $\mathbf{k}$ is restricted to one half of k-space. With the substitutions

$$\begin{aligned} a_{\mathbf{k},\lambda} &= \frac{1}{\sqrt{2kc}}\left(-ip_{\mathbf{k},\lambda} + kcq_{\mathbf{k},\lambda}\right) \\ a^*_{\mathbf{k},\lambda} &= \frac{1}{\sqrt{2kc}}\left(ip_{\mathbf{k},\lambda} + kcq_{\mathbf{k},\lambda}\right) \end{aligned} \tag{10.99}$$

$H_{\rm r}$ becomes

$$H_{\rm r} = \frac{V}{c} \sum_{\mathbf{k}}{}' \sum_{\lambda=1}^{4} \mathcal{N}^2 k \left(p^2_{\mathbf{k},\lambda} + k^2c^2 q^2_{\mathbf{k},\lambda}\right) \tag{10.100}$$

This can be identified as the Hamiltonian of a collection of harmonic oscillators, provided that

$$\mathcal{N} = \sqrt{\frac{c}{2kV}} \tag{10.101}$$

If we make the identification $a^*_{\mathbf{k},\lambda} \to a^\dagger_{\mathbf{k},\lambda}$, and choose the value of $\mathcal{N}$ given by (10.101), then $H_{\rm r}$ takes on the form

$$H_{\rm r} = \frac{1}{2} \sum_{\mathbf{k}}{}' \sum_{\lambda=1}^{4} kc \left(a_{\mathbf{k},\lambda} a^\dagger_{\mathbf{k},\lambda} + a^\dagger_{\mathbf{k},\lambda} a_{\mathbf{k},\lambda}\right) \tag{10.102}$$

This can be recognized as the Hamiltonian of a collection of harmonic oscillators with frequencies $\omega_k = kc$. In the quantum treatment of such a system, $a^\dagger_{\mathbf{k},\lambda}$ and $a_{\mathbf{k},\lambda}$ correspond to creation and annihilation operators, and they obey the commutation relations

$$\begin{aligned} \left[a_{\mathbf{k},\lambda}, a^\dagger_{\mathbf{k},\lambda}\right] &= \delta_{\mathbf{k},\mathbf{k}'}\delta_{\lambda,\lambda'} \\ \left[a^\dagger_{\mathbf{k},\lambda}, a^\dagger_{\mathbf{k},\lambda}\right] &= 0 \\ \left[a_{\mathbf{k},\lambda}, a_{\mathbf{k},\lambda}\right] &= 0 \end{aligned} \tag{10.103}$$

(See equations (A.32)–(A.34).) Expanding the potentials in terms of the normal modes of the system, we have

$$A_\lambda = \sqrt{\frac{c}{2kV}} \sum_{\mathbf{k}}' \left(a_{\mathbf{k},\lambda} e^{i\mathbf{k}\cdot\mathbf{x}} + a^\dagger_{\mathbf{k},\lambda} e^{-i\mathbf{k}\cdot\mathbf{x}} \right) \tag{10.104}$$

The interaction Hamiltonian, expressed in terms of the photon creation and annihilation operators, then becomes:

$$\begin{aligned} H_{\mathrm{i}} &= -\frac{4\pi}{c} \sum_{\lambda=1}^{4} \int d\mathbf{x}\; A_\lambda\; j_\lambda \\ &= -\sum_{\mathbf{k}}' \sum_{\lambda=1}^{4} \frac{4\pi}{\sqrt{2\omega_k V}} \left[a_{\mathbf{k},\lambda} \int d\mathbf{x}\; e^{i\mathbf{k}\cdot\mathbf{x}} j_\lambda + a^\dagger_{\mathbf{k},\lambda} \int d\mathbf{x}\; e^{-i\mathbf{k}\cdot\mathbf{x}} j_\lambda \right] \end{aligned} \tag{10.105}$$

10.8 Metric tensors

Let us consider a coordinate system $x^1, x^2, \cdots, x^d$ labelling the points in a d-dimensional space. We can label the points in a different way by going to a new coordinate system $X^1, X^2, \cdots, X^d$ where the new coordinates are expressed as functions of the old ones.

$$\begin{aligned} X^1 &= X^1(x^1, x^2, \cdots, x^d) \\ X^2 &= X^2(x^1, x^2, \cdots, x^d) \\ &\vdots \\ X^d &= X^d(x^1, x^2, \cdots, x^d) \end{aligned} \tag{10.106}$$

For example, (10.106) might represent a transformation from Cartesian coordinates to spherical polar coordinates. If we have an equation written in terms of the old coordinates, we might ask how to rewrite it in terms of the new ones. More generally, we can try to write a physical equation in such a way that it will look the same in every coordinate system. Suppose that the space is Euclidean (flat), so that in terms of the Cartesian coordinates $x^1, x^2, \cdots, x^d$, the infinitesimal element of length separating two points is given by the Pythagorean rule:

$$ds^2 = \delta_{i,j} dx^i dx^j \equiv g_{i,j} dx^i dx^j \tag{10.107}$$

(In equation (10.107) and in the remainder of this section, we use the Einstein convention, in which a sum over repeated indices is understood, although not written explicitly.) The symbol $g_{i,j}$ which appears in the definition of the infinitesimal length ds^2 is called the *covariant metric tensor*, and for Cartesian coordinates in a Euclidean space, it is just the Kronecker delta function. Using the identity

$$dx^i = \frac{\partial x^i}{\partial X^\mu} dX^\mu \tag{10.108}$$

we can rewrite (10.107) as

$$ds^2 = \delta_{i,j} \frac{\partial x^i}{\partial X^\mu} \frac{\partial x^j}{\partial X^\nu} dX^\mu dX^\nu \equiv G_{\mu,\nu} dX^\mu dX^\nu \tag{10.109}$$

where

$$G_{\mu,\nu} \equiv g_{i,j} \frac{\partial x^i}{\partial X^\mu} \frac{\partial x^j}{\partial X^\nu} \tag{10.110}$$

The quantity $G_{\mu,\nu}$ which appears in equations (10.109) and (10.110) is the covariant metric tensor in the new coordinate system. In any space, whether Euclidean or not, the covariant metric tensor is defined by the expression which yields ds^2, the square of the infinitesimal distance between two points, as in equation (10.107) or (10.109). The word *tensor* refers to the way in which a quantity transforms under changes in the coordinate system. The *rank* of a tensor is the number of indices. The covariant metric tensor is the prototype of a covariant tensor of second rank. Any physical quantity which must be transformed according to the rule

$$A_{\mu,\nu} = a_{i,j} \frac{\partial x^i}{\partial X^\mu} \frac{\partial x^j}{\partial X^\nu} \tag{10.111}$$

under the coordinate transformation $x^1, x^2, \cdots, x^d \rightarrow X^1, X^2, \cdots, X^d$ is said to be a covariant tensor of second rank. The d-component entity

$$dX^\mu = \frac{\partial X^\mu}{\partial x^i} dx^i \tag{10.112}$$

is the prototype of a contravariant tensor of first rank. Any quantity that transforms according to the rule

$$A^\mu = \frac{\partial X^\mu}{\partial x^i} a^i \tag{10.113}$$

is said to be a contravariant tensor of first rank (or contravariant vector). The distance element ds is the prototype of an invariant or *scalar*. Any quantity ϕ which is invariant under coordinate transformations is said to be a scalar. The gradient of a scalar

$$\frac{\partial \phi}{\partial X^\mu} = \frac{\partial x^i}{\partial X^\mu} \frac{\partial \phi}{\partial x^i} \tag{10.114}$$

is the prototype of a covariant tensor of first rank, or covariant vector. Any quantity which transforms according to the rule

$$A_\mu = \frac{\partial x^i}{\partial X^\mu} a_i \tag{10.115}$$

is said to be a covariant vector. We can also define tensors of higher rank. For example,

$$A^{\mu\nu\sigma} = \frac{\partial X^\mu}{\partial x^i}\frac{\partial X^\nu}{\partial x^j}\frac{\partial X^\sigma}{\partial x^k} a^{ijk} \tag{10.116}$$

is said to be a contravariant tensor of third rank. A covariant vector and a contravariant vector can be *contracted* into a scalar:

$$A_\mu B^\mu = \frac{\partial x^i}{\partial X^\mu}\frac{\partial X^\mu}{\partial x^j} a_i b^j = \delta^i_j a_i b^j = a_i b^i \tag{10.117}$$

Similarly, if we contract a contravariant vector with the covariant metric tensor, we obtain a covariant vector:

$$\begin{aligned} G_{\mu\nu} A^\nu &= A_\mu \\ g_{ij} a^i &= a_i \end{aligned} \tag{10.118}$$

It is useful to define a quantity called the *contravariant metric tensor*, which gives the Kronecker δ-function when it is contracted with the covariant metric tensor:

$$\begin{aligned} G^{\mu\nu} G_{\nu\sigma} &= \delta^\mu_\sigma \\ g^{ij} g_{jk} &= \delta^i_k \\ G^{\mu\nu} &= \frac{\partial X^\mu}{\partial x^i}\frac{\partial X^\nu}{\partial x^j} g^{ij} \end{aligned} \tag{10.119}$$

If we contract a covariant vector with the contravariant metric tensor, we obtain a contravariant vector:

$$G^{\mu\nu} A_\nu = A^\mu \tag{10.120}$$

In a similar way, we can raise or lower the indices of a tensor of higher rank. For example, it is easy to show that

$$G_{\mu\nu} A^{\nu\sigma\rho} = A_\mu^{\sigma\rho} \tag{10.121}$$

In a Cartesian coordinate system with unit metric we are accustomed to writing the volume element as

$$dv = dx^1 dx^2 \cdots dx^d \tag{10.122}$$

This is obviously unsatisfactory from the standpoint of tensor analysis, since the right-hand side of equation (10.122) appears to be a contravariant tensor of rank d (or rather a particular component of such a tensor), while the left-hand side has no indices at all. In order to write the volume element in an invariant way, the Italian mathematician Tulio Levi-Civita (1873–1941) introduced a totally antisymmetric covariant tensor of rank d. In a Cartesian coordinate system, for a flat space, the Levi-Civita tensor is given by

$$e_{ijkl\cdots} = \begin{cases} (-1)^\sigma \text{ if } ijkl\cdots = \sigma(1234\cdots) \\ \\ 0 \quad \text{otherwise} \end{cases} \tag{10.123}$$

In other words, the Levi-Civita tensor is ± 1 if $ijkl\cdots$ is a permutation of $1234\cdots$, with the sign depending on whether the permutation is even or odd, and it is zero otherwise. In terms of this tensor, the volume element of equation (10.122) becomes

$$dv = \frac{1}{d!} e_{ijkl\cdots} dx^i dx^j dx^k dx^l \cdots \tag{10.124}$$

while in a transformed coordinate system it is

$$dV = \frac{1}{d!} E_{\mu\nu\sigma\rho\cdots} dX^\mu dX^\nu dX^\sigma dX^\rho \cdots \tag{10.125}$$

where

$$E_{\mu\nu\sigma\cdots} = e_{ijk\cdots} \frac{\partial x^i}{\partial X^\mu} \frac{\partial x^j}{\partial X^\nu} \frac{\partial x^k}{\partial X^\sigma} \cdots \tag{10.126}$$

In this way, Levi-Civita used the formalism of tensor calculus to re-derive the previous result of the German mathematician Carl Gustav Jacobi (1804–1851), who had shown that in a curvilinear coordinate system, the volume element is given by

$$dV = \left| \frac{\partial x^i}{\partial X^\mu} \right| dX^1 dX^2 \cdots dX^d \tag{10.127}$$

where $|\partial x^i / \partial X^\mu|$ is the determinant of the $d \times d$ square matrix of transformation coefficients from Cartesian coordinates to curvilinear coordinates. This determinant is called the *Jacobian* of the transformation. From the relationship

$$G_{\mu,\nu} = \frac{\partial x^i}{\partial X^\mu} \delta_{ij} \frac{\partial x^j}{\partial X^\nu} \tag{10.128}$$

one can show that the Jacobian

$$\sqrt{|G_{\mu\nu}|} \equiv \left| \frac{\partial x^i}{\partial X^\mu} \right| \equiv \sqrt{|G|} \tag{10.129}$$

is the square root of the determinant of the covariant metric tensor. The Jacobian is usually represented by the symbol $\sqrt{|G|}$. Levi-Civita's book *Absolute Differential Calculus* has been translated into many languages. It is still in print, and it remains one of the best textbooks in the field, along with Schrödinger's *Space-Time Structure*, Brillouin's *Les Tenseurs* and Landau and Lifshitz's *The Classical Theory of Fields.*

The Jacobian, $\sqrt{|G|}$, is the prototype of a *scalar density.* We can construct *tensor densities* by multiplying tensors by the Jacobian appropriate for the coordinate system. When a tensor density is transformed to another coordinate system, the Jacobian has to be recalculated from the transformed covariant metric tensor. *Tensor capacities* can be constructed by dividing tensors by the Jacobian. Now consider a scalar function ψ. Its gradient is a covariant vector, and therefore

$$G^{\mu\nu}\frac{\partial\psi}{\partial X^\mu}\frac{\partial\psi}{\partial X^\nu} = \text{scalar} \tag{10.130}$$

It follows that if we let

$$\mathcal{L} = \sqrt{|G|}\left[G^{\mu\nu}\frac{\partial\psi}{\partial X^\mu}\frac{\partial\psi}{\partial X^\nu} + \kappa\psi^2\right] \tag{10.131}$$

where κ is a constant, then the variational principle

$$\delta\int\int\cdots\int\mathcal{L}\ dX^1dX^2\cdots dX^d = 0 \tag{10.132}$$

will be invariant under a curvilinear coordinate transformation. As we saw above, the Euler-Lagrange equations that follow from this variational principle are

$$\frac{\partial}{\partial X^\mu}\frac{\partial\mathcal{L}}{\partial\left(\partial\psi/\partial X^\mu\right)} - \frac{\partial\mathcal{L}}{\partial\psi} = 0 \tag{10.133}$$

With the Lagrangian density of equation (10.131), this becomes

$$\frac{1}{\sqrt{|G|}}\frac{\partial}{\partial X^\mu}\sqrt{|G|}\ G^{\mu\nu}\frac{\partial\psi}{\partial X^\nu} = \kappa\ \psi \tag{10.134}$$

10.9 The Laplace-Beltrami operator

The operator

$$\Delta = \sum_{\mu=1}^{d}\sum_{\nu=1}^{d}\frac{1}{\sqrt{|G|}}\frac{\partial}{\partial X^\mu}\sqrt{|G|}\ G^{\mu\nu}\frac{\partial}{\partial X^\nu} \tag{10.135}$$

is the generalized Laplacian operator, which plays such an important role in the theory of hyperspherical harmonics, but here it is written in a form due to Eugenio Beltrami (1835–1899), which is invariant under coordinate transformations. (In equation (10.135), we have abandoned the Einstein convention, and have re-introduced explicit sums.) To illustrate this equation, let us consider some examples. In a d-dimensional space, we can let

$$\begin{aligned} x^1 &= r\ \sin\theta_1 \sin\theta_2 \cdots \sin\theta_{d-2} \cos\theta_{d-1} \\ x^2 &= r\ \sin\theta_1 \sin\theta_2 \cdots \sin\theta_{d-2} \sin\theta_{d-1} \\ x^3 &= r\ \sin\theta_1 \sin\theta_2 \cdots \cos\theta_{d-2} \\ \vdots\ &\ \vdots\ \vdots \\ x^{d-1} &= r\ \sin\theta_1 \cos\theta_2 \\ x^d &= r\ \cos\theta_1 \end{aligned} \tag{10.136}$$

while

$$\begin{aligned} X^1 &= r \\ X^2 &= \theta_1 \\ X^3 &= \theta_2 \\ \vdots\ &\ \vdots\ \vdots \\ X^{d-1} &= \theta_{d-2} \\ X^d &= \theta_{d-1} \end{aligned} \tag{10.137}$$

Then the Jacobi-ans for various values of d are

$$\begin{aligned} d = 3 &\qquad \sqrt{|G|} = r^2 \sin\theta_1 \\ d = 4 &\qquad \sqrt{|G|} = r^3 \sin^2\theta_1 \sin\theta_2 \\ d = 5 &\qquad \sqrt{|G|} = r^4 \sin^3\theta_1 \sin^2\theta_2 \sin\theta_3 \\ d = 6 &\qquad \sqrt{|G|} = r^5 \sin^4\theta_1 \sin^3\theta_2 \sin^2\theta_3 \sin\theta_4 \\ \vdots\ \vdots &\qquad\qquad \vdots \\ d = d &\qquad \sqrt{|G|} = r^{d-1} \sin^{d-2}\theta_1 \sin^{d-3}\theta_2 \cdots \sin^2\theta_{d-3} \sin\theta_{d-2} \end{aligned} \tag{10.138}$$

The covariant metric tensor for $d = 3$ is

$$G_{\mu,\nu} = \begin{pmatrix} 1 & 0 & 0 \\ 0 & r^2 & 0 \\ 0 & 0 & r^2 \sin^2\theta_1 \end{pmatrix} \tag{10.139}$$

while for $d = 4$

$$G_{\mu,\nu} = \begin{pmatrix} 1 & 0 & 0 & 0 \\ 0 & r^2 & 0 & 0 \\ 0 & 0 & r^2 \sin^2\theta_1 & 0 \\ 0 & 0 & 0 & r^2 \sin^2\theta_1 \sin^2\theta_2 \end{pmatrix} \tag{10.140}$$

and for $d = 5$

$$G_{\mu,\nu} = \begin{pmatrix} 1 & 0 & 0 & 0 & 0 \\ 0 & r^2 & 0 & 0 & 0 \\ 0 & 0 & r^2\sin^2\theta_1 & 0 & 0 \\ 0 & 0 & 0 & r^2\sin^2\theta_1\sin^2\theta_2 & 0 \\ 0 & 0 & 0 & 0 & r^2\sin^2\theta_1\sin^2\theta_2\sin^2\theta_3 \end{pmatrix} \tag{10.141}$$

The contravariant metric tensors are just their reciprocals.

$$G^{\mu,\nu} = \begin{pmatrix} 1 & 0 & 0 \\ 0 & \frac{1}{r^2} & 0 \\ 0 & 0 & \frac{1}{r^2\sin^2\theta_1} \end{pmatrix} \tag{10.142}$$

$$G^{\mu,\nu} = \begin{pmatrix} 1 & 0 & 0 & 0 \\ 0 & \frac{1}{r^2} & 0 & 0 \\ 0 & 0 & \frac{1}{r^2\sin^2\theta_1} & 0 \\ 0 & 0 & 0 & \frac{1}{r^2\sin^2\theta_1\sin^2\theta_2} \end{pmatrix} \tag{10.143}$$

$$G^{\mu,\nu} = \begin{pmatrix} 1 & 0 & 0 & 0 & 0 \\ 0 & \frac{1}{r^2} & 0 & 0 & 0 \\ 0 & 0 & \frac{1}{r^2\sin^2\theta_1} & 0 & 0 \\ 0 & 0 & 0 & \frac{1}{r^2\sin^2\theta_1\sin^2\theta_2} & 0 \\ 0 & 0 & 0 & 0 & \frac{1}{r^2\sin^2\theta_1\sin^2\theta_2\sin^2\theta_3} \end{pmatrix} \tag{10.144}$$

and so on. Combining these results, we obtain the Laplace-Beltrami operators:

$$\sum_\nu G^{\mu,\nu}\frac{\partial}{\partial X^\nu}$$
$$= \left(\frac{\partial}{\partial r}, \frac{1}{r^2}\frac{\partial}{\partial\theta_1}, \frac{1}{r^2\sin^2\theta_1}\frac{\partial}{\partial\theta_2}, \frac{1}{r^2\sin^2\theta_1\sin^2\theta_2}\frac{\partial}{\partial\theta_3}, \ldots\right) \tag{10.145}$$

For $d = 3$,

$$\sum_{\nu=1}^{3} \sqrt{|G|}\, G^{\mu,\nu} \frac{\partial}{\partial X^\nu} = r^2 \sin\theta_1 \left(\frac{\partial}{\partial r}, \frac{1}{r^2} \frac{\partial}{\partial \theta_1}, \frac{1}{r^2 \sin^2\theta_1} \frac{\partial}{\partial \theta_2} \right) \tag{10.146}$$

$$\sum_{\mu=1}^{3}\sum_{\nu=1}^{3} \frac{1}{\sqrt{|G|}} \frac{\partial}{\partial X^\mu} \sqrt{|G|}\, G^{\mu,\nu} \frac{\partial}{\partial X^\nu} = \frac{1}{r^2}\frac{\partial}{\partial r} r^2 \frac{\partial}{\partial r} + \frac{1}{r^2 \sin\theta_1} \frac{\partial}{\partial \theta_1} \sin\theta_1 \frac{\partial}{\partial \theta_1} + \frac{1}{r^2 \sin^2\theta_1} \frac{\partial^2}{\partial \theta_2^2} \tag{10.147}$$

For $d = 4$,

$$\sum_{\nu=1}^{4} \sqrt{|G|}\, G^{\mu,\nu} \frac{\partial}{\partial X^\nu} = r^3 \sin^2\theta_1 \sin\theta_2 \left(\frac{\partial}{\partial r}, \frac{1}{r^2} \frac{\partial}{\partial \theta_1}, \frac{1}{r^2 \sin^2\theta_1} \frac{\partial}{\partial \theta_2}, \frac{1}{r^2 \sin^2\theta_1 \sin^2\theta_2} \frac{\partial}{\partial \theta_3} \right) \tag{10.148}$$

$$\begin{aligned}\sum_{\mu=1}^{4}\sum_{\nu=1}^{4} \frac{1}{\sqrt{|G|}} \frac{\partial}{\partial X^\mu} \sqrt{|G|}\, G^{\mu,\nu} \frac{\partial}{\partial X^\nu} &= \frac{1}{r^3}\frac{\partial}{\partial r} r^3 \frac{\partial}{\partial r} + \frac{1}{r^2 \sin^2\theta_1} \frac{\partial}{\partial \theta_1} \sin^2\theta_1 \frac{\partial}{\partial \theta_1} \\ &+ \frac{1}{r^2 \sin^2\theta_1 \sin\theta_2} \frac{\partial}{\partial \theta_2} \sin\theta_2 \frac{\partial}{\partial \theta_2} + \frac{1}{r^2 \sin^2\theta_1 \sin^2\theta_2} \frac{\partial^2}{\partial \theta_3^2}\end{aligned} \tag{10.149}$$

For $d = 5$,

$$\begin{aligned}\sum_{\mu=1}^{5}\sum_{\nu=1}^{5} \frac{1}{\sqrt{|G|}} \frac{\partial}{\partial X^\mu} \sqrt{|G|}\, G^{\mu,\nu} \frac{\partial}{\partial X^\nu} &= \frac{1}{r^4}\frac{\partial}{\partial r} r^4 \frac{\partial}{\partial r} \\ &+ \frac{1}{r^2 \sin^3\theta_1} \frac{\partial}{\partial \theta_1} \sin^3\theta_1 \frac{\partial}{\partial \theta_1} + \frac{1}{r^2 \sin^2\theta_1 \sin^2\theta_2} \frac{\partial}{\partial \theta_2} \sin^2\theta_2 \frac{\partial}{\partial \theta_2} \\ &+ \frac{1}{r^2 \sin^2\theta_1 \sin^2\theta_2 \sin\theta_3} \frac{\partial}{\partial \theta_3} \sin\theta_3 \frac{\partial}{\partial \theta_3} + \frac{1}{r^2 \sin^2\theta_1 \sin^2\theta_2 \sin^2\theta_3} \frac{\partial^2}{\partial \theta_4^2}\end{aligned} \tag{10.150}$$

For general values of d,

$$\sum_{\nu=1}^{d} \sqrt{|G|}\, G^{\mu,\nu} \frac{\partial}{\partial X^{\nu}}$$
$$= r^{d-1} \sin^{d-2}\theta_1 \sin^{d-3}\theta_2 \cdots \sin\theta_{d-2}$$
$$\times \left(\frac{\partial}{\partial r}, \frac{1}{r^2}\frac{\partial}{\partial\theta_1}, \frac{1}{r^2\sin^2\theta_1}\frac{\partial}{\partial\theta_2}, \dots, \frac{1}{r^2\sin^2\theta_1\sin^2\theta_2\cdots\sin^2\theta_{d-2}}\frac{\partial}{\partial\theta_{d-1}} \right) \tag{10.151}$$

$$\sum_{\mu=1}^{d}\sum_{\nu=1}^{d} \frac{1}{\sqrt{|G|}}\frac{\partial}{\partial X^{\mu}} \sqrt{|G|}\, G^{\mu,\nu} \frac{\partial}{\partial X^{\nu}}$$
$$= \frac{1}{r^{d-1}}\frac{\partial}{\partial r} r^{d-1}\frac{\partial}{\partial r} + \frac{1}{r^2\sin^{d-2}\theta_1}\frac{\partial}{\partial\theta_1}\sin^{d-2}\theta_1\frac{\partial}{\partial\theta_1}$$
$$+ \frac{1}{r^2\sin^2\theta_1\sin^{d-3}\theta_2}\frac{\partial}{\partial\theta_2}\sin^{d-3}\theta_2\frac{\partial}{\partial\theta_2} + \cdots$$
$$+ \frac{1}{r^2\sin^2\theta_1\sin^2\theta_2\sin^2\theta_3\cdots\sin\theta_{d-2}}\frac{\partial}{\partial\theta_{d-2}}\sin\theta_{d-2}\frac{\partial}{\partial\theta_{d-2}}$$
$$+ \frac{1}{r^2\sin^2\theta_1\sin^2\theta_2\cdots\sin^2\theta_{d-2}}\frac{\partial^2}{\partial\theta_{d-1}^2} \tag{10.152}$$

As we saw in equation (2.1), the Laplace-Beltrami operator in hyperspherical coordinates can be written as

$$\Delta = \frac{1}{r^{d-1}}\frac{\partial}{\partial r} r^{d-1}\frac{\partial}{\partial r} - \frac{\Lambda^2}{r^2} \tag{10.153}$$

where r is the hyperradius and where Λ^2 is the generalized angular momentum operator. Comparing this with the results that we have just been discussing, we can see that for $d = 3$,

$$-\Lambda^2 = \frac{1}{\sin\theta_1}\frac{\partial}{\partial\theta_1}\sin\theta_1\frac{\partial}{\partial\theta_1} + \frac{1}{\sin^2\theta_1}\frac{\partial^2}{\partial\theta_2^2} \tag{10.154}$$

while for $d = 4$,

$$-\Lambda^2 = \frac{1}{\sin^2\theta_1}\frac{\partial}{\partial\theta_1}\sin^2\theta_1\frac{\partial}{\partial\theta_1}$$
$$+ \frac{1}{\sin^2\theta_1\sin\theta_2}\frac{\partial}{\partial\theta_2}\sin\theta_2\frac{\partial}{\partial\theta_2} + \frac{1}{\sin^2\theta_1\sin^2\theta_2}\frac{\partial^2}{\partial\theta_3^2} \tag{10.155}$$

and for $d = 5$,

$$-\Lambda^2 = \frac{1}{\sin^3\theta_1}\frac{\partial}{\partial\theta_1}\sin^3\theta_1\frac{\partial}{\partial\theta_1} + \frac{1}{\sin^2\theta_1\sin^2\theta_2}\frac{\partial}{\partial\theta_2}\sin^2\theta_2\frac{\partial}{\partial\theta_2}$$
$$+ \frac{1}{\sin^2\theta_1\sin^2\theta_2\sin\theta_3}\frac{\partial}{\partial\theta_3}\sin\theta_3\frac{\partial}{\partial\theta_3} + \frac{1}{\sin^2\theta_1\sin^2\theta_2\sin^2\theta_3}\frac{\partial^2}{\partial\theta_4^2} \tag{10.156}$$

For general values of d, we have

$$\begin{aligned}
-\Lambda^2 &= \frac{1}{\sin^{d-2}\theta_1}\frac{\partial}{\partial\theta_1}\sin^{d-2}\theta_1\frac{\partial}{\partial\theta_1} \\
&+ \frac{1}{\sin^2\theta_1\sin^{d-3}\theta_2}\frac{\partial}{\partial\theta_2}\sin^{d-3}\theta_2\frac{\partial}{\partial\theta_2} + \cdots \\
&+ \frac{1}{\sin^2\theta_1\sin^2\theta_2\sin^2\theta_3\cdots\sin\theta_{d-2}}\frac{\partial}{\partial\theta_{d-2}}\sin\theta_{d-2}\frac{\partial}{\partial\theta_{d-2}} \\
&+ \frac{1}{\sin^2\theta_1\sin^2\theta_2\cdots\sin^2\theta_{d-2}}\frac{\partial^2}{\partial\theta_{d-1}^2} \qquad (10.157)
\end{aligned}$$

We have until now been considering spaces that are intrinsically flat, but a d-dimensional hyperspherical surface embedded in a $d+1$-dimensional space has intrinsic curvature. If the hyperradius r is regarded as a constant, then the Laplace-Beltrami operator for such a surface is given by

$$\Delta = -\frac{\Lambda^2}{r^2} \qquad (10.158)$$

while the covariant metric tensor on the surface is

$$G_{\mu,\nu} = \begin{pmatrix} r^2 & 0 & 0 & 0 & \cdots \\ 0 & r^2\sin^2\theta_1 & 0 & 0 & \cdots \\ 0 & 0 & r^2\sin^2\theta_1\sin^2\theta_2 & 0 & \cdots \\ 0 & 0 & 0 & r^2\sin^2\theta_1\sin^2\theta_2\sin^2\theta_3 & \cdots \\ \vdots & \vdots & \vdots & \vdots & \end{pmatrix} \qquad (10.159)$$

The infinitesimal element of length on the surface, ds is given by

$$\begin{aligned}
ds^2 &= \sum_{\mu=1}^{d}\sum_{\nu=1}^{d} G_{\mu,\nu}dX^\mu dX^\nu \\
&= r^2\left(d\theta_1^2 + \sin^2\theta_1 d\theta_2^2 + \sin^2\theta_1\sin^2\theta_2 d\theta_3^2 + \cdots\right) \qquad (10.160)
\end{aligned}$$

10.10 Geodesics

In the geometry of curved spaces, *geodesics* play the role that straight lines play in Euclidean geometry. The geodesic curves are local minima of path length. The *minimal geodesics* between two points are the shortest paths through the curved space, and play an important role when analyzing physical systems in curved space. They can be determined by the variational principle

$$s = \int ds = \int \sqrt{\sum_{\mu=1}^{d}\sum_{\nu=1}^{d} G_{\mu,\nu}\frac{dX^\mu}{ds}\frac{dX^\nu}{ds}}\, ds$$

$$= \int \sum_{\mu=1}^{d}\sum_{\nu=1}^{d} G_{\mu,\nu}\frac{dX^\mu}{ds}\frac{dX^\nu}{ds}\, ds = \text{minimum} \tag{10.161}$$

The Euler-Lagrange equations which follow from this variational principle are

$$\frac{d}{ds}\frac{\partial L}{\partial(dX^\mu/ds)} - \frac{\partial L}{\partial X^\mu} = 0 \qquad \mu = 1, 2, \cdots, d \tag{10.162}$$

with

$$L = \sum_{\mu=1}^{d}\sum_{\nu=1}^{d} G_{\mu,\nu}(\mathbf{X})\frac{dX^\mu}{ds}\frac{dX^\nu}{ds} \tag{10.163}$$

The Euler-Lagrange equations for geodesics can be written in the form

$$\frac{d^2X^\sigma}{ds} = \Gamma^\sigma_{\mu\nu}\frac{dX^\mu}{ds}\frac{dX^\sigma}{ds} \tag{10.164}$$

Here $\Gamma^\sigma_{\mu\nu}$ is a Christoffel symbol, which is related to the metric tensors by

$$\Gamma^\sigma_{\mu\nu} = \frac{1}{2}G^{\sigma\rho}\left(\frac{\partial G_{\rho\mu}}{\partial X^\nu} + \frac{\partial G_{\rho\nu}}{\partial X^\mu} - \frac{\partial G_{\mu\nu}}{\partial X^\rho}\right) \tag{10.165}$$

In general relativity theory, the trajectories of particles are geodesics in a space-time continuum, whose metric is affected by the presence of other masses.

10.11 Huygens' principle

The Dutch mathematician, physicist and astronomer Christian Huygens (1629–1695) proposed a principle, according to which each point on a wave front can be regarded as the source of a small circular wavelet propagating outward with a velocity characteristic of that point. The new wave front, an instant later, can be regarded as the sum of the contributions of all the small wavelets. This principle might furnish a numerical method for constructing Hamilton's eikonal surfaces, Hamilton's characteristic function and also the surfaces perpendicular to geodesics on curved surfaces. For example we can pick a point $\mathbf{x}_0$ on a curved surface and construct a small hypersphere around it with radius

$$\delta s = \sqrt{\sum_{\mu=1}^{d}\sum_{\nu=1}^{d} G_{\mu,\nu}(\mathbf{x}_0)\delta X^\mu \delta X^\nu} \tag{10.166}$$

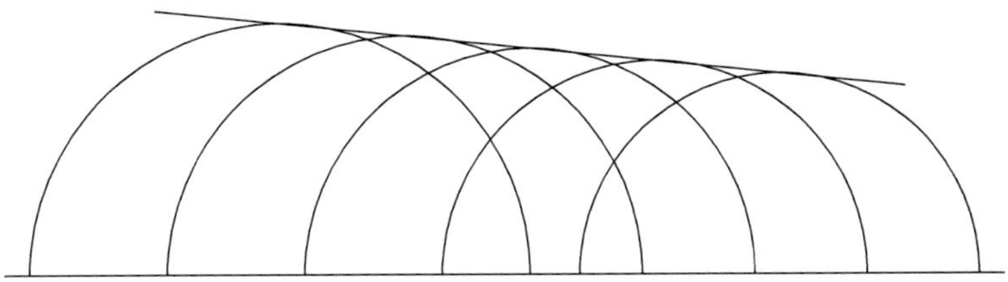

Fig. 10.4: This figure illustrates Huygens' principle. Wavelets propagating outward from the old wave front with velocities appropriate to their positions collectively form the new wave front. Huygens was able to show that his principle led to the Snell-Descartes laws of refraction.

Having done this, we can move to a variety of points on the surface of this small hypersphere and construct similar small hyperspheres on each of the points, using the values of the covariant metric tensor which are appropriate to those points. Continuing in this way, we can construct a set of hypersurfaces to which the geodesics are perpendicular.

In a similar way, Huygens' principle could be used in optics to construct the surfaces corresponding to Hamilton's eikonal function, to which light rays are perpendicular. In the case of optics the radius of the small sphere surrounding a point on the wave front is given by

$$\delta S = \sqrt{\sum_{i=1}^{3} n(\mathbf{x}_0)(\delta x^i)^2} \tag{10.167}$$

where $n(\mathbf{x}_0)$ is the index of refraction.

10.12 Exercises

(1) Consider the system whose Lagrangian is given by equations (10.60) and (10.61). With $\kappa = 2$, diagonalize the potential energy matrix V_{ij}, and verify that the solutions correspond to those given in equations (10.63)–(10.65).

(2) Show that the Schrödinger equation of the system discussed in Exercise 1 is separable. Use the methods discussed in Appendix A to write down solutions to the separated equations.

(3) In a 4-dimensional space, the Cartesian coordinates are related to the

hyperspherical polar coordinates by

$$\begin{aligned} x^1 &= r \ \sin\theta_1 \sin\theta_2 \cos\theta_3 \\ x^2 &= r \ \sin\theta_1 \sin\theta_2 \sin\theta_3 \\ x^3 &= r \ \sin\theta_1 \cos\theta_2 \\ x^4 &= r \ \cos\theta_1 \end{aligned}$$

Calculate the covariant metric tensor, the contravariant metric tensor, the Jacobian, the volume element, and the Laplace-Beltrami operator in the 4-dimensional hyperspherical polar coordinate system.

(4) The spheroidal coordinates in a 3-dimensional space are related to Cartesian coordinates by

$$\begin{aligned} x^1 &= C\zeta\eta \\ x^2 &= C\sqrt{(\zeta^2-1)(1-\eta^2)}\cos\phi \\ x^2 &= C\sqrt{(\zeta^2-1)(1-\eta^2)}\sin\phi \end{aligned}$$

Calculate the covariant metric tensor, the contravariant metric tensor, the Jacobian, the volume element, and the Laplace-Beltrami operator in the spheroidal coordinate system.

Chapter 11

COORDINATE TRANSFORMATIONS FOR N BODIES

11.1 Transforming the kinetic energy operator

The kinetic energy operator can be transformed from the original Cartesian coordinate system $x^1, x^2, \cdots, x^{3N}$ to a new set of coordinates, $X^1, X^2, \cdots, X^{3N}$ by means of the chain rule for partial differentiation. Let us start with the 2-particle case, where the center of mass and relative coordinates are related to the original Cartesian coordinates by

$$\begin{aligned} X^1 &= \frac{M_1 x^1 + M_2 x^4}{M_1 + M_2} \\ X^2 &= \frac{M_1 x^2 + M_2 x^5}{M_1 + M_2} \\ X^3 &= \frac{M_1 x^3 + M_2 x^6}{M_1 + M_2} \\ X^4 &= x^1 - x^4 \\ X^5 &= x^2 - x^5 \\ X^6 &= x^3 - x^6 \end{aligned} \tag{11.1}$$

Then

$$\frac{\partial X^\mu}{\partial x^i} = \begin{pmatrix} \frac{M_1}{M_1+M_2} & 0 & 0 & 1 & 0 & 0 \\ 0 & \frac{M_1}{M_1+M_2} & 0 & 0 & 1 & 0 \\ 0 & 0 & \frac{M_1}{M_1+M_2} & 0 & 0 & 1 \\ \frac{M_2}{M_1+M_2} & 0 & 0 & -1 & 0 & 0 \\ 0 & \frac{M_2}{M_1+M_2} & 0 & 0 & -1 & 0 \\ 0 & 0 & \frac{M_2}{M_1+M_2} & 0 & 0 & -1 \end{pmatrix} \tag{11.2}$$

From (11.2), we can see that

$$\frac{\partial^2 X^\mu}{\partial x^i \partial x^i} = 0 \tag{11.3}$$

Therefore, the chain rule for partial differentiation:

$$\frac{\partial}{\partial x^i} = \sum_{\mu} \frac{\partial X^\mu}{\partial x^i} \frac{\partial}{\partial X^\mu} \tag{11.4}$$

applied twice, yields

$$\begin{aligned}
\frac{\partial^2}{\partial x^i \partial x^i} &= \frac{\partial}{\partial x^i}\left(\sum_{\mu} \frac{\partial X^\mu}{\partial x^i} \frac{\partial}{\partial X^\mu}\right) \\
&= \sum_{\nu} \frac{\partial X^\nu}{\partial x^i} \frac{\partial}{\partial X^\nu}\left(\sum_{\mu} \frac{\partial X^\mu}{\partial x^i} \frac{\partial}{\partial X^\mu}\right) \\
&= \sum_{\mu}\sum_{\nu} \frac{\partial X^\mu}{\partial x^i} \frac{\partial X^\nu}{\partial x^i} \frac{\partial^2}{\partial X^\mu \partial X^\nu}
\end{aligned} \tag{11.5}$$

Then

$$\begin{aligned}
\mathsf{T}_1 &= -\frac{1}{2M_1}\left[\frac{\partial^2}{\partial x^1 \partial x^1} + \frac{\partial^2}{\partial x^2 \partial x^2} + \frac{\partial^2}{\partial x^3 \partial x^3}\right] \\
&= -\frac{1}{2M_1} \sum_{\mu=1}^{6}\sum_{\nu=1}^{6}\sum_{i=1}^{3} \left[\frac{\partial X^\mu}{\partial x^i} \frac{\partial X^\nu}{\partial x^i}\right] \frac{\partial^2}{\partial X^\mu \partial X^\nu}
\end{aligned} \tag{11.6}$$

and similarly

$$\begin{aligned}
\mathsf{T}_2 &= -\frac{1}{2M_2}\left[\frac{\partial^2}{\partial x^4 \partial x^4} + \frac{\partial^2}{\partial x^5 \partial x^5} + \frac{\partial^2}{\partial x^6 \partial x^6}\right] \\
&= -\frac{1}{2M_2} \sum_{\mu=1}^{6}\sum_{\nu=1}^{6}\sum_{i=4}^{6} \left[\frac{\partial X^\mu}{\partial x^i} \frac{\partial X^\nu}{\partial x^i}\right] \frac{\partial^2}{\partial X^\mu \partial X^\nu}
\end{aligned} \tag{11.7}$$

Thus

$$\begin{aligned}
&\mathsf{T} = \mathsf{T}_1 + \mathsf{T}_2 \\
&= -\frac{1}{2} \sum_{\mu=1}^{6}\sum_{\nu=1}^{6}\sum_{i=1}^{6} \left[\frac{\partial X^\mu}{\partial x^i} \frac{1}{M_i} \frac{\partial X^\nu}{\partial x^i}\right] \frac{\partial^2}{\partial X^\mu \partial X^\nu} \\
&\equiv -\frac{1}{2} \sum_{\mu=1}^{6}\sum_{\nu=1}^{6} G^{\mu\nu} \frac{\partial^2}{\partial X^\mu \partial X^\nu}
\end{aligned} \tag{11.8}$$

where

$$
\begin{aligned}
M_i &\equiv (M_1, M_1, M_1, M_2, M_2, M_2) \\
G^{\mu\nu} &\equiv \sum_{i=1}^{6} \left[\frac{\partial X^\mu}{\partial x^i} \frac{1}{M_i} \frac{\partial X^\nu}{\partial x^i} \right] \\
&= \begin{pmatrix}
\frac{1}{M_1+M_2} & 0 & 0 & 0 & 0 & 0 \\
0 & \frac{1}{M_1+M_2} & 0 & 0 & 0 & 0 \\
0 & 0 & \frac{1}{M_1+M_2} & 0 & 0 & 0 \\
0 & 0 & 0 & \frac{1}{M_1}+\frac{1}{M_2} & 0 & 0 \\
0 & 0 & 0 & 0 & \frac{1}{M_1}+\frac{1}{M_2} & 0 \\
0 & 0 & 0 & 0 & 0 & \frac{1}{M_1}+\frac{1}{M_2}
\end{pmatrix}
\end{aligned}
\tag{11.9}
$$

We next consider the 3-particle case where

$$
\begin{aligned}
X^1 &= \frac{M_1 x^1 + M_2 x^4 + M_3 x^7}{M_1 + M_2 + M_3} \\
X^2 &= \frac{M_1 x^2 + M_2 x^5 + M_3 x^8}{M_1 + M_2 + M_3} \\
X^3 &= \frac{M_1 x^3 + M_2 x^6 + M_3 x^9}{M_1 + M_2 + M_3} \\
X^4 &= x^1 - x^4 \\
X^5 &= x^2 - x^5 \\
X^6 &= x^3 - x^6 \\
X^7 &= x^4 - x^7 \\
X^8 &= x^5 - x^8 \\
X^9 &= x^6 - x^9
\end{aligned}
\tag{11.10}
$$

and

$$
\frac{\partial X^\mu}{\partial x^i} = \begin{pmatrix}
\frac{M_1}{M_1+M_2+M_3} & 0 & 0 & 1 & 0 & 0 & 0 & 0 & 0 \\
0 & \frac{M_1}{M_1+M_2+M_3} & 0 & 0 & 1 & 0 & 0 & 0 & 0 \\
0 & 0 & \frac{M_1}{M_1+M_2+M_3} & 0 & 0 & 1 & 0 & 0 & 0 \\
\frac{M_2}{M_1+M_2+M_3} & 0 & 0 & -1 & 0 & 0 & 1 & 0 & 0 \\
0 & \frac{M_2}{M_1+M_2+M_3} & 0 & 0 & -1 & 0 & 0 & 1 & 0 \\
0 & 0 & \frac{M_2}{M_1+M_2+M_3} & 0 & 0 & -1 & 0 & 0 & 1 \\
\frac{M_3}{M_1+M_2+M_3} & 0 & 0 & 0 & 0 & 0 & -1 & 0 & 0 \\
0 & \frac{M_3}{M_1+M_2+M_3} & 0 & 0 & 0 & 0 & 0 & -1 & 0 \\
0 & 0 & \frac{M_3}{M_1+M_2+M_3} & 0 & 0 & 0 & 0 & 0 & -1
\end{pmatrix}
\tag{11.11}
$$

In this case, the total kinetic energy operator is

$$\begin{aligned} \mathsf{T} &= \mathsf{T}_1 + \mathsf{T}_2 + \mathsf{T}_3 \\ &= -\frac{1}{2}\sum_{\mu=1}^{9}\sum_{\nu=1}^{9}\sum_{i=1}^{9}\left[\frac{\partial X^\mu}{\partial x^i}\frac{1}{M_i}\frac{\partial X^\nu}{\partial x^i}\right]\frac{\partial^2}{\partial X^\mu \partial X^\nu} \\ &\equiv -\frac{1}{2}\sum_{\mu=1}^{9}\sum_{\nu=1}^{9} G^{\mu\nu}\frac{\partial^2}{\partial X^\mu \partial X^\nu} \end{aligned} \tag{11.12}$$

where

$$\begin{aligned} M_i &\equiv (M_1, M_1, M_1, M_2, M_2, M_2, M_3, M_3, M_3) \\ G^{\mu\nu} &\equiv \sum_{i=1}^{9}\left[\frac{\partial X^\mu}{\partial x^i}\frac{1}{M_i}\frac{\partial X^\nu}{\partial x^i}\right] \end{aligned}$$

$$= \begin{pmatrix} \frac{1}{M_{tot}} & 0 & 0 & 0 & 0 & 0 & 0 & 0 & 0 \\ 0 & \frac{1}{M_{tot}} & 0 & 0 & 0 & 0 & 0 & 0 & 0 \\ 0 & 0 & \frac{1}{M_{tot}} & 0 & 0 & 0 & 0 & 0 & 0 \\ 0 & 0 & 0 & \frac{1}{\mu_{12}} & 0 & 0 & -\frac{1}{M_2} & 0 & 0 \\ 0 & 0 & 0 & 0 & \frac{1}{\mu_{12}} & 0 & 0 & -\frac{1}{M_2} & 0 \\ 0 & 0 & 0 & 0 & 0 & \frac{1}{\mu_{12}} & 0 & 0 & -\frac{1}{M_2} \\ 0 & 0 & 0 & -\frac{1}{M_2} & 0 & 0 & \frac{1}{\mu_{23}} & 0 & 0 \\ 0 & 0 & 0 & 0 & -\frac{1}{M_2} & 0 & 0 & \frac{1}{\mu_{23}} & 0 \\ 0 & 0 & 0 & 0 & 0 & -\frac{1}{M_2} & 0 & 0 & \frac{1}{\mu_{23}} \end{pmatrix} \tag{11.13}$$

and where

$$\begin{aligned} M_{tot} &\equiv M_1 + M_2 + M_3 \\ \mu_{12} &\equiv \frac{M_1 M_2}{M_1 + M_2} \\ \mu_{23} &\equiv \frac{M_2 M_3}{M_2 + M_3} \end{aligned} \tag{11.14}$$

In general, for all transformations for which equation (11.4) holds, the transformed kinetic energy operator can be written in the form

$$\mathsf{T} = -\frac{1}{2}\sum_{\mu=1}^{3N}\sum_{\nu=1}^{3N} G^{\mu\nu}\frac{\partial^2}{\partial X^\mu \partial X^\nu} \tag{11.15}$$

where

$$\begin{aligned} M_i &\equiv (M_1, M_1, M_1, M_2, M_2, M_2, M_3, M_3, M_3, M_4, \cdots) \\ G^{\mu\nu} &\equiv \sum_{i=1}^{3N}\left[\frac{\partial X^\mu}{\partial x^i}\frac{1}{M_i}\frac{\partial X^\nu}{\partial x^i}\right] \\ &\equiv \sum_{i=1}^{3N}\sum_{j=1}^{3N}\left[\frac{\partial X^\mu}{\partial x^i} g^{ij}\frac{\partial X^\nu}{\partial x^i}\right] \end{aligned} \tag{11.16}$$

where g^{ij} is a diagonal matrix whose diagonal elements are given by $1/M_i$.

But what are we to do in cases where equation (11.4) does not hold? The form of equation (11.16) suggests that in this more general case, we can make use of the transformation properties of the Laplace-Beltrami operator discussed in Chapter 10, provided that we assign to our original coordinate system a contravariant metric tensor which is a diagonal matrix whose diagonal elements are given by $1/M_i$:

$$g^{ij} = \begin{pmatrix} \frac{1}{M_1} & 0 & 0 & 0 & 0 & 0 & 0 & 0 & \cdots \\ 0 & \frac{1}{M_1} & 0 & 0 & 0 & 0 & 0 & 0 & \cdots \\ 0 & 0 & \frac{1}{M_1} & 0 & 0 & 0 & 0 & 0 & \cdots \\ 0 & 0 & 0 & \frac{1}{M_2} & 0 & 0 & 0 & 0 & \cdots \\ 0 & 0 & 0 & 0 & \frac{1}{M_2} & 0 & 0 & 0 & \cdots \\ 0 & 0 & 0 & 0 & 0 & \frac{1}{M_2} & 0 & 0 & \cdots \\ 0 & 0 & 0 & 0 & 0 & 0 & \frac{1}{M_3} & 0 & \cdots \\ 0 & 0 & 0 & 0 & 0 & 0 & 0 & \frac{1}{M_3} & \cdots \\ \vdots & \vdots & \vdots & \vdots & \vdots & \vdots & \vdots & \vdots & \ddots \end{pmatrix} \tag{11.17}$$

11.2 The Laplace-Beltrami operator

As we discussed in Chapter 10, the generalized Laplacian operator Δ can be written in a way that is formally invariant under general curvilinear coordinate transformations. In a d-dimensional space whose contravariant metric tensor is g^{ij}, and whose Jacobian determinant is $\sqrt{|g|}$, Δ can be written as

$$\Delta = \sum_{i=1}^{d} \sum_{j=1}^{d} \frac{1}{\sqrt{|g|}} \frac{\partial}{\partial x^i} \sqrt{|g|} g^{ij} \frac{\partial}{\partial x^j} \tag{11.18}$$

In the space represented in terms of the coordinates $x^1, x^2, \cdots, x^d$, the covariant and contravariant metric tensors are matrix inverses of each other,

$$\sum_j g_{ij} g^{jk} = \delta_i^k \tag{11.19}$$

and the Jacobian is the square root of the absolute value of the determinant of the covariant metric tensor:

$$\sqrt{|g|} = \sqrt{|g_{ij}|} \tag{11.20}$$

Let us now represent the same space in terms of a new coordinate system, $X^1, X^2, \cdots, X^d$. In the new system, the Laplace-Beltrami operator Δ has the same form

$$\Delta = \sum_{\mu=1}^{d} \sum_{\nu=1}^{d} \frac{1}{\sqrt{|G|}} \frac{\partial}{\partial X^\mu} \sqrt{|G|} G^{\mu\nu} \frac{\partial}{\partial X^\nu} \tag{11.21}$$

but now it is represented in terms of a new contravariant metric tensor $G^{\mu\nu}$, where

$$G^{\mu\nu} = \sum_{i=1}^{d}\sum_{j=1}^{d} \frac{\partial X^\mu}{\partial x^i} g^{ij} \frac{\partial X^\nu}{\partial x^j} \tag{11.22}$$

and a new Jacobian determinant, which is the square root of the determinant of the new covariant metric tensor:

$$\sqrt{|G|} = \sqrt{|G_{\mu\nu}|} \tag{11.23}$$

where $G_{\mu\nu}$ is the inverse of $G^{\mu\nu}$:

$$\sum_{\nu} G_{\mu\nu} G^{\nu\sigma} = \delta_\mu^\sigma \tag{11.24}$$

11.3 Transformation of the kinetic energy operator

It is convenient, when solving quantum mechanical systems, to start with mass-weighted Cartesian coordinates, in which the coordinates of particle number s is scaled by the square root of the particle mass, $\sqrt{M_s}$. In this setting, the kinetic energy operator acts identically on all particles regardless of mass, and the particle masses disappear from the physical equations, appearing only in the metric tensors. In mass-weighted Cartesian coordinates, the covariant metric tensor becomes:

$$\begin{aligned} g_{ij} &= \delta_{ij} M_i \\ M_i &= M_1 \qquad i = 1,2,3 \\ M_i &= M_2 \qquad i = 4,5,6 \\ M_i &= M_3 \qquad i = 7,8,9 \\ \vdots \quad & \vdots \quad \vdots \qquad \quad \vdots \end{aligned} \tag{11.25}$$

Then, in the mass-weighted coordinate system x^i, the kinetic energy operator T of a system of N particles is given by

$$\mathsf{T} = -\frac{1}{2}\Delta = -\frac{1}{2}\sum_{i=1}^{3N}\sum_{j=1}^{3N} \frac{1}{\sqrt{|g|}} \frac{\partial}{\partial x^i} \sqrt{|g|} g^{ij} \frac{\partial}{\partial x^j} \tag{11.26}$$

When transforming to a second coordinate system, the transformed kinetic energy operator in terms of the new coordinates X^μ is

$$\mathsf{T} = -\frac{1}{2}\Delta = -\frac{1}{2}\sum_{\mu=1}^{3N}\sum_{\nu=1}^{3N} \frac{1}{\sqrt{|G|}} \frac{\partial}{\partial X^\mu} \sqrt{|G|} G^{\mu\nu} \frac{\partial}{\partial X^\nu} \tag{11.27}$$

where the new and old metrics are related to each other in the manner described above.

11.4 A simple example

In the case of a 2-particle system, the block of the contravariant metric tensor corresponding to the internal coordinates has the form

$$G^{\mu\nu}_{rel} \rightarrow g^{ij} = \begin{pmatrix} \frac{1}{\mu_{12}} & 0 & 0 \\ 0 & \frac{1}{\mu_{12}} & 0 \\ 0 & 0 & \frac{1}{\mu_{12}} \end{pmatrix} \tag{11.28}$$

We have renamed the block because we wish to use it as a starting point for a transformation to spherical polar coordinates. The corresponding covariant metric tensor is given by

$$g_{ij} = \begin{pmatrix} \mu_{12} & 0 & 0 \\ 0 & \mu_{12} & 0 \\ 0 & 0 & \mu_{12} \end{pmatrix} \tag{11.29}$$

We now introduce a new system: the spherical polar coordinates of the relative coordinates of our 2-particle system:

$$\begin{aligned} X^1 &= r_{12} \\ X^2 &= \theta_{12} \\ X^3 &= \phi_{12} \end{aligned} \tag{11.30}$$

which are related to the relative coordinates in Cartesian form by

$$\begin{aligned} x_{12} &= r_{12} \sin\theta_{12} \cos\phi_{12} \\ y_{12} &= r_{12} \sin\theta_{12} \sin\phi_{12} \\ z_{12} &= r_{12} cos\theta_{12} \end{aligned} \tag{11.31}$$

If we rewrite this relationship as

$$\begin{aligned} x^1 &= X_1 \sin X^2 \cos X^3 \\ x^2 &= X^1 \sin X^2 \sin X^3 \\ x^3 &= X^1 \cos X^2 \end{aligned} \tag{11.32}$$

then the Jacobian matrix for the transformation is given by

$$\frac{\partial x^j}{\partial X^\mu} = \begin{pmatrix} \cos(\phi_{12})\sin(\theta_{12}) & \sin(\phi_{12})\sin(\theta_{12}) & \cos(\theta_{12}) \\ r_{12}\cos(\phi_{12})\cos(\theta_{12}) & r_{12}\cos(\theta_{12})\sin(\phi_{12}) & -r_{12}\sin(\theta_{12}) \\ -r_{12}\sin(\phi_{12})\sin(\theta_{12}) & r_{12}\cos(\phi_{12})\sin(\theta_{12}) & 0 \end{pmatrix} \tag{11.33}$$

The covariant metric tensor is then found to be

$$G_{\mu\nu} = \sum_{i=1}^{3}\sum_{j=1}^{3} \frac{\partial x^i}{\partial X^\mu} g_{ij} \frac{\partial x^j}{\partial X^\mu} = \begin{pmatrix} \mu_{12} & 0 & 0 \\ 0 & \mu_{12} r_{12}^2 & 0 \\ 0 & 0 & \mu_{12} r_{12}^2 \sin^2(\theta_{12}) \end{pmatrix} \tag{11.34}$$

The contravariant metric tensor is the reciprocal of $G_{\mu\nu}$:

$$G^{\mu\nu} = \begin{pmatrix} \frac{1}{\mu_{12}} & 0 & 0 \\ 0 & \frac{1}{\mu_{12} r_{12}^2} & 0 \\ 0 & 0 & \frac{1}{\mu_{12} r_{12}^2 \sin^2(\theta_{12})} \end{pmatrix} \tag{11.35}$$

while the Jacobian for the transformation is

$$\sqrt{|G|} = \mu_{12}^{3/2} r_{12}^2 \sin\theta_{12} \tag{11.36}$$

The kinetic energy operator for the internal coordinates then becomes

$$\begin{aligned} \mathsf{T}_{\text{rel}} &= -\frac{1}{2}\sum_{\mu=1}^{3}\sum_{\nu=1}^{3} \frac{1}{\sqrt{|G|}} \frac{\partial}{\partial X^\mu} \sqrt{|G|} G^{\mu\nu} \frac{\partial}{\partial X^\nu} \\ &= \frac{1}{2\mu_{12}} \left(\frac{1}{r_{12}^2} \frac{\partial}{\partial r_{12}} r_{12}^2 \frac{\partial}{\partial r_{12}} \right) \\ &+ \frac{1}{2\mu_{12}} \left(\frac{1}{r_{12}^2 \sin\theta_1} \frac{\partial}{\partial \theta_{12}} \sin\theta_{12} \frac{\partial}{\partial \theta_1} + \frac{1}{r_{12}^2 \sin^2\theta_{12}} \frac{\partial^2}{\partial \phi_{12}^2} \right) \\ &\equiv \frac{1}{2\mu_{12}} \left(\frac{1}{r_{12}^2} \frac{\partial}{\partial r_{12}} r_{12}^2 \frac{\partial}{\partial r_{12}} - \frac{\mathsf{L}_{12}^2}{r_{12}^2} \right) \end{aligned} \tag{11.37}$$

11.5 Jacobi coordinates of a 3-body system with equal masses

As another example of transformation of the kinetic energy operator, let us consider the coordinates defined by

$$\begin{aligned} \mathbf{X}_1 &= \mathbf{x}_1 - \mathbf{x}_2 \\ \mathbf{X}_2 &= \frac{2}{\sqrt{3}} \left(\mathbf{x}_3 - \frac{1}{2}(\mathbf{x}_1 + \mathbf{x}_2) \right) \\ \mathbf{X}_3 &= \frac{1}{3} (\mathbf{x}_1 + \mathbf{x}_2 + \mathbf{x}_3) \end{aligned} \tag{11.38}$$

Then

$$\frac{\partial X^\mu}{\partial x^j} = \begin{pmatrix} 1 & 0 & 0 & -1 & 0 & 0 & 0 & 0 & 0 \\ 0 & 1 & 0 & 0 & -1 & 0 & 0 & 0 & 0 \\ 0 & 0 & 1 & 0 & 0 & -1 & 0 & 0 & 0 \\ -\frac{1}{\sqrt{3}} & 0 & 0 & -\frac{1}{\sqrt{3}} & 0 & 0 & \frac{2}{\sqrt{3}} & 0 & 0 \\ 0 & -\frac{1}{\sqrt{3}} & 0 & 0 & -\frac{1}{\sqrt{3}} & 0 & 0 & \frac{2}{\sqrt{3}} & 0 \\ 0 & 0 & -\frac{1}{\sqrt{3}} & 0 & 0 & -\frac{1}{\sqrt{3}} & 0 & 0 & \frac{2}{\sqrt{3}} \\ \frac{1}{3} & 0 & 0 & \frac{1}{3} & 0 & 0 & \frac{1}{3} & 0 & 0 \\ 0 & \frac{1}{3} & 0 & 0 & \frac{1}{3} & 0 & 0 & \frac{1}{3} & 0 \\ 0 & 0 & \frac{1}{3} & 0 & 0 & \frac{1}{3} & 0 & 0 & \frac{1}{3} \end{pmatrix} \tag{11.39}$$

and the contravariant metric tensor in the new coordinate system becomes:

$$G^{\mu\nu} = \frac{1}{M}\begin{pmatrix} 2&0&0&0&0&0&0&0&0\\ 0&2&0&0&0&0&0&0&0\\ 0&0&2&0&0&0&0&0&0\\ 0&0&0&2&0&0&0&0&0\\ 0&0&0&0&2&0&0&0&0\\ 0&0&0&0&0&2&0&0&0\\ 0&0&0&0&0&0&\frac{1}{3}&0&0\\ 0&0&0&0&0&0&0&\frac{1}{3}&0\\ 0&0&0&0&0&0&0&0&\frac{1}{3} \end{pmatrix} \tag{11.40}$$

and

$$\mathsf{T} = -\frac{1}{2}\Delta = -\frac{1}{M}\left(\nabla^2_{\mathbf{X}_1} + \nabla^2_{\mathbf{X}_2} + \frac{1}{6}\nabla^2_{\mathbf{X}_3}\right) \tag{11.41}$$

11.6 Normal mode transformations

Let us now consider a molecule or cluster whose classical kinetic energy and potential energy are given respectively by

$$T = \frac{1}{2}\sum_{i=1}^{d}\sum_{j=1}^{d}\delta_{i,j}\frac{dx^i}{dt}\frac{dx^j}{dt} \tag{11.42}$$

and

$$V \approx v_0 + \frac{1}{2}\sum_{i=1}^{d}\sum_{j=1}^{d}\frac{v_{i,j}}{\sqrt{M_i M_j}}x^i x^j + \cdots \tag{11.43}$$

Here the coordinates $x_1, x_2, \cdots, x_d$ represent mass-weighted displacements from a local minimum of the potential energy, while M_j is the mass appropriate for the coordinate x_j. The normal modes of the system are found by solving the secular equation

$$\sum_{j=1}^{d}\left(\frac{v_{i,j}}{\sqrt{M_i M_j}} - \mathcal{V}_\mu \delta_{i,j}\right)U_{j,\mu} = 0 \tag{11.44}$$

The normal mode coordinates are then given by

$$X_\mu = \sum_{j=1}^{d} x_j U_{j,\mu} \tag{11.45}$$

where $U_{j,\mu}$ is a unitary matrix with $d = 3N$, and where N is the number of atoms in the molecule or cluster. The quantum mechanical kinetic energy operator is

$$\begin{aligned} \mathsf{T} &= -\frac{1}{2}\sum_{j=1}^{d}\frac{\partial^2}{\partial x_j^2} = -\frac{1}{2}\sum_{j=1}^{d}\sum_{\mu=1}^{d}\sum_{\nu=1}^{d}\frac{\partial X_\mu}{\partial x_j}\frac{\partial X_\nu}{\partial x_j}\frac{\partial^2}{\partial X_\mu X_\nu} \\ &= -\frac{1}{2}\sum_{j=1}^{d}\sum_{\mu=1}^{d}\sum_{\nu=1}^{d}U_{j,\mu}U_{j,\nu}\frac{\partial^2}{\partial X_\mu X_\nu} = -\frac{1}{2}\sum_{\nu=1}^{d}\frac{\partial^2}{\partial X_\mu^2} \end{aligned} \tag{11.46}$$

In the last step of equation (11.46), we have made use of the fact that $U_{j,\mu}$ is unitary and real.

11.7 A simple model for interatomic forces in molecules

In an earlier chapter, equations (10.68)–(10.72), we saw that in the simplest approximation, a molecule or cluster in a non-melted state could be regarded as a collection of atoms held together by springlike bonds with a potential of the form

$$V = \frac{1}{2}\sum_{t>s}^{N}\sum_{s=1}^{N} k_{st}\left(|\mathbf{x}_s + \mathbf{R}_s - \mathbf{x}_t - \mathbf{R}_t| - |\mathbf{R}_s - \mathbf{R}_t|\right)^2 \tag{11.47}$$

where k_{st} represents the force constant of the "spring" which connects atom s with atom t. We introduced the notation

$$\begin{aligned}\mathbf{R}_{st} &\equiv \mathbf{R}_s - \mathbf{R}_t \\ \mathbf{x}_{st} &\equiv \mathbf{x}_s - \mathbf{x}_t\end{aligned} \tag{11.48}$$

and when we expanded the potential about its equilibrium position with the assumption that the vibrations were very small, we obtained as the leading term in the Taylor series expansion of the potential

$$\begin{aligned}V &\approx \sum_{t>s}^{N}\sum_{s=1}^{N} \frac{k_{st}}{|\mathbf{R}_{st}|^2}(\mathbf{R}_{st}\cdot\mathbf{x}_{st})^2 \\ &\equiv \frac{1}{2}\sum_{t>s}^{N}\sum_{s=1}^{N}\sum_{\mu=1}^{3}\sum_{\nu=1}^{3} V_{s,\mu;t,\nu}x_{s,\mu}x_{t,\nu}\end{aligned} \tag{11.49}$$

where $\mathbf{x}_s \equiv (x_{s1}, x_{s2}, x_{s3})$. In order to find the normal modes of the system, one then diagonalizes the mass-weighted potential energy matrix $\frac{V_{s,\mu;t,\nu}}{\sqrt{m_s m_t}}$.

11.8 A simple example

In order to illustrate this procedure, let us consider the simple case of three equal masses joined by equal springs with force constants κ, with their equilibrium positions forming an equilateral triangle. Let us denote the original mass-weighted displacements from equilibrium as

$$\mathbf{x} = (x_1, y_1, z_1, x_2, y_2, z_2, x_3, y_3, z_3) \tag{11.50}$$

while the normal modes are written as

$$\mathbf{X} = (X_1, X_2, X_3, X_4, X_5, X_6, X_7, X_8, X_9) \tag{11.51}$$

By diagonalizing the mass-weighted potential we obtain a unitary transformation matrix relating the normal modes to the original coordinates:

$$X_\mu = \sum_{j=1}^{9} x_j U_{j,\mu} \tag{11.52}$$

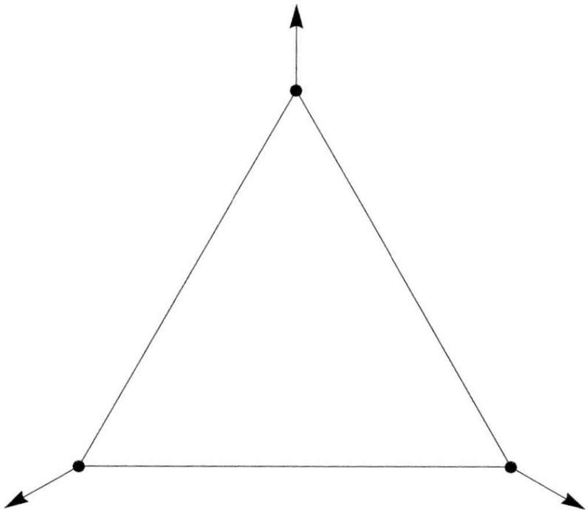

Fig. 11.1: This figure shows one of the normal modes of a system of elastically-joined point masses (the "breathing mode"). In this example, the masses are all equal, the springs have equal force constants κ, and their equilibrium positions are at the corners of an equilateral triangle. The breathing mode is related to the coordinates representing small displacements from equilibrium by X_1 in equation (11.53). When the normal modes are calculated by diagonalization of the mass-weighted potential energy matrix, one also finds six zero-frequency modes corresponding to translations and rotations of the entire system

As predicted by theory, six of the X_μ are zero-frequency modes, corresponding to rotation and translation of the system as a whole. There are however, three modes whose vibrational frequency does not vanish. These are

$$\begin{aligned}
X_1 &= \frac{x_1}{2} + \frac{y_1}{2\sqrt{3}} - \frac{x_2}{2} + \frac{y_2}{2\sqrt{3}} - \frac{y_3}{\sqrt{3}} \\
X_2 &= -\frac{x_1}{2} + \frac{y_1}{2\sqrt{3}} + \frac{x_2}{2} + \frac{y_2}{2\sqrt{3}} - \frac{y_3}{\sqrt{3}} \\
X_3 &= -\frac{x_1}{2\sqrt{3}} - \frac{y_1}{2} - \frac{x_2}{2\sqrt{3}} + \frac{y_2}{2} + \frac{x_3}{\sqrt{3}}
\end{aligned} \tag{11.53}$$

The next three normal modes are zero-frequency modes corresponding to translations of the system as a whole. These are related to the original coordinates by

$$\begin{aligned}
X_4 &= \frac{x_1}{\sqrt{3}} + \frac{x_2}{\sqrt{3}} + \frac{x_3}{\sqrt{3}} \\
X_5 &= \frac{y_1}{\sqrt{3}} + \frac{y_2}{\sqrt{3}} + \frac{y_3}{\sqrt{3}} \\
X_6 &= \frac{z_1}{\sqrt{3}} + \frac{z_2}{\sqrt{3}} + \frac{z_3}{\sqrt{3}}
\end{aligned} \tag{11.54}$$

Finally there are three zero-frequency normal modes which correspond to rotations of the system as a whole:

$$\begin{aligned}
X_7 &= \frac{x_1}{2\sqrt{3}} - \frac{y_1}{2} + \frac{x_2}{2\sqrt{3}} + \frac{y_2}{2} - \frac{x_3}{\sqrt{3}} \\
X_8 &= -\frac{z_1}{\sqrt{2}} + \frac{z_2}{\sqrt{2}} \\
X_9 &= -\frac{z_1}{\sqrt{6}} - \frac{z_2}{\sqrt{6}} + \sqrt{\frac{2}{3}}z_3
\end{aligned} \tag{11.55}$$

Since the transformation matrix $U_{j,\mu}$ which relates the original coordinates to the normal mode coordinates is unitary, and since the original coordinates are Cartesian, the covariant and contravariant metric tensors are unaffected by the transformation, and hence the Laplace-Beltrami operator is also unaffected. In terms of the normal mode coordinates, the Schrödinger equation of the system is separable. Let us look first at the degrees of freedom X_1, X_2 and X_3, which are associated with non-zero vibrational frequencies. The wave equations associated with these modes have the form:

$$\left(-\frac{1}{2}\frac{\partial^2}{\partial X_\mu^2} + \mathcal{V}_\mu - \epsilon_\mu\right)\psi_\mu(X_\mu) = 0 \qquad \mu = 1,2,3 \tag{11.56}$$

where the three $\mathcal{V}_\mu$ are eigenvalues of the potential energy matrix corresponding to first three normal modes. The solutions to the wave equations in (11.56) harmonic oscillator wave functions corresponding to the frequencies

$$\omega_1 = \sqrt{\frac{2\kappa}{\mathcal{M}}} \qquad \omega_2 = \sqrt{\frac{\kappa}{\mathcal{M}}} \qquad \omega_3 = \sqrt{\frac{\kappa}{\mathcal{M}}} \tag{11.57}$$

The inverse Jacobian matrix for the transformation shown in equations (11.53)–(11.55) is

$$\frac{\partial X^\mu}{\partial x^j} = \begin{pmatrix} \frac{1}{2} & -\frac{1}{2} & -\frac{1}{2\sqrt{3}} & \frac{1}{\sqrt{3}} & 0 & 0 & \frac{1}{2\sqrt{3}} & 0 & 0 \\ \frac{1}{2\sqrt{3}} & \frac{1}{2\sqrt{3}} & -\frac{1}{2} & 0 & \frac{1}{\sqrt{3}} & 0 & -\frac{1}{2} & 0 & 0 \\ 0 & 0 & 0 & 0 & 0 & \frac{1}{\sqrt{3}} & 0 & -\frac{1}{\sqrt{2}} & -\frac{1}{\sqrt{6}} \\ -\frac{1}{2} & \frac{1}{2} & -\frac{1}{2\sqrt{3}} & \frac{1}{\sqrt{3}} & 0 & 0 & \frac{1}{2\sqrt{3}} & 0 & 0 \\ \frac{1}{2\sqrt{3}} & \frac{1}{2\sqrt{3}} & \frac{1}{2} & 0 & \frac{1}{\sqrt{3}} & 0 & \frac{1}{2} & 0 & 0 \\ 0 & 0 & 0 & 0 & 0 & \frac{1}{\sqrt{3}} & 0 & \frac{1}{\sqrt{2}} & -\frac{1}{\sqrt{6}} \\ 0 & 0 & \frac{1}{\sqrt{3}} & \frac{1}{\sqrt{3}} & 0 & 0 & -\frac{1}{\sqrt{3}} & 0 & 0 \\ -\frac{1}{\sqrt{3}} & -\frac{1}{\sqrt{3}} & 0 & 0 & \frac{1}{\sqrt{3}} & 0 & 0 & 0 & 0 \\ 0 & 0 & 0 & 0 & 0 & \frac{1}{\sqrt{3}} & 0 & 0 & \sqrt{\frac{2}{3}} \end{pmatrix} \tag{11.58}$$

while its inverse, the Jacobian matrix, is

$$\frac{\partial x^j}{\partial X^\mu} = \begin{pmatrix} \frac{1}{2} & \frac{1}{2\sqrt{3}} & 0 & -\frac{1}{2} & \frac{1}{2\sqrt{3}} & 0 & 0 & -\frac{1}{\sqrt{3}} & 0 \\ -\frac{1}{2} & \frac{1}{2\sqrt{3}} & 0 & \frac{1}{2} & \frac{1}{2\sqrt{3}} & 0 & 0 & -\frac{1}{\sqrt{3}} & 0 \\ -\frac{1}{2\sqrt{3}} & -\frac{1}{2} & 0 & -\frac{1}{2\sqrt{3}} & \frac{1}{2} & 0 & \frac{1}{\sqrt{3}} & 0 & 0 \\ \frac{1}{\sqrt{3}} & 0 & 0 & \frac{1}{\sqrt{3}} & 0 & 0 & \frac{1}{\sqrt{3}} & 0 & 0 \\ 0 & \frac{1}{\sqrt{3}} & 0 & 0 & \frac{1}{\sqrt{3}} & 0 & 0 & \frac{1}{\sqrt{3}} & 0 \\ 0 & 0 & \frac{1}{\sqrt{3}} & 0 & 0 & \frac{1}{\sqrt{3}} & 0 & 0 & \frac{1}{\sqrt{3}} \\ \frac{1}{2\sqrt{3}} & -\frac{1}{2} & 0 & \frac{1}{2\sqrt{3}} & \frac{1}{2} & 0 & -\frac{1}{\sqrt{3}} & 0 & 0 \\ 0 & 0 & -\frac{1}{\sqrt{2}} & 0 & 0 & \frac{1}{\sqrt{2}} & 0 & 0 & 0 \\ 0 & 0 & -\frac{1}{\sqrt{6}} & 0 & 0 & -\frac{1}{\sqrt{6}} & 0 & 0 & \sqrt{\frac{2}{3}} \end{pmatrix} \tag{11.59}$$

One can verify that both these matrices are unitary, and that each is the transpose of the other. Furthermore, the determinants of both matrices are unity, so that

$$\sqrt{|G|} \equiv \left| \frac{\partial x^j}{\partial X^\mu} \right| = 1 \tag{11.60}$$

More generally, we can consider a molecule or cluster whose potential V is a function of $d = 3N$ coordinates. Suppose that this potential has a local minimum at a point given by the $d = 3N$-dimensional vector $\mathbf{R}$, while $\mathbf{x}$ represents small displacements of the particles from the minimum. Then the leading term in a Taylor series expansion of the potential in terms of the small displacements will be

$$V \approx V(\mathbf{R}) + \frac{1}{2} \sum_{i=1}^{d} \sum_{j=1}^{d} V_{i,j} x_i x_j \tag{11.61}$$

where

$$V_{i,j} = \left[\frac{\partial^2 V}{\partial x_i x_j} \right]_{\mathbf{x}=0} \tag{11.62}$$

By diagonalizing the mass-weighted potential energy matrix $V_{i,j}/\sqrt{M_i M_j}$ we obtain the normal modes of the system, and in terms of these its Schrödinger equation is approximately separable, but only approximately so. At high excitation energies this approximate separability breaks down, and eventually, with the particles free to move between local minima, reference to a particular minimum becomes meaningless. At such excitation energies, the molecule or cluster can be said to be melted.

11.9 Separability in the harmonic approximation

In the simple example discussed above, the interatomic potential is taken to be exactly harmonic, and the 9-dimensional Schrödinger equation separates exactly into nine 1-dimensional equations. In the later sections, we will discuss more general systems, where there are anharmonic terms in the interaction potentials. In such systems, the Schrödinger equation is only approximately separable, but solutions obtained in the harmonic approximation can still be used as basis functions.

If the vibrational and rotational energies of a molecule or cluster are not too great, and if the change in equilibrium for different rotational states is neglected, the harmonic approximation discussed in equations (11.47)–(11.62) may be used, and in this approximation, the Schrödinger equation of the system is separable. The normal modes are given by

$$X_\mu = \sum_{j=1}^{d} x_j U_{j,\mu} \tag{11.63}$$

Since the transformation is unitary, the covariant and contravariant metric tensors are unaffected by it, and hence the Laplace-Beltrami operator has the same form in in the old and new coordinate :

$$\Delta = \sum_{j=1}^{d} \frac{\partial^2}{\partial x_j^2} = \sum_{\mu=1}^{d} \frac{\partial^2}{\partial X_\mu^2} \tag{11.64}$$

In the harmonic approximation, the potential energy is factorable when expressed in terms of the normal mode coordinates:

$$V(\mathbf{X}) \approx V_0 + \prod_{\mu=1}^{d-6} \left[\frac{1}{2}\kappa_\mu X_\mu^2\right] \tag{11.65}$$

The approximate Schrödinger equation then becomes

$$\left\{-\frac{1}{2}\sum_{\mu=1}^{d} \frac{\partial^2}{\partial X_\mu^2} + V_0 + \prod_{\mu=1}^{d-6} \left[\frac{1}{2}\kappa_\mu X_\mu^2\right] - E\right\} \Psi(\mathbf{X}) = 0 \tag{11.66}$$

This equation is separable, and it has solutions of the form

$$\Psi(\mathbf{X}) = \psi_{rot}(X_d, X_{d-1}, X_{d-2})\psi_{trans}(X_{d-3}, X_{d-4}, X_{d-5}) \prod_{\mu=1}^{d-6} \psi_{n_\mu}(X_\mu) \tag{11.67}$$

The translational part of the wave function is given by

$$\psi_{trans}(X_{d-3}, X_{d-4}, X_{d-5}) = \prod_{\mu=d-5}^{d-3} e^{ik_\mu X_\mu} \tag{11.68}$$

11.10 The Morse potential

Let us consider a system of N identical atoms, each with mass M, interacting through a Morse potential:

$$V(\mathbf{x}) = D_e \sum_{i>j}^{N} \sum_{j=1}^{N} \left[e^{-2\alpha(r_{i,j}-r_e)} - 2e^{-\alpha(r_{i,j}-r_e)} \right] \tag{11.69}$$

In equation (11.69), D_e is the dissociation energy, $r_{i,j}$ is the interatomic distance, r_e is the equilibrium distance and α is an adjustable parameter. Expanding this potential in a Taylor series about its equilibrium positions, we obtain the mass-weighted potential energy matrix, equation (10.72). The unitary transformation which diagonalizes this matrix takes us from coordinates representing small displacements of the atoms from their equilibrium positions to normal mode coordinates.

$$X_\mu = \sum_{j=1}^{3N} x_j U_{j,\mu} \tag{11.70}$$

Had we approximated the system by the model of particles joined by springs, the same unitary transformation would bring us to the normal modes. This is because, close to the equilibrium positions, the Morse potential can be approximated by an harmonic potential, as is shown in Figure 12.2. Of the normal modes, six will be zero-frequency modes corresponding to translations and rotations of the system as a whole. As a simple example, we can consider a system of three identical atoms interacting with each other through the Morse potential. Their equilibrium positions will form an equilateral triangle, just as in Figure 12.1, and the unitary transformation shown in equations (11.53)–(11.55) will diagonalize the mass-weighted potential energy matrix. The non-zero frequencies are found to be

$$\omega_1 = \sqrt{\frac{6\alpha^2}{M}} \qquad \omega_2 = \sqrt{\frac{3\alpha^2}{M}} \qquad \omega_3 = \sqrt{\frac{3\alpha^2}{M}} \tag{11.71}$$

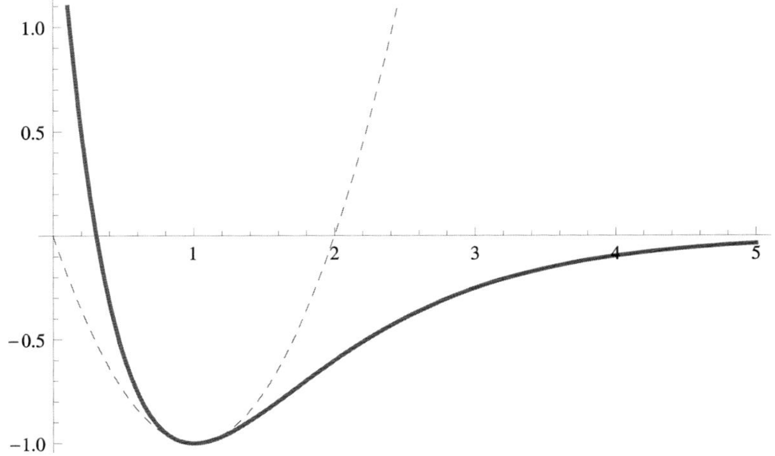

Fig. 11.2: This figure shows the Morse potential, which is sometimes used as a model for interatomic interactions (see equation (11.62)). The potential is shown as a function of the interatomic distance $r_{i,j}$, and the parameters are chosen to be $D_e = 1$, $r_e = 1$ and $\alpha = 1$. Thus the potential has the form $e^{-2((r_{i,j}-1)} - 2e^{-(r_{i,j}-1)}$ (thick curve). The dashed curve in the figure shows the leading terms in a Taylor series expansion of the potential about its equilibrium position. These leading terms have the form $-1+(r_{i,j}-1)^2$.

As long as the oscillations of the system about its equilibrium position are small, the Schrödinger equation describing it is separable when expressed in terms of the normal coordinates. However, as the amount of excitation energy becomes greater, this approximate separability gradually breaks down. Finally, when the system has enough energy, the separability breaks down completely, the cluster can be regarded as melted, and the hyperspherical approach of the previous chapter approach must be used.

11.11 Exact solutions for the Morse oscillator

If we introduce the new variables

$$\rho \equiv \alpha r \qquad \rho_e \equiv \alpha r_e \qquad C \equiv \frac{\sqrt{2MD_e}}{\alpha} \qquad \epsilon \equiv \frac{2ME}{\alpha^2} \tag{11.72}$$

the Schrödinger equation of a single Morse oscillator can be written in the form

$$\left[-\frac{d^2}{d\rho^2} + V_m(\rho) - \epsilon_n\right] \psi_n(\rho) = 0 \tag{11.73}$$

where

$$V_m(\rho) = C^2 \left[e^{-2(\rho-\rho_e)} - 2e^{-(\rho-\rho_e)} \right] \tag{11.74}$$

The eigenvalues are then given by

$$\epsilon_n = -\left(C - n - \frac{1}{2} \right)^2 \qquad n = 0, 1, 2, 3, \cdots \tag{11.75}$$

The eigenfunctions can be expressed in terms of associated Laguerre polynomials:

$$\psi_n(\rho) = N_n \, z^{C-n-1/2} e^{-z/2} \, L_n^{2C-2n-1}(z) \tag{11.76}$$

where

$$z \equiv 2Ce^{-(\rho-\rho_e)} \tag{11.77}$$

and where the normalization constant is

$$N_n = n! \left[\frac{2C - 2n - 1}{\Gamma(n+1)\Gamma(2C-n)} \right]^{1/2} \tag{11.78}$$

The Morse oscillator has only a finite number of bound states. These start out being evenly spaced near the bottom of the well, where the potential is approximately harmonic. The higher excited states become gradually more closely spaced. The size of the constant C can be found from equation (11.72). Since we are using atomic units, the mass of an atom is given in units of the electron mass. Thus, for example, a carbon atom is $1836 \times 12.07 = 22162$ times the mass of an electron. With $D_e = 1$ and $\alpha = 1$, this corresponds to $C = 210.4$. In this example, the Morse oscillator has 210 bound states.

11.12 Morse oscillator eigenfunctions as a basis

If the function $\psi_n(\rho)$ is given by equations (11.76)–(11.78), so that it satisfies equation (11.73), then it obeys the orthonormality relation:

$$\int_0^\infty d\rho \; \psi_{n'}(\rho)\psi_n(\rho) = \delta_{n',n} \tag{11.79}$$

Since $\psi_n(\rho)$ is a solution to equation (11.73), we can infer that

$$\left[-\frac{d^2}{d\rho^2} \right] \psi_n(\rho) = \left[-V_m(\rho) + \epsilon_n \right] \psi_n(\rho) \tag{11.80}$$

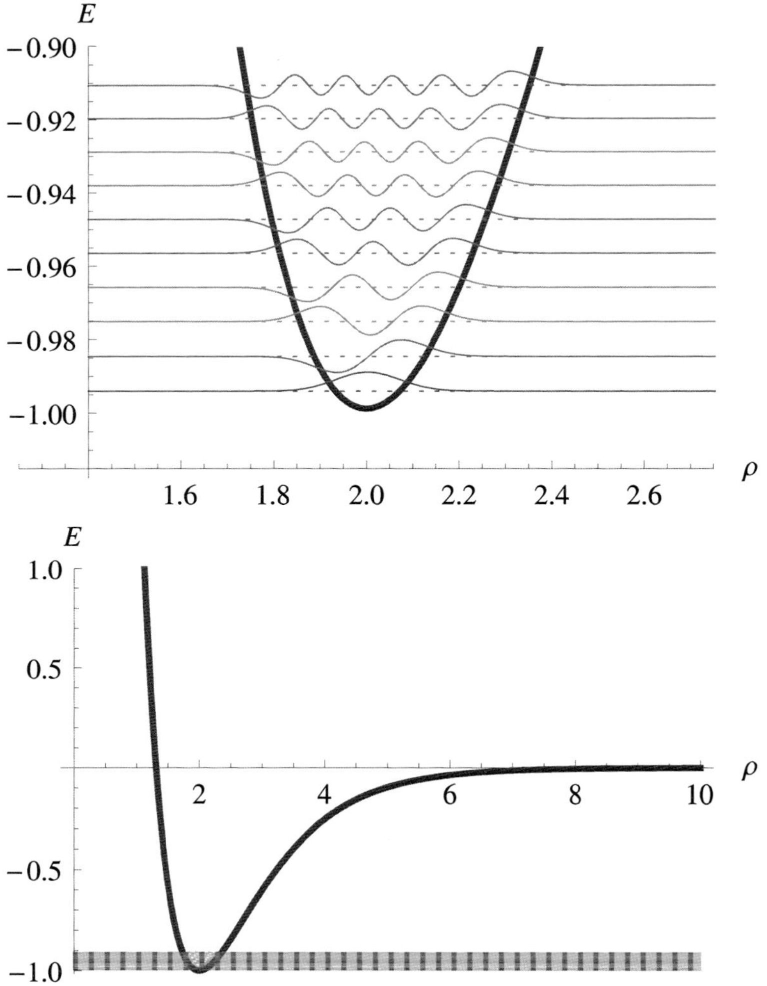

Fig. 11.3: The exact solutions to the Morse potential for $C = 210.4$, *with dissociation energy* $D_e = 1$, *and* $\alpha = 1$. *The first few solutions near to the bottom of the well are evenly spaced, because the potential is approximately harmonic. As we go up in energy, one can see that the energy levels become more closely spaced. Of the 210 bound states, only the first 10 are shown. The figure on the left shows the bottom of the potential well, while the figure on the right shows the full potential.*

Now suppose that we want to solve the equation

$$\left[-\frac{d^2}{d\rho^2} + V(\rho) - E_\kappa\right] \Phi_\kappa(\rho) = 0 \tag{11.81}$$

where $V(\rho) \neq V_m(\rho)$. We can build up solutions to this equation in terms of Morse oscillator wave functions by letting

$$\Phi_\kappa(\rho) = \sum_n \psi_n(\rho)\langle n|\kappa\rangle \tag{11.82}$$

Substituting this expansion into (11.81), multiplying on the left by $\psi_{n'}(\rho)$ and integrating over ρ, we obtain the secular equations:

$$\begin{aligned}
&\sum_n \int_0^\infty d\rho\, \psi_{n'}(\rho)\left[-\frac{d^2}{d\rho^2} + V(\rho) - E_\kappa\right] \psi_n(\rho)\langle n|\kappa\rangle \\
&= \sum_n \int_0^\infty d\rho\, \psi_{n'}(\rho)\left[-\frac{d^2}{d\rho^2} + V(\rho) - E_\kappa\right] \psi_n(\rho)\langle n|\kappa\rangle \\
&= \sum_n \int_0^\infty d\rho\, \psi_{n'}(\rho)\left[V(\rho) - V_m(\rho) + \epsilon_n - E_\kappa\right] \psi_n(\rho)\langle n|\kappa\rangle \\
&= \sum_n \left[V'_{n',n} + \delta_{n',n}(\epsilon_n - E_\kappa)\right]\langle n|\kappa\rangle = 0
\end{aligned} \tag{11.83}$$

where

$$V'_{n',n} \equiv \int_0^\infty d\rho\, \psi_{n'}(\rho)\left[V(\rho) - V_m(\rho)\right]\psi_{n'}(\rho) \tag{11.84}$$

Thus, we can obtain solutions to the secular equations by diagonalizing the matrix $V'_{n',n} + \delta_{n',n}\epsilon_n$

11.13 Rotations and vibrations of diatomic molecules

As a simple example of the use of the Morse oscillator eigenfunctions as basis sets, we can consider the rotational and vibrational spectra of diatomic molecules. Let us denote the masses of the two atoms in the dimer by M_1 and M_2, while $\mathbf{R} \equiv \mathbf{R}_1 - \mathbf{R}_2$ represents the distance between them. Then, after the center of mass motion has been separated out from the problem, the Hamiltonian for the molecule's internal dynamics will be given by

$$\mathsf{H}(\mathbf{R}) = -\frac{1}{2\mu}\nabla^2_{\mathbf{R}} + V(R) \tag{11.85}$$

where $R \equiv |\mathbf{R}_1 - \mathbf{R}_2|$. and where

$$\mu = \frac{M_1 M_2}{M_1 + M_2} \tag{11.86}$$

is the reduced mass. The eigenfunctions of $\mathsf{H}(\mathbf{R})$ have the form

$$\Psi_{\nu,J,M}(\mathbf{R}) = \frac{\chi_{\nu,J}(R)}{R} Y_{J,M}(\theta_R, \phi_R) \tag{11.87}$$

Here $Y_{J,M}(\theta_R, \phi_R)$ is a spherical harmonic representing the rotational motion of the diatomic molecule about its center of mass, while $\chi_{\nu,J}(R)$ obeys the equation

$$\left[-\frac{1}{2\mu}\frac{d^2}{dR^2} + V(R) + \frac{J(J+1)}{2\mu R^2} - E_{\nu,J}\right]\chi_{\nu,J}(R) = 0 \tag{11.88}$$

This can be rewritten as

$$\left[-\frac{d^2}{dR^2} + V_{eff}(R) - 2\mu E_{\nu,J}\right]\chi_{\nu,J}(R) = 0 \tag{11.89}$$

where

$$V_{eff}(R) = 2\mu V(R) + \frac{J(J+1)}{R^2} \tag{11.90}$$

Comparing equation (11.88) with (11.73), we can see that Morse eigenfunctions will be an appropriate basis for solving the differential equation provided that $V_{eff}(R)$ has a minimum, and provided that we can find a Morse potential with parameters such that it fits $V_{eff}(R)$ as closely as possible in the neighborhood of the minimum.

Although we have been discussing the rotations and vibrations of a diatomic molecule, the method used here is more general. For example it could be applied to the dynamics of a system of two nucleons. In the next section, we will discuss the dynamics of clusters of atoms. Once again, the methods are more general, and can be applied to other problems.

11.14 The rotational and vibrational modes of molecules and clusters

Let us next consider a molecule or cluster with N atoms of mass M_j, and let $\mathbf{R}_j$, $j = 1, 2, 3, \cdots, N$, denote the positions of the atoms. The moment of inertia tensor for the system is then given by

$$I_{s,t} = \sum_{j=1}^{N} M_j \left[\delta_{s,t}|\mathbf{R}_j|^2 - R_{j,s}R_{j,t}\right] \qquad s = 1, 2, 3 \qquad t = 1, 2, 3 \tag{11.91}$$

where $R_{j,s}$ is the projection of the vector $\mathbf{R}_j$ onto the s-axis of the coordinate system. The position of the center of mass of the system is given by

$$\mathbf{R}_{cm} = \frac{\sum_{j=1}^{N} M_j \mathbf{R}_j}{\sum_{j=1}^{N} M_j} \tag{11.92}$$

Let us denote the angles between the center of mass vector $\mathbf{R}_{cm}$ and the three axes of our coordinate system by θ_1, θ_2 and θ_3. Then the classical rotational kinetic energy of the system will be given by

$$T_{rot} = \frac{1}{2} \sum_{s=1}^{3} \sum_{t=1}^{3} I_{s,t} \frac{d\theta_s}{dt} \frac{d\theta_t}{dt} \tag{11.93}$$

Before passing over to a quantum treatment of the system, we need to make a coordinate transformation which diagonalizes the moment of inertia tensor $I_{s,t}$. We solve the secular equations

$$\sum_{t=1}^{3} \left[I_{s,t} - \delta_{s,t} I_\kappa\right] c_{t,\kappa} = 0 \tag{11.94}$$

The three roots I_κ of the secular equation (11.94) are called the *principal moments of inertia.* In terms of them, the kinetic energy operator for rotational motion is given by

$$\mathsf{T}_{rot} = \frac{1}{2} \left[\frac{J_x^2}{I_x} + \frac{J_y^2}{I_y} + \frac{J_x^2}{I_z} \right] \tag{11.95}$$

Now let $|J, M\rangle$ be an angular momentum eigenfunction, i.e a solution to

$$\left[\mathsf{J}^2 - J(J+1)\right] |J, M\rangle = 0 \tag{11.96}$$

Then, using the usual quantum treatment of angular momentum eigenfunctions, one can show that (in atomic units)

$$\begin{aligned} &\langle J, M \pm 2 | \mathsf{T}_{rot} | J, M \rangle \\ &= \frac{1}{8} \left[\frac{1}{I_x} - \frac{1}{I_y} \right] \sqrt{[J(J+1) - (M \pm 1)(M \pm 2)][J(J+1) - M(M \pm 1)]} \end{aligned} \tag{11.97}$$

while

$$\langle J, M | \mathsf{T}_{rot} | J, M \rangle = \frac{1}{4} \left[\frac{1}{I_x} + \frac{1}{I_y} \right] \left[J(J+1) - M^2\right] + \frac{M^2}{2I_z} \tag{11.98}$$

and all the other elements are zero. Equations (11.97) and (11.98) give us a matrix representation of T_{rot} based on the solutions to equation (11.96). By diagonalizing this matrix, we obtain the rotational eigenfunctions of the system.

11.15 Normal modes

Assuming that the system is in a particular rotational eigenstate, we can next turn to the problem of finding its classical normal modes, and then treating these quantum mechanically. Let us assume that the atomic positions $\mathbf{R}_j$ correspond to equilibrium positions of the atoms for some conformation of the system, i.e. a local potential energy minimum in the particular rotational eigenstate that we are examining. If this is not the case, the equilibrium positions need to be recalculated, perhaps iteratively. We now let $\mathbf{x} = (\mathbf{x}_1, \mathbf{x}_2, \cdots, \mathbf{x}_N) = (x_1, x_2, \cdots, x_{3N})$ represent small displacements from the equilibrium positions, and we make a Taylor series expansion of the potential about the local minimum.

$$V(\mathbf{x}) = v_0 + \sum_{i=1}^{3N}\sum_{j=1}^{3N} v_{i,j} x_i x_j + \cdots \tag{11.99}$$

The normal modes of the system can then be found by solving secular equations of the form

$$\sum_{j=1}^{3N} \left[\frac{v_{i,j}}{\sqrt{M_i M_j}} - \delta_{i,j} \mathcal{V}_\kappa \right] U_{j,\kappa} = 0 \tag{11.100}$$

where $U_{j,\kappa}$ is a unitary matrix, and where M_j is the mass appropriate to the coordinate x_j. In order to study eigenstates of the system near to the local potential minimum, we make the normal mode coordinate transformation:

$$X_\kappa = \sum_{j=1}^{3N} x_j U_{j,\kappa} \tag{11.101}$$

The zero-frequency normal modes are cyclic, while vibrations of those modes whose whose frequencies are non-zero can be studied using a basis set consisting of a product of Morse oscillator eigenfunctions.

11.16 A simple example

As a simple example of the rotations and vibrations of a cluster, let us consider three identical atoms, each of mass $\mathcal{M}$, with equilibrium positions at the corners of an equilateral triangle, as shown in Figure 12.1. If the origin of the coordinate system is taken to be the center of mass, then the equilibrium positions of the three atoms are given by

$$\begin{aligned} R_1 &= (0, a, 0) \\ R_2 &= (a\cos[\pi/6], -a\sin[\pi/6], 0) \\ R_3 &= (-a\cos[\pi/6], -a\sin[\pi/6], 0) \end{aligned} \tag{11.102}$$

where a is the distance from the center of mass to the equilibrium positions of the atoms. If we use equation (11.91) to calculate the moment of inertia tensor for the system, we obtain a result which is already diagonal:

$$I = \begin{pmatrix} 3a^2\mathcal{M} & 0 & 0 \\ 0 & 3a^2\mathcal{M} & 0 \\ 0 & 0 & 3a^2\mathcal{M} \end{pmatrix} \tag{11.103}$$

so that the step described in equation (11.94) can be omitted. Since all of the diagonal elements of the moment of inertia are equal, the term shown in equation (11.97) vanishes, and in this simple example, the matrix representation of H_{rot} based on eigenfunctions of the operator J^2 is already diagonal, the diagonal elements being give by

$$\begin{aligned} \langle J, M|\mathsf{H}_{rot}|J, M\rangle &= \frac{1}{4}\left[\frac{1}{I_x} + \frac{1}{I_y}\right]\left[J(J+1) - M^2\right] + \frac{M^2}{2I_z} \\ &= \frac{1}{6a^2\mathcal{M}}\left[J(J+1) - M^2\right] + \frac{M^2}{6a^2\mathcal{M}} \\ &= \frac{1}{6a^2\mathcal{M}}\, J(J+1) \end{aligned} \tag{11.104}$$

We can see that because of the symmetry of our simple example, the rotational energy levels are $(2J+1)$-fold degenerate.

In our simple example, the transformation from the coordinate system $x_1, y_1, z_1, x_2, y_2, z_2, x_3, y_3, z_3$, which represent mass-weighted displacements of the three particles from their equilibrium positions to the normal mode coordinates is given by the unitary transformation

$$\frac{1}{\sqrt{3}}\begin{pmatrix} \frac{\sqrt{3}}{2} & \frac{1}{2} & 0 & \frac{-\sqrt{3}}{2} & \frac{1}{2} & 0 & 0 & -1 & 0 \\ \frac{-\sqrt{3}}{2} & \frac{1}{2} & 0 & \frac{\sqrt{3}}{2} & \frac{1}{2} & 0 & 0 & -1 & 0 \\ \frac{-1}{2} & \frac{-\sqrt{3}}{2} & 0 & \frac{-1}{2} & \frac{\sqrt{3}}{2} & 0 & 1 & 0 & 0 \\ 1 & 0 & 0 & 1 & 0 & 0 & 1 & 0 & 0 \\ 0 & 1 & 0 & 0 & 1 & 0 & 0 & 1 & 0 \\ 0 & 0 & 1 & 0 & 0 & 1 & 0 & 0 & 1 \\ \frac{1}{2} & \frac{-\sqrt{3}}{2} & 0 & \frac{1}{2} & \frac{\sqrt{3}}{2} & 0 & -1 & 0 & 0 \\ 0 & 0 & -\sqrt{\frac{3}{2}} & 0 & 0 & \sqrt{\frac{3}{2}} & 0 & 0 & 0 \\ 0 & 0 & \frac{-1}{\sqrt{2}} & 0 & 0 & \frac{-1}{\sqrt{2}} & 0 & 0 & \sqrt{2} \end{pmatrix}\begin{pmatrix} x_1 \\ y_1 \\ z_1 \\ x_2 \\ y_2 \\ z_2 \\ x_3 \\ y_3 \\ z_3 \end{pmatrix} = \begin{pmatrix} X_1 \\ X_2 \\ X_3 \\ X_4 \\ X_5 \\ X_6 \\ X_7 \\ X_8 \\ X_9 \end{pmatrix} \tag{11.105}$$

Because the transformation to normal mode coordinates is orthogonal, i.e. both unitary and real, the kinetic energy operator is unaffected by the transformation, and

$$\mathsf{T} = -\frac{1}{2}\sum_{\mu=1}^{9} \frac{\partial^2}{\partial X_\mu^2} \tag{11.106}$$

Notice that since the original coordinates represented mass-weighted displacements from the equilibrium positions of the atoms, the kinetic energy operator is independent of mass. In the harmonic approximation, the energy associated with the three vibrational normal modes is given by

$$E_{vib} = \sum_{i=1}^{3}\left(n_i + \frac{1}{2}\right)\omega_i \qquad n_i = 1, 2, 3, \cdots$$

$$\omega_1 = \sqrt{\frac{2\kappa}{\mathcal{M}}} \qquad \omega_2 = \sqrt{\frac{\kappa}{\mathcal{M}}} \qquad \omega_3 = \sqrt{\frac{\kappa}{\mathcal{M}}} \tag{11.107}$$

where the effective force constant κ depends on the rotational quantum number J. In calculating the total energy,

$$E_{tot} = \frac{1}{6a^2\mathcal{M}}J(J+1) + E_{vib} \tag{11.108}$$

we have to remember that both κ and a depend on the value of the rotational quantum number J, and also on the exact form of the interatomic interaction potential.

11.17 Exercises

(1) Repeat for yourself the calculation of the kinetic energy of a 2-body system in center-of-mass and relative coordinates, as shown in equations (11.1)–(11.9). Discuss the physical meaning of the two terms, T_1 and T_2. How is the quantity $1/M_1 + 1/M_2$ related to the reduced mass?

(2) Transform the kinetic energy operator of a 3-body system from Cartesian coordinates to center-of-mass coordinates and the relative coordinates shown in equation (11.10), using method shown in equations (11.10)–(11.14).

(3) Transform the kinetic energy operator of a 4-body system to center-of-mass coordinates and to the chainlike relative coordinate system analogous to that of the previous exercise.

(4) Show that the coordinate transformation of equations (11.53)–(11.55) is unitary and that it leads to the Jacobian matrices shown in equations (11.58) and (11.59).

Chapter 12

SOME ILLUSTRATIVE EXAMPLES

12.1 Matrix elements of the kinetic energy operator

If we consider (for example) a 3-body problem, then, once the center of mass motion has been separated out, we are left with a 6-dimensional problem where the kinetic energy and the internal potential energy matrices must be evaluated using appropriate basis sets. Our choice of basis sets will depend on the problem being treated. For the case where $d = 6$, the basis function $\Phi_\nu(\mathbf{X})$, expressed in terms of its Fourier transform, is given by

$$\Phi_\nu(\mathbf{X}) = \frac{1}{(2\pi)^3} \int d\mathbf{p}\ e^{i\mathbf{p}\cdot\mathbf{X}} \Phi_\nu^t(\mathbf{p}) \tag{12.1}$$

Then

$$\frac{\partial^2}{\partial X_i \partial X_j} \Phi_\nu(\mathbf{X}) = \frac{1}{(2\pi)^3} \int d\mathbf{p}\ p_i p_j e^{i\mathbf{p}\cdot\mathbf{X}} \Phi_\nu^t(\mathbf{p}) \tag{12.2}$$

and

$$\begin{aligned} \int d\mathbf{X}\ \Phi_{\nu'}^*(\mathbf{X}) \frac{\partial^2}{\partial X_i \partial X_j} \Phi_\nu(\mathbf{X}) &= \int d\mathbf{p}\ p_i p_j \Phi_\nu^t(\mathbf{p})\ \frac{1}{(2\pi)^3} \int d\mathbf{X}\ e^{i\mathbf{p}\cdot\mathbf{X}}\ \Phi_{\nu'}^*(\mathbf{X}) \\ &= \int d\mathbf{p}\ p_i p_j \Phi_\nu^t(\mathbf{p})\ \Phi_{\nu'}^{*t}(-\mathbf{p}) \end{aligned} \tag{12.3}$$

12.2 Matrix elements of the potential energy

If there are no external forces acting on the system, then the potential energy V is independent of the center of mass coordinate $\mathbf{X}_1$. As a first step towards evaluating matrix elements of V, we let

$$v(r) = \frac{1}{(2\pi)^{3/2}} \int d\mathbf{p}\ e^{-i\mathbf{p}\cdot\mathbf{x}} v^t(p) \tag{12.4}$$

where

$$v^t(p) = \frac{1}{(2\pi)^{3/2}} \int d\mathbf{x}\ e^{i\mathbf{p}\cdot\mathbf{x}} v(r)$$
$$= \sqrt{\frac{2}{\pi}} \int_0^\infty dr\ r^2 j_0(pr) v(r) \tag{12.5}$$

For example, when

$$v(r) = \frac{e^{-\gamma r}}{r} \tag{12.6}$$

Then

$$v^t(p) = \sqrt{\frac{2}{\pi}} \int_0^\infty dr\ r\ j_0(pr)\ e^{-\gamma r}$$
$$= \frac{\sqrt{\frac{2}{\pi}}}{\gamma^2 + p^2} \tag{12.7}$$

Notice that in these relationships, $\mathbf{p} = (p_1, p_2, p_3)$ is a 3-dimensional vector. Thus we can write the internal potential in the form

$$v_{12}(|\mathbf{X}_2|) = \frac{1}{(2\pi)^{3/2}} \int d\mathbf{p}\ e^{-i\mathbf{p}\cdot(\mathbf{X}_2} v_{12}^t(p)$$
$$v_{23}(|\mathbf{X}_3|) = \frac{1}{(2\pi)^{3/2}} \int d\mathbf{p}\ e^{-i\mathbf{p}\cdot\mathbf{X}_3} v_{23}^t(p)$$
$$v_{13}(|\mathbf{X}_2 + \mathbf{X}_3|) = \frac{1}{(2\pi)^{3/2}} \int d\mathbf{p}\ e^{-i\mathbf{p}\cdot(\mathbf{X}_2+\mathbf{X}_3)} v_{13}^t(p) \tag{12.8}$$

We now introduce the 6-dimensional vector

$$\mathbf{X} \equiv (\mathbf{X}_2, \mathbf{X}_3) \tag{12.9}$$

If we then let

$$\mathbf{p}_{12} = (p_1, p_2.p_3, 0, 0, 0)$$
$$\mathbf{p}_{23} = (0, 0, 0, p_1, p_2.p_3)$$
$$\mathbf{p}_{13} = (p_1, p_2, p_3, p_1, p_2.p_3) \tag{12.10}$$

then we can express the internal interaction potentials in the form:

$$v_{ij}(\mathbf{X}) = \frac{1}{(2\pi)^{3/2}} \int d\mathbf{p}\ v_{ij}^t(p)\ e^{-i\mathbf{p}_{ij}\cdot\mathbf{X}} \tag{12.11}$$

It follows that for any function $f(\mathbf{X})$,

$$\int d\mathbf{X} f(\mathbf{X}) v_{ij}(\mathbf{X}) = \frac{1}{(2\pi)^{3/2}} \int d\mathbf{p}\ v_{ij}^t(p) \int d\mathbf{X}\ f(\mathbf{X})\ e^{-i\mathbf{p}_{ij}\cdot\mathbf{X}}$$
$$= (2\pi)^{3/2} \int d\mathbf{p}\ v_{ij}^t(p)\ f^t(\mathbf{p}_{ij}) \tag{12.12}$$

Now suppose that we have chosen to express the internal wave function of the system in terms of some basis set $\Phi_\nu(\mathbf{X})$. We can let

$$f_{\nu',\nu}(\mathbf{X}) = \Phi^*_{\nu'}(\mathbf{X})\Phi_\nu(\mathbf{X})$$
$$f^t_{\nu',\nu}(\mathbf{p}_{ij}) = \frac{1}{(2\pi)^3}\int d\mathbf{X}\; e^{-i\mathbf{p}_{ij}\cdot\mathbf{X}}\;\Phi^*_{\nu'}(\mathbf{X})\Phi_\nu(\mathbf{X}) \tag{12.13}$$

Then

$$\int d\mathbf{X}\;\Phi^*_{\nu'}(\mathbf{X})v_{ij}(\mathbf{X})\Phi_\nu(\mathbf{X}) = (2\pi)^{3/2}\int d\mathbf{p}\; v^t_{ij}(p)\; f^t_{\nu',\nu}(\mathbf{p}_{ij}) \tag{12.14}$$

12.3 Products of basis functions

Suppose that we are in possession of a set of coefficients $\mathcal{C}^{\nu''}_{\nu',\nu}$ such that

$$\Phi^*_{\nu'}(\mathbf{X})\Phi_\nu(\mathbf{X}) = \sum_{\nu''}\mathcal{C}^{\nu''}_{\nu',\nu}\; W_{\nu''}(\mathbf{X}) \tag{12.15}$$

Then

$$\begin{aligned} f^t_{\nu',\nu}(\mathbf{p}_{ij}) &= \frac{1}{(2\pi)^3}\int d\mathbf{X}\; e^{-i\mathbf{p}_{ij}\cdot\mathbf{X}}\;\Phi^*_{\nu'}(\mathbf{X})\Phi_\nu(\mathbf{X}) \\ &= \sum_{\nu''}\mathcal{C}^{\nu''}_{\nu',\nu}\; W^t_{\nu''}(\mathbf{p}_{ij}) \end{aligned} \tag{12.16}$$

12.4 Examples of basis sets

There are a number of cases where the coefficients $\mathcal{C}^{\nu''}_{\nu',\nu}$ are easy to calculate, and where $W^t_{\nu''}(\mathbf{p}_{ij})$ can also be evaluated exactly by Mathematica. In these cases, expansions of the type shown in equations (12.15) and (12.16) are helpful in evaluating matrix elements of the potential. For example , suppose that d-dimensional Coulomb Sturmians are used as a basis. Then

$$\begin{aligned} &\chi^*_{n',\lambda',\mu'}(\mathbf{X})\chi_{n,\lambda,\mu}(\mathbf{X}) \\ &= \sum_{n'',\lambda',\mu''}\chi_{n'',\lambda',\mu''}(2k,\mathbf{X})b^{n'',\lambda''}_{n',\lambda';n,\lambda}\int d\Omega_d Y^*_{\lambda'',\mu''}(\mathbf{u_X})Y^*_{\lambda',\mu'}(\mathbf{u_X})Y_{\lambda,\mu}(\mathbf{u_X}) \end{aligned} \tag{12.17}$$

where

$$b^{n'',\lambda''}_{n',\lambda';n,\lambda} = \frac{2n+d-3}{4k}\int_0^\infty dr\; r^{d-2}\; R_{n''.\lambda''}(2k,r)R_{n'.\lambda'}(r)R_{n.\lambda}(r) \tag{12.18}$$

The expansion of a product of d-dimensional Coulomb Sturmian basis functions shown in equation (12.17) is exact, and it involves only a limited number of coefficients. The Fourier transform of $\chi_{n'',\lambda',\mu''}(2k,\mathbf{X})$ is easy to find by means of the d-dimensional generalization of Fock's relationship.

$$\chi^t_{n'',\lambda',\mu''}(2k,\mathbf{p}) = \frac{(4k)^{d/2+1}}{\sqrt{2}(4k^2+p^2)^{d/2+1/2}} Y_{n-1,\lambda,\mu}(2k,\Omega_{d+1}) \quad (12.19)$$

As a second example, let us consider a basis set consisting of d-dimensional harmonic oscillator wave functions. As discussed in Appendix A, we then have:

$$\chi^*_{\nu_1}(\mathbf{x})\chi_{\nu_2}(\mathbf{x}) = \sum_{\nu'} \chi_{\nu'}(2\omega,\mathbf{x})\ C^{\nu'}_{\nu_1,\nu_2} \quad (12.20)$$

where

$$\begin{aligned} C^{\nu}_{\nu_1,\nu_2} &= \int dx\ \chi^*_{\nu}(2\omega,\mathbf{x})\chi^*_{\nu_1}(\mathbf{x})\chi_{\nu_2}(\mathbf{x}) \\ &= \int_0^\infty dr\ r^{d-1} R_{n,\lambda_1+\lambda_2}(2\omega,r) R_{n_1,\lambda_1}(r) R_{n_2,\lambda_2}(r) \\ &\times \int d\Omega_d\ Y^*_{\lambda,\mu}(\mathbf{u}) Y^*_{\lambda_1,\mu_1}(\mathbf{u}) Y_{\lambda_2,\mu_2}(\mathbf{u}) \end{aligned} \quad (12.21)$$

and where the hyperradial part of the wave function, expressed in terms of associated Legendre polynomials, is given by

$$R_{n,\lambda}(r) = \sqrt{\frac{2n!\omega^{d/2}}{\Gamma(\lambda+n+d/2)}}\ (\sqrt{\omega}r)^\lambda e^{-\omega r^2/2} L_n^{\lambda+(d-2)/2}(\omega r^2) \quad (12.22)$$

This radial wave function is normalized in such a way that

$$\int_0^\infty dr\ r^{d-1}|R_{n,\lambda}(r)|^2 = 1 \quad (12.23)$$

It satisfies equation (21) with

$$E_{n,\lambda} = \omega(2n+\lambda+d/2) \quad (12.24)$$

The Fourier transform of the 2ω basis functions can be evaluated exactly by Mathematica.

As a third example illustrating the composition properties of basis sets, let us think of the case where the basis set has the form:

$$\Phi_{\mathbf{n}}(\mathbf{X}) = \prod_{j=1}^{d} e^{-aX_j^2} X_j^{n_j} \quad (12.25)$$

A basis set of this kind has the very simple composition property

$$\Phi_{\mathbf{n}'}(\mathbf{X})\Phi_{\mathbf{n}}(\mathbf{X}) = \prod_{j=1}^{d} e^{-2aX_j^2} X_j^{n'_j+n_j} \equiv \Phi_{\mathbf{n}'+\mathbf{n}}(2a, \mathbf{X}) \tag{12.26}$$

The Fourier transform of a function of this type is easy to evaluate. For example, using Mathematica, we find that

$$\begin{aligned}
\int_{-\infty}^{\infty} dX_j\ e^{-2aX_j^2 - ipX_j} &= \frac{\sqrt{\frac{\pi}{2}}e^{-\frac{p^2}{8a}}}{\sqrt{a}} \\
\int_{-\infty}^{\infty} dX_j\ e^{-2aX_j^2 - ipX_j} X_j &= -\frac{i\sqrt{\frac{\pi}{2}}pe^{-\frac{p^2}{8a}}}{4a^{3/2}} \\
\int_{-\infty}^{\infty} dX_j\ e^{-2aX_j^2 - ipX_j} X_j^2 &= \frac{\sqrt{\frac{\pi}{2}}e^{-\frac{p^2}{8a}}\left(4a - p^2\right)}{16a^{5/2}} \\
\int_{-\infty}^{\infty} dX_j\ e^{-2aX_j^2 - ipX_j} X_j^3 &= \frac{i\sqrt{\frac{\pi}{2}}pe^{-\frac{p^2}{8a}}\left(p^2 - 12a\right)}{64a^{7/2}} \\
\int_{-\infty}^{\infty} dX_j\ e^{-2aX_j^2 - ipX_j} X_j^4 &= \frac{\sqrt{\frac{\pi}{2}}e^{-\frac{p^2}{8a}}\left(48a^2 - 24ap^2 + p^4\right)}{256a^{9/2}}
\end{aligned} \tag{12.27}$$

Notice that by differentiating the first integral in equation (12.27) with respect to p we can bring down higher and higher powers of X_j under the integral sign, so that

$$\int_{-\infty}^{\infty} dX_j\ e^{-2aX_j^2 - ipX_j} X_j^n = \left(i\frac{\partial}{\partial p}\right)^n \frac{\sqrt{\frac{\pi}{2}}e^{-\frac{p^2}{8a}}}{\sqrt{a}} \tag{12.28}$$

Basis sets of other forms may be better than the ones just discussed, depending on the potential that is to be treated.

12.5 The problem of slow convergence

As has been discussed by Professor T.K. Das [Das, 2016], hyperspherical expansions tend to converge slowly when used to treat potentials that produce wave functions with cusps. In order to achieve more rapid convergence, one can initially use a limited basis set and then find additional basis functions of appropriate symmetry using the iteration methods discussed in Chapter 7. The potential harmonic methods of Aquilanti and Avery [Aquilanti and Avery, 1997] or Das [Das, 2016] can also be used.

12.6 2-electron atoms: the results of Das et al.

In Chapter 6 of his recent book entitled *Hyperspherical Expansion Techniques: Applications to problems in Physics* [Das, 2016], [Das et al., 1994], Professor T.K. Das discusses the application of hyperspherical harmonics to 2-electron atoms. He lists results of calculations on the first few atoms of the heliumlike isoelectronic series, and on positronium. In obtaining these results, Professor Das and his coworkers make use of a large-scale calculational technique which involves extrapolation. The results of these calculations are shown in Table 12.1, compared with experimental values.

Table 12.1: Results of Das et al. (Positronium, Ps)

Atom	State	Calculated	Experiment
H^-	1^1S	0.52668	0.52777
He	1^1S	2.90368	2.9038
Li^+	1^1S	7.28007	7.2804
Be^{++}	1^1S	13.65600	13.6572
He	2^1S	2.13895	2.14606
Ps^-	1^1S	0.2616689	

Our own calculations on some other atoms in the heliumlike isoelectronic series are shown in Tables 12.2 and 12.3. These calculations are made by two methods, both of which are extremely rapid and computationally undemanding. The first method makes use of the Goscinskian configurations discussed in Chapter 7 of this book, while the second method makes use of configurations based on Coulomb Sturmians. For large values of the nuclear charge Z, the results of the two methods approach each other. However, as Z increases, relativistic effects become more important, and a rough relativistic correction is needed before the calculational results agree well with experimental results.

Table 12.2: Results of Avery and Avery (1^1S states). Rapid calculations for heavier members of the heliumlike isoelectronic series using Goscinskians and Coulomb Sturmians are compared with experimental results. As the nuclear charge increases, the Goscinskian and Coulomb Sturmian binding energies agree more and more closely with each other. They would also agree more and more closely with the experimental values, were it not for relativistic effects which become more and more important as the nuclear charge increased. When a rough relativistic correction is included, the calculated values agree much better with experiment, as is shown in the final table.

Atom	Goscinskians	Coul. Sturm.	Experiment
B^{3+}	22.0218	22.0291	22.0349
C^{4+}	32.3970	32.4043	32.4158
N^{5+}	44.7721	44.7795	44.8018
O^{6+}	59.1471	59.1546	59.1922
F^{7+}	75.5222	75.5297	75.5942
Ne^{8+}	93.8972	93.9048	94.0055
Na^{9+}	114.272	114.280	114.431
Mg^{10+}	136.647	136.655	136.872
Al^{11+}	161.022	161.030	161.334
Si^{12+}	187.397	187.405	187.819
P^{13+}	215.772	215.780	216.334
S^{14+}	246.147	246.155	246.881
Cl^{15+}	278.522	278.530	279.470

Table 12.3: Here the results of the previous table are repeated, but the calculated results are corrected to take relativistic effects into account. It can be seen that for very large values of Z, the calculated results slightly overestimate the binding energy. This may be due to effects such as vacuum polarization and the Lamb shift, which become more important with increasing nuclear charge, but which are not included in these calculations.

Atom	Goscinskians	Coul. Sturm.	Experiment
B^{3+}	22.0291	22.0364	22.0349
C^{4+}	32.4126	32.4199	32.4158
N^{5+}	44.8012	44.8086	44.8018
O^{6+}	59.1974	59.2049	59.1922
F^{7+}	75.6038	75.6113	75.5942
Ne^{8+}	94.0221	94.0297	94.0055
Na^{9+}	114.457	114.465	114.431
Mg^{10+}	136.909	136.917	136.872
Al^{11+}	161.386	161.394	161.334
Si^{12+}	187.888	187.896	187.819
P^{13+}	216.421	216.429	216.334
S^{14+}	246.991	246.999	246.881
Cl^{15+}	279.603	279.611	279.470

12.7 Approximate relativistic corrections

As we discussed in Chapter 7, it is possible to make a rough correction for relativistic effects, a correction which becomes progressively more accurate in an isoelectronic series as the nuclear charge Z increases. In this correction method, the calculated results are multiplied with a correction factor which is the relativistic energy of a configuration, divided by the configuration's non-relativistic energy, completely neglecting interelectron repulsion in both cases. It clear that such a correction for relativistic effects will become more accurate as Z increases in an isoelectronic series, since the effects of the nuclear attraction potential will increasingly preponderate over the effects of interelectron repulsion. When interelectron repulsion is entirely neglected, the total energy of a configuration is just the sum of the energies of single electrons moving around the nucleus. In the non-relativistic case the energy of a single electron in the field of a nucleus is given by the familiar formula

$$E_\mu = -\frac{Z^2}{2n^2} \tag{12.29}$$

while in the relativistic case, where it also can be calculated exactly, the single-electron energy is found to be

$$\epsilon_\mu = c^2 \left[1 + \left(\frac{Z}{c(\gamma + n - |j + 1/2|)}\right)^2\right]^{-1/2} - c^2 \tag{12.30}$$

$$\gamma \equiv \sqrt{\left(j + \frac{1}{2}\right)^2 - \left(\frac{Z}{c}\right)^2} \qquad c = 137.036 \tag{12.31}$$

Thus the rough correction factor has the form

$$F = \frac{\sum_\mu \epsilon_\mu}{\sum_\mu E_\mu} \tag{12.32}$$

where the sums run over all the single-electron indices in the configuration. Here are the calculated relativistic correction factors listed after the Z values to which they correspond:

$$F = \begin{pmatrix} 2 & 1.00005 \\ 3 & 1.00012 \\ 4 & 1.00021 \\ 5 & 1.00033 \\ 6 & 1.00048 \\ 7 & 1.00065 \\ 8 & 1.00085 \\ 9 & 1.00108 \\ 10 & 1.00133 \\ 11 & 1.00162 \\ 12 & 1.00192 \\ 13 & 1.00226 \\ 14 & 1.00262 \\ 15 & 1.00301 \\ 16 & 1.00343 \\ 17 & 1.00388 \\ 18 & 1.00435 \end{pmatrix} \tag{12.33}$$

12.8 Exercises

(1) Calculate the Fourier transforms of 6-dimensional Coulomb Sturmian basis functions of the type shown in equation (6.23).
(2) Repeat for 6-dimensional harmonic oscillator basis sets.
(3) Repeat for basis sets of the form $\Phi_n u(\mathbf{X}) = \prod_{j=1}^{6} e^{-aX_j^2} X_j^{n_j}$.
(4) Show that when the crude corrections of equation (12.33) are applied to the results shown in Table 12.2, they yield the corrected results of Table 12.3.

Appendix A

THE D-DIMENSIONAL HARMONIC OSCILLATOR

A.1 Harmonic oscillators in one dimension

We begin by reviewing the theory of the simple harmonic oscillator in one dimension. We wish to find solutions to the equation

$$\left[-\frac{1}{2}\frac{d^2}{dq^2} + \frac{1}{2}\omega^2 q^2 - \epsilon_n\right]\psi_n(q) = 0 \tag{A.1}$$

where ω^2 is the force constant, and q is the mass-weighted coordinate. It is convenient to introduce the dimensionless parameter $\zeta = \sqrt{\omega}q$. Then the solutions to (A.1) have the form

$$\psi_n(\zeta) = N_n e^{-\zeta^2/2} H_n(\zeta) \tag{A.2}$$

Here, $H_n(\zeta)$ is a Hermite polynomial, named after the French mathematician Charles Hermite (1822–1901), whose name is also associated with self-adjointness. The first few Hermite polynomials are

$$\begin{aligned}
H_0(\zeta) &= 1 \\
H_1(\zeta) &= 2\zeta \\
H_2(\zeta) &= 4\zeta^2 - 2 \\
H_3(\zeta) &= 8\zeta^3 - 12\zeta \\
\vdots \quad & \vdots \quad \vdots
\end{aligned} \tag{A.3}$$

If the solutions are normalized in such a way that

$$\int_{-\infty}^{\infty} dq\ \psi_{n'}(\sqrt{\omega}q)\psi_n(\sqrt{\omega}q) = \delta_{n',n} \tag{A.4}$$

then the normalization constant is

$$N_n = \sqrt{\frac{\sqrt{m\omega}}{2^n n!\sqrt{\pi}}} \tag{A.5}$$

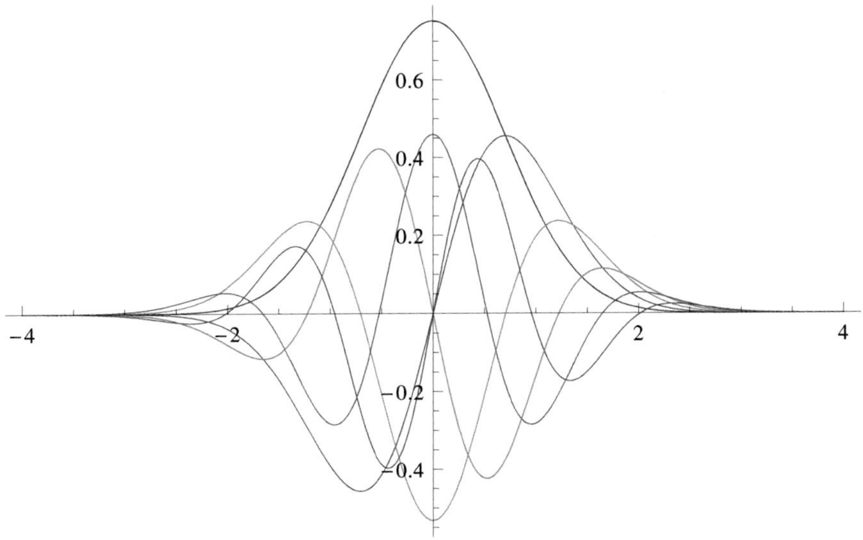

Fig. A.1: This figure shows the first few wave functions for a 1-dimensional harmonic oscillator. The functions with quantum numbers n have n nodes. When n is even the functions are symmetric with respect to inversion, while when n is odd, they are odd.

In atomic units, where $\hbar = 1$, the corresponding energies are

$$\epsilon_n = \omega\left(n + \frac{1}{2}\right) \tag{A.6}$$

A.2 Creation and annihilation operators for harmonic oscillators

Equation (A.1) can be rewritten in the form

$$H|n\rangle = \epsilon_n|n\rangle \tag{A.7}$$

where

$$H = \frac{1}{2}\left(-\frac{\partial^2}{\partial q^2} + \omega^2 q^2\right) \tag{A.8}$$

If we let

$$p \equiv \frac{1}{i}\frac{\partial}{\partial q} \tag{A.9}$$

then

$$H = \frac{1}{2}\left(p^2 + \omega^2 q^2\right) \tag{A.10}$$

Also we can see the p and q obey the commutation relations

$$[p,q] = -i \qquad [p,p] = 0 \qquad [q,q] = 0 \tag{A.11}$$

From these commutations relations it follows that

$$\begin{aligned}[H,p] &= \frac{1}{2}\omega^2[q^2,p] \\ &= \frac{1}{2}\omega^2(q[q,p] + [q,p]q) \\ &= i\omega^2 q\end{aligned} \tag{A.12}$$

and

$$\begin{aligned}[H,q] &= \frac{1}{2}[p^2,q] \\ &= \frac{1}{2}(p[p,q] + [p,q]p) \\ &= -ip\end{aligned} \tag{A.13}$$

Now suppose that we have found an eigenfunction of H, so that

$$H|n\rangle = \epsilon_n|n\rangle \tag{A.14}$$

We can show by means of the commutation relations (A.13) and (A.14) that when the operator $p \pm i\omega q$ acts on $|n\rangle$, the resulting function is also an eigenfunction of H:

$$\begin{aligned}H(p \pm i\omega q)|n\rangle &= \{[H,p] \pm [H,q] + (p \pm i\omega q)H\}\,|n\rangle \\ &= \left\{i\omega^2 q \pm i\omega(-ip) + (p \pm i\omega q)\epsilon_n\right\}|n\rangle \\ &= (\epsilon_n \pm \omega)(p \pm i\omega q)|n\rangle\end{aligned} \tag{A.15}$$

Equation (A.15) shows that the function $(p \pm i\omega q)|n\rangle$ is an eigenfunction of H corresponding to the eigenvalue $\epsilon_n \pm \omega$. The operator $p + i\omega q$ is thus a "raising operator". When it acts on $|n\rangle$, it produces a new eigenfunction, whose eigenvalue is raised by an amount ω. Similarly, $p - i\omega q$ is a lowering operator. When it acts on $|n\rangle$, it produces a new eigenfunction, whose eigenvalue is lowered by an amount ω.

If we continue to act on $|n\rangle$ with the lowering operator $p - i\omega q$, we must eventually come to the ground state of the harmonic oscillator, a state of minimum energy beyond which it is impossible to lower the energy eigenvalue. Let us represent the ground state by the symbol $|0\rangle$. The lowering operator, acting on the ground state, must give zero, since it cannot give an eigenfunction corresponding to a lower energy. Therefore we have the relation:

$$(p - i\omega q)|0\rangle = 0 \tag{A.16}$$

Acting on (A.16) with $p + i\omega q$, we obtain

$$\begin{aligned}(p + i\omega q)(p - i\omega q)|0\rangle &= \left(p^2 + \omega^2 q^2 + i\omega[p, q]\right)|0\rangle \\ &= (2H - \omega)|0\rangle \\ &= (2\epsilon_0 - \omega)|0\rangle = 0\end{aligned} \tag{A.17}$$

Thus the energy of the ground state is given by

$$\epsilon_0 = \frac{\omega}{2} \tag{A.18}$$

Combining (A.18) and (A.15), we can see that the energy of a general state $|n\rangle$ is given by

$$\epsilon_n = \omega\left(n + \frac{1}{2}\right) \tag{A.19}$$

It is convenient to define a normalized raising operator, which we will call a "creation operator",

$$a^\dagger \equiv \mathcal{N}(-ip + \omega q) \tag{A.20}$$

and a normalized lowering operator, which we will call an "annihilation operator".

$$a \equiv \mathcal{N}(ip + \omega q) \tag{A.21}$$

The constant of normalization is chosen in such a way that

$$a^\dagger|n\rangle = \sqrt{n+1}\;|n+1\rangle \tag{A.22}$$

and

$$a|n+1\rangle = \sqrt{n+1}\;|n\rangle \tag{A.23}$$

Then

$$\begin{aligned}aa^\dagger|n\rangle &= a\sqrt{n+1}\;|n+1\rangle \\ &= (n+1)|n\rangle \\ &= \mathcal{N}^2(p - i\omega q)(p + i\omega q)|n\rangle \\ &= \mathcal{N}^2\left(2H + i\omega[p, q]\right)|n\rangle \\ &= \mathcal{N}^2 2\omega(n+1)|n\rangle\end{aligned} \tag{A.24}$$

Solving (A.24) for $\mathcal{N}$, we obtain

$$\mathcal{N} = \frac{1}{\sqrt{2\omega}} \tag{A.25}$$

Thus

$$a^\dagger = \frac{1}{\sqrt{2\omega}}(-ip + \omega q) \tag{A.26}$$

and

$$a = \frac{1}{\sqrt{2\omega}}(ip + \omega q) \tag{A.27}$$

We can also solve for p and q in terms of the creation and annihilation operators:

$$p = i\sqrt{\frac{\omega}{2}}(a^\dagger - a) \tag{A.28}$$

while

$$q = i\frac{1}{\sqrt{2\omega}}(a^\dagger + a) \tag{A.29}$$

Then, making use of (A.10) we have

$$H = \omega\left(a^\dagger a + \frac{1}{2}\right) \tag{A.30}$$

From (A.28), (A.29) and (A.11), it follows that $a^\dagger$ and a obey the commutation relations

$$\begin{aligned} [a, a^\dagger] &= 1 \\ [a^\dagger, a^\dagger] &= 0 \\ [a, a] &= 0 \end{aligned} \tag{A.31}$$

A.3 A collection of harmonic oscillators

Let us now consider a system whose Hamiltonian can be represented by a sum of simple harmonic oscillator Hamiltonians:

$$H = \sum_{k=1}^{d} \omega_k \left(a_k^\dagger a_k + \frac{1}{2}\right) \tag{A.32}$$

Then the commutation relations corresponding to (A.31) will be

$$\begin{aligned} [a_{k'}, a_k^\dagger] &= \delta_{k',k} \\ \left[a_{k'}^\dagger, a_k^\dagger\right] &= 0 \\ [a_{k'}, a_k] &= 0 \end{aligned} \tag{A.33}$$

The eigenfunctions of the Hamiltonian are just products of simple harmonic oscillator eigenfunctions, and they can be labelled by a set of numbers

$n_1, n_2, \cdots, n_d$, one quantum number for each normal mode of the system. If we use the symbol $|n_1, n_2, \cdots, n_d\rangle$ to denote such a state, then we have:

$$H|n_1, n_2, \cdots, n_d\rangle = \sum_{k=1}^{d} \omega_k \left(a_k^\dagger a_k + \frac{1}{2} \right) |n_1, n_2, \cdots, n_d\rangle$$
$$= \sum_{k=1}^{d} \omega_k \left(n_k + \frac{1}{2} \right) |n_1, n_2, \cdots, n_d\rangle \qquad (A.34)$$

The operator $a_k^\dagger a_k$ is called the "number operator", because its eigenvalues correspond to the quantum number n_k.

A.4 *d*-dimensional isotropic harmonic oscillators

In terms of the mass-weighted coordinates $\mathbf{q} = \sqrt{m}\ \mathbf{x}$, the Schrödinger equation of a d-dimensional isotropic harmonic oscillator can be written in the form

$$\sum_{i=1}^{d} \frac{1}{2} \left[-\frac{\partial^2}{\partial q_i^2} + \omega^2 q_i^2 \right] \Psi_{\mathbf{n}}(\mathbf{q}) = E_{\mathbf{n}} \Psi_{\mathbf{n}}(\mathbf{q}) \qquad (A.35)$$

If we let

$$\Psi_{\mathbf{n}}(\mathbf{q}) = \prod_{i=1}^{d} \psi_{n_i}(q_i)$$
$$E_{\mathbf{n}} = \sum_{i=1}^{d} \epsilon_{n_i} \qquad (A.36)$$

Then (A.35) separates into d independent equations of the form

$$\frac{1}{2} \left[-\frac{\partial^2}{\partial q_i^2} + \omega^2 q_i^2 \right] \psi_{n_i}(q_i) = \epsilon_{n_i} \psi_{n_i}(q_i) \qquad (A.37)$$

In other words, the isotropic d-dimensional harmonic oscillator can be treated as a system of independent simple harmonic oscillators, all having the same frequency. so that

$$\sum_{i=1}^{d} \epsilon_{n_i} = \omega \sum_{i=1}^{d} \left(n_i + \frac{1}{2} \right) \qquad (A.38)$$

Alternatively we can write the Schrödinger equation in the form:

$$\left[-\frac{1}{2m} \Delta + \frac{1}{2} m \omega^2 r^2 - E_{n,\lambda} \right] \chi_{n,\lambda,\mu}(\mathbf{x}) = 0 \qquad (A.39)$$

where Δ is the generalized Laplacian operator. We can try to find solutions which are functions of the hyperradius multiplied by hyperspherical harmonics:

$$\chi_{n,\lambda,\mu}(\mathbf{x}) = R_{n,\lambda}(r) Y_{\lambda,\mu}(\mathbf{u}) \tag{A.40}$$

Substituting this into equation (A.39), we find that the hyperradial part of the solution must obey an ordinary differential equation. If we use units for which $m = 1$ and make use of equations (2.1) and (2.8), we have

$$\begin{aligned}
&\left[-\Delta + \omega^2 r^2 - 2E_{n,\lambda}\right] R_{n,\lambda}(r) \\
&= \left[-\frac{1}{r^{d-1}}\frac{d}{dr} r^{d-1}\frac{d}{dr} - \frac{\Lambda^2}{r^2} + \omega^2 r^2 - 2E_{n,\lambda}\right] R_{n,\lambda}(r) \\
&= \left[-\frac{1}{r^{d-1}}\frac{d}{dr} r^{d-1}\frac{d}{dr} - \frac{\lambda(\lambda+d-2)}{r^2} + \omega^2 r^2 - 2E_{n,\lambda}\right] R_{n,\lambda}(r) = 0
\end{aligned} \tag{A.41}$$

With the substitution $\zeta = \sqrt{\omega} r$, equation (21) becomes

$$\left[\frac{d^2}{d\zeta^2} + \frac{d-1}{\zeta}\frac{d}{d\zeta} + \frac{\lambda(\lambda+d-2)}{\zeta^2} - \zeta^2 + 2E_{n,\lambda}\right] R_{n,\lambda}(\zeta) = 0 \tag{A.42}$$

According to Valliéres et al. [Valliéres et al., 1975], this equation has a solution of the form

$$R_{n,\lambda}(\zeta) = \sqrt{\frac{2n!}{\Gamma(\lambda+n+d/2)}}\, \zeta^\lambda e^{-\zeta^2/2} L_n^{\lambda+(d-2)/2}(\zeta^2) \tag{A.43}$$

where L is an associated Legendre polynomial and

$$E_{n,\lambda} = 2n + \lambda + d/2 \tag{A.44}$$

The normalization is chosen so that

$$\int_0^\infty d\zeta\, \zeta^{d-1} |R_{n,\lambda}(\zeta)|^2 = 1 \tag{A.45}$$

Written in terms of the hyperradius r, rather than $\zeta = \sqrt{\omega} r$, we have

$$R_{n,\lambda}(r) = \sqrt{\frac{2n!\omega^{d/2}}{\Gamma(\lambda+n+d/2)}}\, (\sqrt{\omega} r)^\lambda e^{-\omega r^2/2} L_n^{\lambda+(d-2)/2}(\omega r^2) \tag{A.46}$$

This radial wave function is normalized in such a way that

$$\int_0^\infty dr\, r^{d-1} |R_{n,\lambda}(r)|^2 = 1 \tag{A.47}$$

It satisfies equation (A.42) with

$$E_{n,\lambda} = \omega(2n + \lambda + d/2) \tag{A.48}$$

Comparing this with equation (A.38), we can see that we must make the identification

$$2n + \lambda = \sum_{i=1}^{d} n_i \tag{A.49}$$

For $d = 3$ this becomes

$$\left[\frac{d^2}{dr^2} + \frac{2}{r}\frac{d}{dr} + \frac{l(l+1)}{r^2} - \omega^2 r^2 + 2E\right] R_{n,l}(r) = 0 \tag{A.50}$$

which has solutions of the form

$$R_{n,l}(r) = N_{n,l}\, r^l\, e^{-\omega r^2/2} L_n^{l+1/2}(\omega r^2) \tag{A.51}$$

where

$$N_{n,l} = \sqrt{\sqrt{\frac{\omega^3}{4\pi}} \frac{2^{n+l+3} n! \omega^l}{(2n+2l+1)!!}} \tag{A.52}$$

is a normalizing constant, and $L_n^{l+1/2}$ is an associated Legendre polynomial. Analogous solutions can be found for higher values of d. The 3-dimensional isotropic harmonic oscillator radial wave functions shown in equations (A.51) and (A.52) obey the orthonormality relation:

$$\int_0^\infty dr\; r^2 R_{n',l}(r) R_{n,l}(r) = \delta_{n',n} \tag{A.53}$$

Since they are solutions to the same differential equation (differently expressed) it must be possible to expand these functions in terms of those solutions to (A.35) which correspond to the same energy.

$$R_{n,\lambda}(r) Y_{\lambda,\mu}(\hat{\mathbf{u}}) = {\sum_{\mathbf{n}}}' \Psi_{\mathbf{n}}(\mathbf{q})\; U_{\mathbf{n};n,\lambda,\mu} \tag{A.54}$$

The prime over the sum in (A.54) indicates that it includes only those values of $\mathbf{n}$ that fulfill equation (A.49).

A.5 Fourier transforms of 3-dimensional harmonic oscillator wave functions

Let

$$\chi_{n,l,m}(\mathbf{x}) = R_{n,l}(r) Y_{l,m}(\mathbf{u}) \tag{A.55}$$

be a 3-dimensional harmonic oscillator wave function expressed in spherical polar coordinates. Its Fourier transform is given by

$$\begin{aligned}\chi^t_{n,l,m}(\mathbf{p}) &= \frac{1}{(2\pi)^{3/2}}\int d\mathbf{x}\ \chi_{n,l,m}(\mathbf{x})\ e^{-i\mathbf{p}\cdot\mathbf{x}} \\ &= \frac{1}{(2\pi)^{3/2}}\int_0^\infty dr\ r^2\ R_{n,l}(r)\int d\Omega_3 Y_{l,m}(\mathbf{u})\ e^{-i\mathbf{p}\cdot\mathbf{x}} \\ &= \frac{4\pi}{(2\pi)^{3/2}}(-i)^l Y_{l,m}(\mathbf{u}_p)\int_0^\infty dr\ r^2\ R_{n,l}(r)j_l(pr) \\ &= (-i)^l Y_{l,m}(\mathbf{u}_p)R^t_{n,l}(p)\end{aligned} \tag{A.56}$$

where we have made use of the expansion

$$e^{-i\mathbf{p}\cdot\mathbf{x}} = 4\pi\sum_{l=0}^{\infty}(-i)^l\ j_l(pr)\sum_{m'=-l}^{l} Y_{l,m'}(\mathbf{u}_p)Y^*_{l,m'}(\mathbf{u}) \tag{A.57}$$

and where

$$R^t_{n,l}(p) = \sqrt{\frac{2}{\pi}}\int dr\ r^2\ R^t_{n,l}(p)j_l(pr) \tag{A.58}$$

Using equation (A.51), we can write this transform as

$$R^t_{n,l}(p) = N_{n,l}\sqrt{\frac{2}{\pi}}\int_0^\infty dr\ r^2\ r^l e^{-\omega r^2/2}L_n^{l+1/2}(\omega r^2)j_l(pr) \tag{A.59}$$

Mathematica is not able to evaluate this integral for general n or l, but it can do so if we give it particular values of n and l. Looking at sufficiently many particular cases, we can make the generalization:

$$R^t_{n,l}(p) = N_{n,l}\ \frac{(-1)^n}{\omega^{(l+3)/2}}\ e^{-p^2/(2\omega)}(p^2/\omega)^{l/2}L_n^{l+1/2}(p^2/\omega) \tag{A.60}$$

The Bessel transformed radial functions obey the orthonormality relation

$$\int_0^\infty dp\ p^2 R^t_{n',l}(p)R^t_{n,l}(p) = \delta_{n',n} \tag{A.61}$$

A.6 The hyperspherical Bessel transform of the radial function

By analogy with equations (A.46) and (A.60), we make the guess

$$\begin{aligned}&R^t_{n,\lambda}(p) \\ &= \sqrt{\frac{2n!}{\Gamma(\lambda+n+d/2)\omega^{d/2}}}\ (-1)^n\ e^{-p^2/(2\omega)}(p^2/\omega)^{\lambda/2}L_n^{\lambda+(d-2)/2}(p^2/\omega)\end{aligned} \tag{A.62}$$

We can check that $R^t_{n,\lambda}(p)$ obeys the orthonormality relation

$$\int_0^\infty dp\; p^{d-1} R^t_{n',\lambda}(p) R^t_{n,\lambda}(p) = \delta_{n',n} \tag{A.63}$$

We can also use the expansion of a d-dimensional plane wave in terms of hyperspherical harmonics and hyperspherical Bessel functions

$$e^{-i\mathbf{p}\cdot\mathbf{x}} = (d-2)!!I(0)\sum_{\lambda=0}^{\infty}(-i)^\lambda j^d_\lambda(pr)\sum_{\mu'} Y^*_{\lambda,\mu}(\mathbf{u})Y_{\lambda,\mu}(\mathbf{u}_p) \tag{A.64}$$

to evaluate the d-dimensional Fourier-Bessel transform:

$$\begin{aligned}
&\frac{1}{(2\pi)^{d/2}}\int_0^\infty dr\; r^{d-1}\int d\Omega_d\; R_{n,\lambda}(r)Y_{\lambda,\mu}(\mathbf{u})e^{-i\mathbf{p}\cdot\mathbf{x}} \\
&= (-i)^\lambda Y_{\lambda,\mu}(\mathbf{u}_p)\frac{(d-2)!!I(0)}{(2\pi)^{d/2}}\int_0^\infty dr\; r^{d-1}\; j^d_\lambda(pr)R_{n,\lambda}(r) \\
&= (-i)^\lambda Y_{\lambda,\mu}(\mathbf{u}_p)R^t_{n,\lambda}(p)
\end{aligned} \tag{A.65}$$

where $I(0)$ is the total solid angle in the d-dimensional space. Mathematica is unable to perform the hyperradial integral of equation (A.65) for general values of n and λ, but it can do so for particular values, and thus we can check the hypothesis shown in equation (A.62).

A.7 Coupling coefficients for harmonic oscillator wave functions

The d-dimensional harmonic oscillator wave functions obey the orthonormality relation

$$\int d\mathbf{x}\; \chi^*_{\nu'}(\mathbf{x})\chi_\nu(\mathbf{x}) = \delta_{\nu',\nu} \tag{A.66}$$

If we double the value of the frequency ω, the relationship is the same:

$$\int d\mathbf{x}\; \chi^*_{\nu'}(2\omega,\mathbf{x})\chi_\nu(2\omega,\mathbf{x}) = \delta_{\nu',\nu} \tag{A.67}$$

We can use the orthonormality relation (A.67) to express the product of two d-dimensional harmonic oscillator wave functions as a sum of single functions of the same kind, but with double the frequency. Let

$$\chi^*_{\nu_1}(\mathbf{x})\chi_{\nu_2}(\mathbf{x}) = \sum_{\nu'}\chi_{\nu'}(2\omega,\mathbf{x})\; C^{\nu'}_{\nu_1,\nu_2} \tag{A.68}$$

Then, making use of (A.67), we have

$$\begin{aligned}\int d\mathbf{x}\ \chi_\nu^*(2\omega,\mathbf{x})\chi_{\mu_1}^*(\mathbf{x})\chi_{\mu_2}(\mathbf{x}) &= \sum_{\nu'}\int d\mathbf{x}\ \chi_\nu^*(2\omega,\mathbf{x})\chi_{\nu'}(2\omega,\mathbf{x})\ C_{\nu_1,\nu_2}^{\nu'} \\ &= \sum_{\nu'}\delta_{\nu,\nu'}C_{\nu_1,\nu_2}^{\nu'} = C_{\nu_1,\nu_2}^{\nu} \qquad \text{(A.69)}\end{aligned}$$

The integral on the left-hand side of (A.69) can be separated into a hyper-radial part and a hyperangular part:

$$\begin{aligned}C_{\nu_1,\nu_2}^{\nu} &= \int d\mathbf{x}\ \chi_\nu^*(2\omega,\mathbf{x})\chi_{\nu_1}^*(\mathbf{x})\chi_{\nu_2}(\mathbf{x}) \\ &= \int_0^\infty dr\ r^{d-1}R_{n,\lambda_1+\lambda_2}(2\omega,r)R_{n_1,\lambda_1}(r)R_{n_2,\lambda_2}(r) \\ &\times \int d\Omega_d\ Y_{\lambda,\mu}^*(\mathbf{u})Y_{\lambda_1,\mu_1}^*(\mathbf{u})Y_{\lambda_2,\mu_2}(\mathbf{u}) \qquad \text{(A.70)}\end{aligned}$$

The hyperangular integral in equation (A.70) can be evaluated rapidly and exactly by means of our general theorem (2.27)–(2.29). The hyperradial integral can also be evaluated exactly, and the sum in equation (A.68) terminates after a finite number of terms.

Appendix B

MOLECULAR INTEGRALS FOR SLATER-TYPE ORBITALS

B.1 Definition of STO's

Slater-Type Orbitals (STO's) are frequently used in quantum chemical calculations. They have a radial part of the form:

$$\mathbb{R}_{n,\zeta}(r) = N_{n,\zeta}\; r^{n-1}e^{-\zeta r} \tag{B.1}$$

where $N_{n,\zeta}$ is a normalizing constant:

$$N_{n,\zeta} = (2\zeta)^n\sqrt{\frac{2\zeta}{(2n)!}} = \frac{(2\zeta)^{n+\frac{1}{2}}}{\sqrt{(2n)!}} \tag{B.2}$$

It is common to use spherical harmonics as the angular parts of STO's. Thus we can write:

$$\Phi_{n,\zeta,l,m}(\mathbf{x}) = \mathbb{R}_{n\zeta}(r)Y_{l,m}(\mathbf{u}) \equiv \Phi_{\bar{\mu}}(\mathbf{x}) \tag{B.3}$$

where

$$\bar{\mu} \equiv (n, \zeta, l, m) \tag{B.4}$$

Although STO's are often denoted by the symbol χ, we denote them here by Φ in order to distinguish them from Coulomb Sturmians, which we have consistently called χ in our books and papers.

B.2 Expansion of an arbitrary function of $s = kr$ in terms of Coulomb Sturmian radial functions

Why is it not trivial to expand STO's in terms of Coulomb Sturmians? Why not just let $k = \zeta$? The reason is that for the hyperspherical method to work, k must be the same for all the Sturmian orbitals involved in the integral. But in general, the molecular integrals that we wish to evaluate involve several values of ζ on different centers. Thus, in general, k will not be equal to ζ, and therefore the expansion is not a trivial one.

Let us first discuss how to expand an arbitrary function of $s = kr$ in terms of Coulomb Sturmian radial functions. If we divide the radial parts of Coulomb Sturmians by a factor of $k^{\frac{3}{2}}$ we will obtain a function

$$\tilde{R}_{n,l}(s) \equiv \frac{1}{k^{\frac{3}{2}}} R_{n,l}(r) \tag{B.5}$$

which is a pure function of $s = kr$. For example,

$$\tilde{R}_{1,0}(s) = 2e^{-s}$$

$$\tilde{R}_{2,0}(s) = 2(1-s)e^{-s}$$

$$\tilde{R}_{2,1}(s) = \frac{2}{\sqrt{3}}\, s\, e^{-s} \tag{B.6}$$

and so on. From the potential-weighted orthonormality relation (6.39), it then follows that

$$n \int_0^\infty ds\ s\ \tilde{R}_{n',l}(s)\tilde{R}_{n,l}(s) = \delta_{n'n} \tag{B.7}$$

Now suppose that we wish to expand a function $f(s)$ in a series of the form:

$$f(s) = \sum_{n=l+1}^{\infty} \tilde{R}_{n,l}(s)\ c_{n,l} \tag{B.8}$$

In the following equations, the symbol $c_{n,l}$ retains this meaning: It is the coefficient corresponding to n in the expansion of some function of s in terms of Coulomb Sturmian radial functions. But the functions being expanded differ in the various equations. Making use of the orthonormality relation obeyed by Coulomb Sturmians, we obtain

$$\begin{aligned} n \int_0^\infty ds\ s\ \tilde{R}_{n,l}(s) f(s) &= \sum_{n'=l+1}^{\infty} \int_0^\infty ds\ s\ \tilde{R}_{n',l}(s)\tilde{R}_{n,l}(s)\ c_{n',l} \\ &= \sum_{n'=l+1}^{\infty} \delta_{n',n} c_{n',l} = c_{n,l} \end{aligned} \tag{B.9}$$

Thus, for example, in the series

$$e^{-\rho s} = \sum_{n=1}^{\infty} \tilde{R}_{n,0}(s)\ c_{n,0}(\rho) \tag{B.10}$$

the coefficients $c_{n,0}$ are given by

$$c_{n,0}(\rho) = n \int_0^\infty ds\ s\ \tilde{R}_{n,0}(s)\ e^{-\rho s} \tag{B.11}$$

We can evaluate this integral exactly, using Mathematica, and the surprisingly simple result is

$$c_{n,0}(\rho) = n \int_0^\infty ds\ s\ \tilde{R}_{n,0}(s)\ e^{-\rho s} = 2n\ \frac{(\rho-1)^{n-1}}{(\rho+1)^{n+1}} \tag{B.12}$$

From (B.9), it follows that the coefficients c_n in the series

$$s^l\ e^{-\rho s} = \sum_{n=l+1}^{\infty} \tilde{R}_{n,l}(s)\ c_{n,l}(\rho) \tag{B.13}$$

are given by

$$c_{n,l}(\rho) = n \int_0^\infty ds\ s^{l+1}\ \tilde{R}_{n,l}(s)\ e^{-\rho s} \tag{B.14}$$

Mathematica is able to evaluate these integrals exactly, and, for the first few values of l, we obtain

$$\begin{aligned}
c_{n,0}(\rho) &= 2n\ \frac{(\rho-1)^{n-1}}{(\rho+1)^{n+1}} \\
c_{n,1}(\rho) &= 4n\sqrt{n^2-1}\ \frac{(\rho-1)^{n-2}}{(\rho+1)^{n+2}} \\
c_{n,2}(\rho) &= 8n\sqrt{n^4-5n^2+4}\ \frac{(\rho-1)^{n-3}}{(\rho+1)^{n+3}} \\
c_{n,3}(\rho) &= 16n\sqrt{\frac{(n+3)!}{n(n-4)!}}\ \frac{(\rho-1)^{n-4}}{(\rho+1)^{n+4}} \\
c_{n,4}(\rho) &= 32n\sqrt{\frac{(n+4)!}{n(n-5)!}}\ \frac{(\rho-1)^{n-5}}{(\rho+1)^{n+5}} \\
c_{n,5}(\rho) &= 64n\sqrt{\frac{(n+5)!}{n(n-6)!}}\ \frac{(\rho-1)^{n-6}}{(\rho+1)^{n+6}} \\
c_{n,6}(\rho) &= 128n\sqrt{\frac{(n+6)!}{n(n-7)!}}\ \frac{(\rho-1)^{n-7}}{(\rho+1)^{n+7}}
\end{aligned} \tag{B.15}$$

In general, if $n > l$, we obtain the result

$$c_{n,l}(\rho) = 2^{l+1} n\ \sqrt{\frac{(n+l)!}{n(n-1-l)!}}\ \frac{(\rho-1)^{n-1-l}}{(\rho+1)^{n+1+l}} \tag{B.16}$$

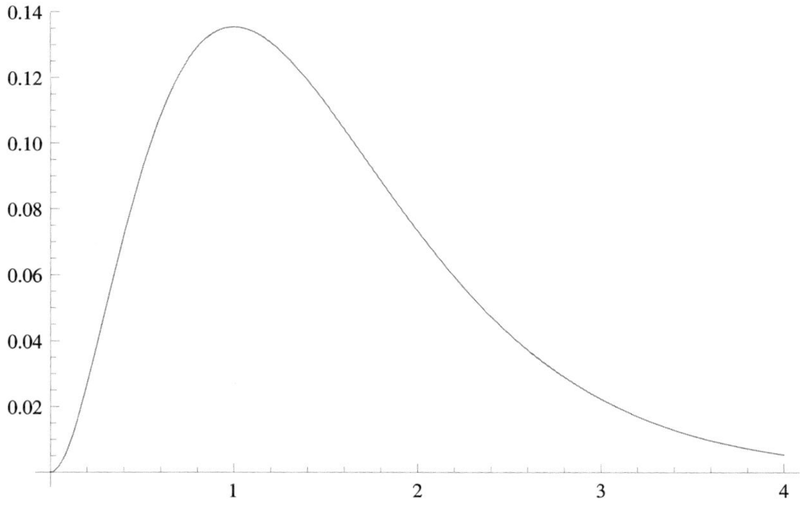

Fig. B.1: This figure shows the curve $s^2\ e^{-\rho s}$ as a function of s compared with its synthesis in terms of Coulomb Sturmian radial functions. The value $\rho = 2$ is chosen. The function and its series representation cannot be distinguished by eye, even when only 10 terms are used in the series.

The series in equation (B.13) converges very rapidly when ρ is close to 1. In the special case where $\rho = 1$, only the first term in the series is non-zero, and it has the value

$$c_{n,n-1}(1) = \frac{1}{2^n}\ \sqrt{n(2n-1)!} \tag{B.17}$$

In this special case, equation (B.13) reduces to a single term:

$$s^{n-1}\ e^{-s} = \tilde{R}_{n,n-1}(s)\ \frac{1}{2^n}\ \sqrt{n(2n-1)!} \tag{B.18}$$

The following figures illustrate the convergence of the series in equation (B.13) when $\rho = 2$.

B.3 Evaluation of molecular integrals

This expansion allows us to apply to STO's our previously-developed methods for the evaluation of molecular integrals by means of the theory of hyperspherical harmonics. We can evaluate the coefficients C_{n,n_1} by writing

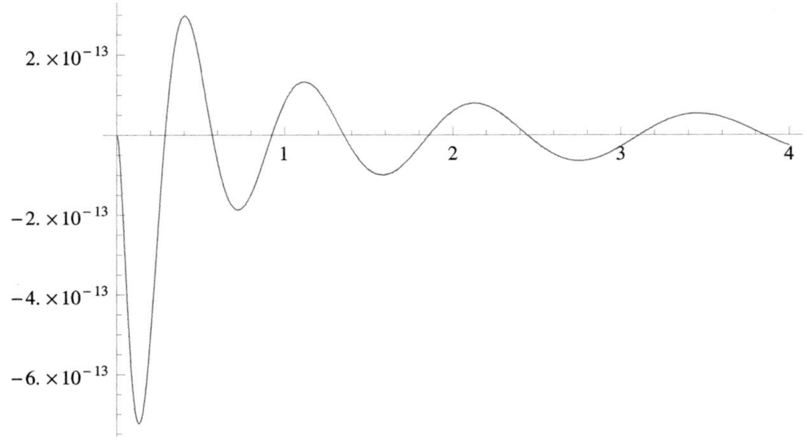

Fig. B.2: Here we see the difference between the curve $s^2\ e^{-\rho s}$ and its series representation when 30 terms are used in the series. The difference is shown as a function of s, for the case where $\rho = 2$. As $\rho \to 1$, the convergence becomes much more rapid, and when $\rho = 1$ only a single term survives.

the STO radial function in the form:

$$\begin{aligned}
\mathbb{R}_{n_1,\zeta_1}(r) &= N_{n_1,\zeta_1}\ r^{n_1-1}e^{-\zeta_1 r} \\
&= \frac{(2\zeta_1)^{n_1+\frac{1}{2}}}{\sqrt{(2n_1)!}}\ r^{n_1-1}e^{-\zeta_1 r} \\
&= \frac{(2\zeta_1)^{n_1+\frac{1}{2}}}{k^{n_1-1}\sqrt{(2n_1)!}}\ s^{n_1-1}e^{-\rho_1 s} \qquad s \equiv kr \\
&= k^{3/2}\frac{(2\rho_1)^{n_1+\frac{1}{2}}}{\sqrt{(2n_1)!}}\ s^{n_1-1}e^{-\rho_1 s} \qquad \rho_1 \equiv \frac{\zeta_1}{k} \\
&= k^{3/2}N_{n_1,\rho_1}\ s^{n_1-1}e^{-\rho_1 s}
\end{aligned} \tag{B.19}$$

When the STO radial function is expanded in terms of Coulomb Sturmian radial functions in a series of the form:

$$\begin{aligned}
\mathbb{R}_{n_1,\zeta_1}(r) &= k^{3/2}N_{n_1,\rho_1}\sum_{n>n_1-1}\tilde{R}_{n,n_1-1}(s)\ c_{n,n_1-1}(\rho_1) \qquad s \equiv kr \\
&= N_{n_1,\rho_1}\sum_{n>n_1-1}R_{n,n_1-1}(r)\ c_{n,n_1-1}(\rho_1)
\end{aligned} \tag{B.20}$$

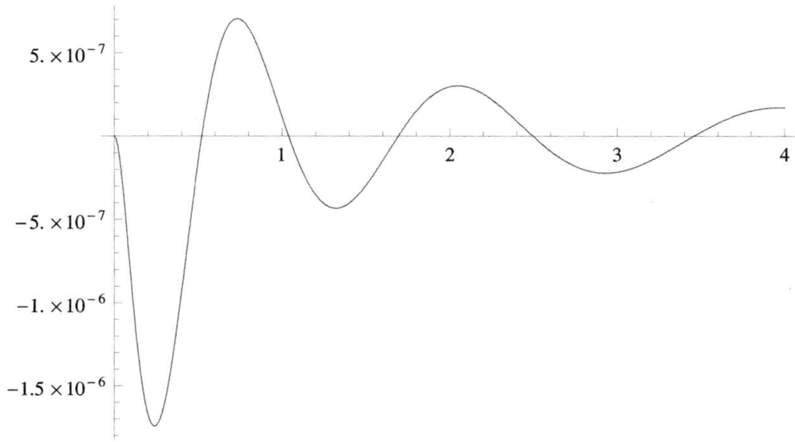

Fig. B.3: This figure is the same as the previous one, except that the series is truncated after 15 terms. This graph corresponds to $\rho = 2$, but the results for $\rho = 1/2$ are very similar.

we can write the expansion as

$$\begin{aligned}
\Phi_{\zeta_j,n_j,l_j,m_j}(\mathbf{x}) &\equiv \mathbb{R}_{n_1,\zeta_j}(r) Y_{l_1,m_1}(\hat{\mathbf{x}}) \\
&= N_{n_j,\rho_j} Y_{l_j,m_j}(\hat{\mathbf{x}}) \sum_{n>n_j-1} R_{n,n_j-1}(r)\; c_{n,n_j-1}(\rho_j) \\
&= N_{n_j,\rho_j} Y_{l_j,m_j}(\hat{\mathbf{x}}) \sum_{n>n_j-1} R_{n,l_j}(r) \left(-\frac{\partial}{\partial \rho_j}\right)^{n_j-1-l_j} c_{n,n_j-1}(\rho_j) \\
&= N_{n_j,\rho_j} \sum_{n>n_j-1} \chi_{n,l_j,m_j}(\mathbf{x}) \left(-\frac{\partial}{\partial \rho_j}\right)^{n_j-1-l_j} c_{n,n_j-1}(\rho_j) \\
&\equiv N_{n_j,\rho_j} \sum_{n>n_j-1} \chi_{n,l_j,m_j}(\mathbf{x})\; \mathcal{B}_{\mu_j}^{(n)}(\rho_j)
\end{aligned} \tag{B.21}$$

where the coefficients $c_{n,l}(\rho)$ are defined in equation (B.14), and where

$$\mathcal{B}_{\mu_j}^{(n)}(\rho_j) \equiv \left(-\frac{\partial}{\partial \rho_j}\right)^{n_j-1-l_j} c_{n,n_j-1}(\rho_j) \tag{B.22}$$

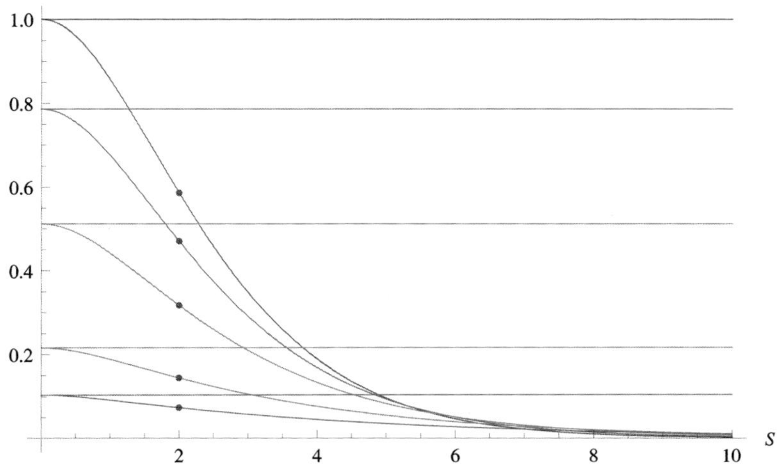

Fig. B.4: Overlap integrals for STO 1s orbitals. The overlap integrals are shown as functions of the parameter $S = \sqrt{\zeta_1\zeta_2}\ |\mathbf{X}_1 - \mathbf{X}_2|$. The top curve shows the case where $\zeta_1 = \zeta_2$, which is exact. The next four curves show the cases where $\sqrt{\zeta_1/\zeta_2}$ is $3/2, 2, 3$, and 4. The dots at $S = 2$ were calculated with three-dimensional numerical integrals. The numerical integrals were calculated with 8-10 significant digits, and agreed with the calculated curves up to that precision.

B.4 STO overlap integrals

From Equations (B.3) and (9.27) we have

$$\begin{aligned}
\Phi_{\bar{\mu}_1}(\mathbf{x} - \mathbf{X}_1) &= N_{n_1,\rho_1} \sum_{n>l_1} \chi_{n,l_1,m_1}(\mathbf{x} - \mathbf{X}_1)\mathcal{B}_{\mu_1}^{(n)}(\rho_1) \\
\Phi_{\bar{\mu}_2}(\mathbf{x} - \mathbf{X}_2) &= N_{n_2,\rho_2} \sum_{n'>l_2} \chi_{n',l_2,m_2}(\mathbf{x} - \mathbf{X}_2)\mathcal{B}_{\mu_2}^{(n')}(\rho_2) \qquad \text{(B.23)}
\end{aligned}$$

Therefore the STO overlap integral can be written as

$$\begin{aligned}
&\int d\mathbf{x}\ \Phi_{\bar{\mu}_1}^*(\mathbf{x} - \mathbf{X}_1)\Phi_{\bar{\mu}_2}(\mathbf{x} - \mathbf{X}_2) \\
&= N_{n_1,\rho_1} N_{n_2,\rho_2} \sum_{n>l_1} \sum_{n'>l_2} \mathcal{B}_{\mu_1}^{(n)}(\rho_1)\mathcal{B}_{\mu_2}^{(n')}(\rho_2) \\
&\times \int d\mathbf{x}\ \chi_{n,l_1,m_1}^*(\mathbf{x} - \mathbf{X}_1)\chi_{n',l_2,m_2}(\mathbf{x} - \mathbf{X}_2) \qquad \text{(B.24)}
\end{aligned}$$

Since we are able to evaluate overlap integrals between Coulomb Sturmians rapidly, and in closed form, by using a method based on the theory of

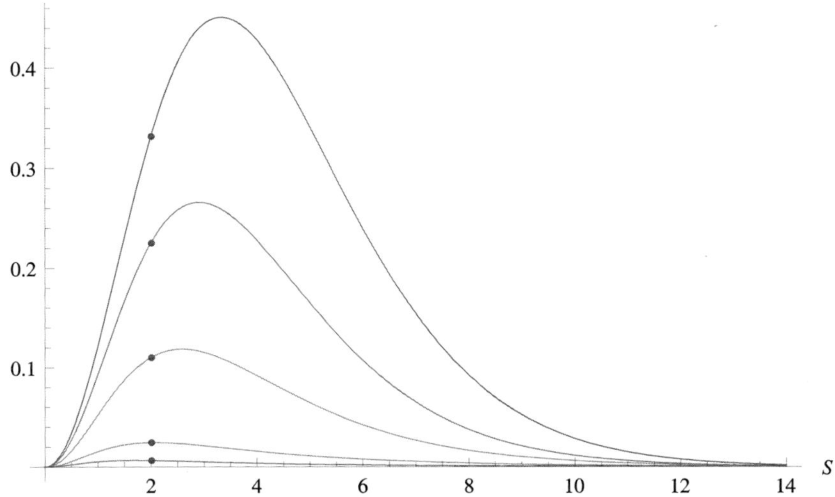

Fig. B.5: STO overlap integrals between 1s and 3d orbitals (with $m = 0$) shown with the same parameters as in Figure B.4. The displacement is in the z-direction. As before, the overlap integrals are correct at least up to 8-10 digits, which is the number of significant digits in the numerically calculated integrals.

hyperspherical harmonics, this gives us a rapid and accurate method for evaluating STO overlap integrals.

B.5 Interelectron repulsion integrals with STO's

In previously published work, we have developed rapid methods for evaluating many-center molecular interelectron repulsion integrals when Coulomb Sturmians located on the atomic centers are used as a basis. Since equation (9.31) allows us to express a 2-center density based on Slater-type orbitals in terms of Coulomb Sturmians, we are in a position to transform many-center interelectron repulsion integrals from a Coulomb Sturmian basis to an STO basis. If we introduce the Fourier transform representation of $1/r_{i,j}$

$$\frac{1}{|\mathbf{x}_i - \mathbf{x}_j|} = \frac{1}{2\pi^2} \int d\mathbf{p} \; \frac{1}{p^2} \; e^{i\mathbf{p}\cdot(\mathbf{x}_i - \mathbf{x}_j)} \tag{B.25}$$

and if $\rho_1(\mathbf{x}_1)$ and $\rho_2(\mathbf{x}_2)$ are 2-center densities, then multicenter STO interelectron repulsion integrals take on the form

$$\begin{aligned}
&\int d\mathbf{x}_1 \int d\mathbf{x}_2 \;\; \Phi^*_{\bar{\mu}_1}(\mathbf{x}_1 - \mathbf{X}_1)\Phi_{\bar{\mu}_2}(\mathbf{x}_1 - \mathbf{X}_2)\; \Phi^*_{\bar{\mu}_3}(\mathbf{x}_2 - \mathbf{X}_3)\Phi_{\bar{\mu}_4}(\mathbf{x}_2 - \mathbf{X}_4) \\
&\times \frac{1}{|\mathbf{x}_1 - \mathbf{x}_2|} \\
&\equiv \int d\mathbf{x}_1 \int d\mathbf{x}_2 \; \rho_i(\mathbf{x}_1)\; \frac{1}{|\mathbf{x}_1 - \mathbf{x}_2|}\; \rho_2(\mathbf{x}_2) \\
&= \frac{1}{2\pi^2} \int d\mathbf{p}\; \frac{1}{p^2} \int d\mathbf{x}_1 \int d\mathbf{x}_2 \; \rho_1(\mathbf{x}_1)\; e^{i\mathbf{p}\cdot(\mathbf{x}_1 - \mathbf{x}_2)}\; \rho_2(\mathbf{x}_2) \\
&= 4\pi \int d\mathbf{p}\; \frac{1}{p^2}\; \rho^t_1(-\mathbf{p})\; \rho^t_2(\mathbf{p})
\end{aligned} \tag{B.26}$$

where

$$\rho^t_j(\mathbf{p}) \equiv \frac{1}{(2\pi)^{3/2}} \int d\mathbf{x}_j \; e^{-i\mathbf{p}\cdot\mathbf{x}_j} \rho_j(\mathbf{x}_j) \tag{B.27}$$

Thus the evaluation of STO multicenter interelectron repulsion integrals requires the we are able to evaluate the Fourier transforms of STO 2-center densities. We can do this when the 2-center densities are based on Coulomb Sturmians, and a transformation very similar to equation (9.32) allows us to calculate the Fourier transforms of 2-center densities based on STO's.

$$\begin{aligned}
\rho^t_1(-\mathbf{p}) &= \frac{1}{(2\pi)^{3/2}} \int d\mathbf{x}_1 \; e^{i\mathbf{p}\cdot\mathbf{x}_1} \Phi^*_{\bar{\mu}_1}(\mathbf{x}_1 - \mathbf{X}_1)\Phi_{\bar{\mu}_2}(\mathbf{x}_1 - \mathbf{X}_2) \\
&= N_{n_1,\rho_1} N_{n_2,\rho_2} \sum_{n>l_1} \sum_{n'>l_2} \mathcal{B}^{(n)}_{\mu_1}(\rho_1)\mathcal{B}^{(n')}_{\mu_2}(\rho_2) \\
&\times \frac{1}{(2\pi)^{3/2}} \int d\mathbf{x}_1 \; e^{i\mathbf{p}\cdot\mathbf{x}_j} \; \chi^*_{n,l_1,m_1}(\mathbf{x}_1 - \mathbf{X}_1)\chi_{n',l_2,m_2}(\mathbf{x}_1 - \mathbf{X}_2) \\
\rho^t_2(\mathbf{p}) &= \frac{1}{(2\pi)^{3/2}} \int d\mathbf{x}_2 \; e^{-i\mathbf{p}\cdot\mathbf{x}_2} \Phi^*_{\bar{\mu}_3}(\mathbf{x}_2 - \mathbf{X}_3)\Phi_{\bar{\mu}_4}(\mathbf{x}_2 - \mathbf{X}_4) \\
&= N_{n_3,\rho_3} N_{n_4,\rho_4} \sum_{n''>l_3} \sum_{n'''>l_4} \mathcal{B}^{(n'')}_{\mu_3}(\rho_3)\mathcal{B}^{(n''')}_{\mu_4}(\rho_4) \\
&\times \frac{1}{(2\pi)^{3/2}} \int d\mathbf{x}_2 \; e^{-i\mathbf{p}\cdot\mathbf{x}_2} \; \chi^*_{n'',l_3,m_3}(\mathbf{x}_2 - \mathbf{X}_3)\chi_{n''',l_4,m_4}(\mathbf{x}_2 - \mathbf{X}_4)
\end{aligned} \tag{B.28}$$

Substituting the Fourier-transformed 2-center STO densities of equation (B.28) into equation (B.26), we obtain 4-center STO interelectron repulsion integrals. Combining these relationships, we obtain

$$\begin{aligned}
&\int d\mathbf{x}_1 \int d\mathbf{x}_2 \;\; \Phi^*_{\bar{\mu}_1}(\mathbf{x}_1 - \mathbf{X}_1)\Phi_{\bar{\mu}_2}(\mathbf{x}_1 - \mathbf{X}_2)\Phi^*_{\bar{\mu}_3}(\mathbf{x}_2 - \mathbf{X}_3)\Phi_{\bar{\mu}_4}(\mathbf{x}_2 - \mathbf{X}_4) \\
&\times \frac{1}{|\mathbf{x}_1 - \mathbf{x}_2|} \\
&= \left(\prod_{j=1}^{4} N_{n_j,\rho_j}\right) \sum_{n>l_1} \sum_{n'>l_2} \sum_{n''>l_3} \sum_{n'''>l_4} \mathcal{B}^{(n)}_{\mu_1}(\rho_1)\mathcal{B}^{(n')}_{\mu_2}(\rho_2)\mathcal{B}^{(n'')}_{\mu_3}(\rho_3)\mathcal{B}^{(n''')}_{\mu_4}(\rho_4) \\
&\times \int d\mathbf{x}_1 \int d\mathbf{x}_2 \; \chi^*_{n',l_1,m_1}(\mathbf{x}_1 - \mathbf{X}_1)\chi_{n,l_2,m_2}(\mathbf{x}_1 - \mathbf{X}_2)\frac{1}{|\mathbf{x}_1 - \mathbf{x}_2|} \\
&\times \chi^*_{n'',l_3,m_3}(\mathbf{x}_2 - \mathbf{X}_3)\chi_{n''',l_4,m_4}(\mathbf{x}_2 - \mathbf{X}_4) \qquad \text{(B.29)}
\end{aligned}$$

We next need to choose an appropriate value of k, in such a way that the convergence of the expansions will be rapid. In general, multicenter repulsion integrals with STO's involve four different values of the Slater exponents, buy for our Coulomb Sturmian methods to work, the value of k should be the same throughout. For evaluating overlap integrals, which involve only two Slater exponents, our experience has shown the convergence is most rapid if we let

$$k = \sqrt{\zeta_1\zeta_2} \qquad \text{(B.30)}$$

This suggests that for 4-center interelectron repulsion integrals we should let

$$k = (\zeta_1\zeta_2\zeta_3\zeta_4)^{1/4} \qquad \text{(B.31)}$$

but of course, other prescriptions can be tried to see which choice of k does indeed give the most rapid convergence.

B.6 Checks in the atomic case

As a check for these equation (B.29), we can see what it reduces to in the atomic case, where there is only a single center. Then equation (B.29) becomes

$$
\begin{aligned}
&\int d\mathbf{x}_1 \int d\mathbf{x}_2 \;\; \Phi^*_{\bar{\mu}_1}(\mathbf{x}_1)\Phi_{\bar{\mu}_2}(\mathbf{x}_1) \, \frac{1}{|\mathbf{x}_1 - \mathbf{x}_2|} \;\; \Phi^*_{\bar{\mu}_3}(\mathbf{x}_2)\Phi_{\bar{\mu}_4}(\mathbf{x}_2) \\
&= \left(\prod_{j=1}^{4} N_{n_j,\rho_j} \right) \sum_{n>l_1} \sum_{n'>l_2} \sum_{n''>l_3} \sum_{n'''>l_4} \mathcal{B}^{(n)}_{\mu_1}(\rho_1)\mathcal{B}^{(n')}_{\mu_2}(\rho_2)\mathcal{B}^{(n'')}_{\mu_3}(\rho_3)\mathcal{B}^{(n''')}_{\mu_4}(\rho_4) \\
&\times \int d\mathbf{x}_1 \int d\mathbf{x}_2 \; \chi^*_{n,l_1,m_1}(\mathbf{x}_1)\chi_{n',l_2,m_2}(\mathbf{x}_1)\chi^*_{n'',l_3,m_3}(\mathbf{x}_2)\chi_{n''',l_4,m_4}(\mathbf{x}_2) \\
&\times \frac{1}{|\mathbf{x}_1 - \mathbf{x}_2|}
\end{aligned}
\qquad \text{(B.32)}
$$

The simplest case is the one for which $l_1 = l_2 = l_3 = l_4 = 0$, where the coefficients $\mathcal{B}^{(n)}_{\mu_j}(\rho_j)$ are

$$
\begin{aligned}
\mathcal{B}^{(n)}_{n_j,0,0}(\rho_j) &= \left(-\frac{\partial}{\partial \rho_j} \right)^{n_j - 1} c_{n,n_j-1}(\rho_j) \\
&= n 2^{n_j} \sqrt{\frac{(n+n_j-1)!}{n(n-n_j)!}} \left(-\frac{\partial}{\partial \rho_j} \right)^{n_j - 1} \frac{(\rho_j - 1)^{n-n_j}}{(\rho_j + 1)^{n+n_j}}
\end{aligned}
\qquad \text{(B.33)}
$$

For example,

$$
\mathcal{B}^{(n)}_{1,0,0}(\rho_j) = c_{n,0}(\rho_j) = 2n \, \frac{(\rho_j - 1)^{n-1}}{(\rho_j + 1)^{n+1}} \qquad \text{(B.34)}
$$

In this very simple case, equation (B.32) becomes:

$$
\begin{aligned}
&\int d\mathbf{x}_1 \int d\mathbf{x}_2 \;\; \Phi^*_{\bar{\mu}_1}(\mathbf{x}_1)\Phi_{\bar{\mu}_2}(\mathbf{x}_1) \, \frac{1}{|\mathbf{x}_1 - \mathbf{x}_2|} \;\; \Phi^*_{\bar{\mu}_3}(\mathbf{x}_2)\Phi_{\bar{\mu}_4}(\mathbf{x}_2) \\
&= \left(\prod_{j=1}^{4} N_{n_j,\rho_j} \right) \sum_{n>0} \sum_{n'>0} \sum_{n''>0} \sum_{n'''>0} c_{n,0}(\rho_1)c_{n',0}(\rho_2)c_{n'',0}(\rho_3)c_{n''',0}(\rho_4) \\
&\times \int d\mathbf{x}_1 \int d\mathbf{x}_2 \; \chi^*_{n,0,0}(\mathbf{x}_1)\chi_{n',0,0}(\mathbf{x}_1) \frac{1}{|\mathbf{x}_1 - \mathbf{x}_2|} \chi^*_{n'',0,0}(\mathbf{x}_2)\chi_{n''',0,0}(\mathbf{x}_2)
\end{aligned}
\qquad \text{(B.35)}
$$

In the 1-center case, both STO interelectron repulsion integrals and Coulomb Sturmian interelectron repulsion integrals can readily be evaluated using methods discussed by us in Appendix A of [Avery and Avery, 2006]. In the STO case, when $(n_j, l_j m_j) = (1, 0, 0)$ for all j, we have:

$$
\begin{aligned}
&\int d\mathbf{x}_1 \int d\mathbf{x}_2 \;\; \Phi^*_{\bar{\mu}_1}(\mathbf{x}_1)\Phi_{\bar{\mu}_2}(\mathbf{x}_1) \; \frac{1}{|\mathbf{x}_1 - \mathbf{x}_2|} \;\; \Phi^*_{\bar{\mu}_3}(\mathbf{x}_2)\Phi_{\bar{\mu}_4}(\mathbf{x}_2) \\
&= \left(\prod_{j=1}^{4} N_{1,\rho_j}\right) \int_0^\infty dr_1 \; r_1^2 \int_0^\infty dr_2 \; r_1^2 \; \frac{1}{r_>} \; e^{-(\zeta_1+\zeta_2)r_1} \; e^{-(\zeta_3+\zeta_4)r_1} \\
&= 16\left(\zeta_1\zeta_2\zeta_3\zeta_4\right)^{3/2} \int_0^\infty dr_1 \; r_1^2 \int_0^\infty dr_2 \; r_1^2 \; \frac{1}{r_>} \; e^{-(\zeta_1+\zeta_2)r_1} \; e^{-(\zeta_3+\zeta_4)r_1} \\
&= \frac{32\left(\zeta_1\zeta_2\zeta_3\zeta_4\right)^{3/2}}{(\zeta_1+\zeta_2+\zeta_3+\zeta_4)^3} \left(\frac{3}{(\zeta_1+\zeta_2)(\zeta_3+\zeta_4)} + \frac{1}{(\zeta_1+\zeta_2)^2} + \frac{1}{(\zeta_3+\zeta_4)^2}\right)
\end{aligned}
\tag{B.36}
$$

while

$$
\begin{aligned}
&J_{n,n',n'',n'''} \\
&\equiv \int d\mathbf{x}_1 \int d\mathbf{x}_2 \; \chi^*_{n,0,0}(\mathbf{x}_1)\chi_{n',0,0}(\mathbf{x}_1) \frac{1}{|\mathbf{x}_1 - \mathbf{x}_2|} \chi^*_{n'',0,0}(\mathbf{x}_2)\chi_{n''',0,0}(\mathbf{x}_2) \\
&= \int_0^\infty dr_1 \; r_1^2 \int_0^\infty dr_2 \; r_2^2 \; \frac{1}{r_>} R_{n,0}(r_1)R_{n',0}(r_1)R_{n'',0}(r_2)R_{n''',0}(r_2)
\end{aligned}
\tag{B.37}
$$

These integrals are also easy to evaluate using the methods described in Appendix A of [Avery and Avery, 2006]. For example, with $n = 1, 2, 3, 4, 5$ and $n' = 1, 2, 3, 4, 5$ while $n'' = b''' = 1$, we obtain the familiar values

$$
J_{n,n',1,1} = \begin{pmatrix}
\frac{5k}{8} & -\frac{5k}{32} & -\frac{k}{16} & -\frac{3k}{128} & -\frac{k}{128} \\
-\frac{5k}{32} & \frac{3k}{8} & -\frac{5k}{64} & -\frac{11k}{256} & -\frac{11k}{512} \\
-\frac{k}{16} & -\frac{5k}{64} & \frac{17k}{64} & -\frac{25k}{512} & -\frac{k}{32} \\
-\frac{3k}{128} & -\frac{11k}{256} & -\frac{25k}{512} & \frac{211k}{1024} & -\frac{35k}{1024} \\
-\frac{k}{128} & -\frac{11k}{512} & -\frac{k}{32} & -\frac{35k}{1024} & \frac{863k}{5120}
\end{pmatrix}
\tag{B.38}
$$

By using the relationship

$$\int d\mathbf{x}_1 \int d\mathbf{x}_2 \;\; \Phi^*_{\bar{\mu}_1}(\mathbf{x}_1)\Phi_{\bar{\mu}_2}(\mathbf{x}_1) \,\frac{1}{|\mathbf{x}_1 - \mathbf{x}_2|}\; \Phi^*_{\bar{\mu}_3}(\mathbf{x}_2)\Phi_{\bar{\mu}_4}(\mathbf{x}_2)$$
$$= \left(\prod_{j=1}^{4} N_{1,\rho_j}\right) \int_0^\infty dr_1\; r_1^2 \int_0^\infty dr_2\; r_1^2 \;\frac{1}{r_>}\; e^{-(\zeta_1+\zeta_2)r_1}\; e^{-(\zeta_3+\zeta_4)r_1}$$
$$= 16\,(\zeta_1\zeta_2\zeta_3\zeta_4)^{3/2} \int_0^\infty dr_1\; r_1^2 \int_0^\infty dr_2\; r_1^2 \;\frac{1}{r_>}\; e^{-(\zeta_1+\zeta_2)r_1}\; e^{-(\zeta_3+\zeta_4)r_1}$$
$$= \frac{32\,(\zeta_1\zeta_2\zeta_3\zeta_4)^{3/2}}{(\zeta_1+\zeta_2+\zeta_3+\zeta_4)^3}\left(\frac{3}{(\zeta_1+\zeta_2)(\zeta_3+\zeta_4)} + \frac{1}{(\zeta_1+\zeta_2)^2} + \frac{1}{(\zeta_3+\zeta_4)^2}\right)$$
$$= \frac{32k\,(\rho_1\rho_2\rho_3\rho_4)^{3/2}}{(\rho_1+\rho_2+\rho_3+\rho_4)^3}\left(\frac{3}{(\rho_1+\rho_2)(\rho_3+\rho_4)} + \frac{1}{(\rho_1+\rho_2)^2} + \frac{1}{(\rho_3+\rho_4)^2}\right)$$
$$\approx \left(\prod_{j=1}^{4} N_{1,\rho_j}\right) \sum_{n=1}^{n_{max}}\sum_{n'=1}^{n_{max}}\sum_{n''=1}^{n_{max}}\sum_{n'''=1}^{n_{max}} c_{n,0}(\rho_1)c_{n',0}(\rho_2)c_{n'',0}(\rho_3)c_{n''',0}(\rho_4)$$
$$\times\, J_{n,n',n'',n'''} \tag{B.39}$$

we can check the convergence of the expansion for various values of the ζ_j's. We can also check that when

$$\rho_1 = \rho_2 = \rho_3 = \rho_4 \to 1 \tag{B.40}$$

the STO interelectron repulsion integral reduces to $J_{1,1,1,1}$:

$$\frac{32k\,(\rho_1\rho_2\rho_3\rho_4)^{3/2}}{(\rho_1+\rho_2+\rho_3+\rho_4)^3}\left(\frac{3}{(\rho_1+\rho_2)(\rho_3+\rho_4)} + \frac{1}{(\rho_1+\rho_2)^2} + \frac{1}{(\rho_3+\rho_4)^2}\right)$$
$$\to \frac{5k}{8} \tag{B.41}$$

Table B.1: This table illustrates the rate of convergence for the synthesis of interelectron repulsion integrals in terms of Coulomb Sturmian results. The error is shown (in units of k) for the synthesis of the exact electron repulsion integral in equation (B.41), for a range of values of ρ_j and $n_{\max}$.

ρ_1	ρ_2	ρ_3	ρ_4	$n_{\max} = 5$	$n_{\max} = 10$	$n_{\max} = 15$
1	1	1	1	0	0	0
2/3	2/3	3/2	3/2	0.0000147699	-9.61191×10^{-10}	1.64313×10^{-14}
1/2	1/2	2	2	0.000620543	-4.98248×10^{-8}	3.40038×10^{-11}
1/3	1/3	3	3	0.012104	0.0000468892	8.88163×10^{-8}
1/4	1/2	3/4	1	-0.000443541	-7.12232×10^{-7}	5.89531×10^{-10}

Bibliography

Abou-Salem, L.I.: Adv. High Energy Phys. **2014, Article ID** 196434, 2014.

Abragam, A. and Bleany B., *Electron Paramagnetic Resonance of Transition Ions*, Clarendon, 1970.

Abramowitz, M., Stegun, I.A., **Handbook of Mathematical Functions**. National Institute of Standards and Technology, USA, 1964.

Adam, R.M., Sofianos, S.A.: Phys. Rev. **A 82**, 053635, 2010.

Afnan, I.R., Tang, Y.C.: Phys. Rev. **175**, 1337, 1968.

Ahlberg, R. and Lindner, P., *The Fermi correlation for electrons in momentum space*, J. Phys. B, **9** (17) 2963-9, 1976.

Akhiezer, A.I. and Berestetskii, V.B., **Quantum Electrodynamics**, Interscience, New York, 1965.

Albeverio, S., Gesztesy, F., Høegh-Krohn, R., Holden, H.: **Solvable Models in Quantum Mechanics**. Springer, Berlin, 1980.

Alijah, A., L. Wolniewicz, and J. Hinze, Molecular Physics **85.6**, 1125-1150, 1996.

Al-Khalili, J., *An Introduction to Halo Nuclei*, in Al-Khalili, J., Roecki, E. (eds.) **The Euroschool Lectures on Physics with Exotic Beams, vol. I**, Lecture Notes in Physics 651. Springer, Berlin, 2004.

Allen, H.C., and Cross, P.C., **Molecular vib-rotors; the theory and interpretation of high resolution infra-red spectra** (324 pages), Wiley, New York, 1963.

Anderson, R.W., Aquilanti, V., Cavalli, S. and Grossi, G., J. Phys. Chem., **95** 8184, 1991.

Alliluev, S.P., Sov. Phys. JETP, **6** 156, 1958.

Amiet, J.-P. et Huguenin, P., **Mécaniques classique et quantiques dans l'espace de phase**, Université de Neuchâtel, 1981.

Amos, A.T. and Hall, G.G., Proc. Roy. Soc. London, **A263**, 483, 1961.

Anderson, R.W., Aquilanti, V., Cavalli, S. and Grossi, G., J. Phys. Chem., **97** 2443, 1993.

Anderson, M.H., Ensher, J.R., Matthews, M.R., Wieman, C.E., Cornell, E.A.: Science **269**, 198, 1995.

Aquilanti, V. and Avery, J., *Sturmian expansions for quantum mechanical many-body problems and hyperspherical harmonics*, Adv. Quant. Chem., **39** 72-101, 2001.

Aquilanti, V. and Cavalli, S., *Coordinates for molecular dynamics*, J. Chem. Phys., **85** 1355-1361, 1986.

Aquilanti, V., Cavalli, S., De Fazio, D., and Grossi, G. *Hyperangular momentum: Applications to atomic and molecular science*, in **New Methods in Quantum Theory**, Tsipis, C.A., Popov, V.S., Herschbach, D.R., and Avery, J.S., Eds., Kluwer, Dordrecht, 1996.

Aquilanti, V., Cavalli, S. and Grossi, G., *Hyperspherical coordinates for molecular dynamics by the method of trees and the mapping of potential-energy surfaces for triatomic systems*, J. Chem. Phys., **85** 1362, 1986.

Aquilanti, V., Grossi, G., Laganá, A., Pelikan, E., and Klar, H., *A decoupling scheme for a 3-body problem treated by expansions into hyperspherical harmonics. The hydrogen molecular ion*, Lett. Nuovo Cimento, **41**, 541, 1984.

Aquilanti, V., Grossi, G., and Laganá, A., *On hyperspherical mapping and harmonic expansions for potential energy surfaces* J. Chem. Phys., **76** 1587-8, 1982.

Aquilanti, V., Laganá, A., and Levine, R.D., Chem. Phys. Lett., **158** 87, 1989.

Aquilanti, V. and Cavalli, S., Chem. Phys. Lett., **141** 309, 1987.

Aquilanti, V., Cavalli, S., Grossi, G., Rosi, M., Pellizzari, V., Sgamellotti, A., and Tarantelli, F., Chem. Phys. Lett., **16** 179, 1989.

Aquilanti, V., Cavalli, S., Grossi, G., and Anderson, R.W., J. Chem. Soc. Faraday Trans., **86** 1681, 1990.

Aquilanti, V., Benevente, L., Grossi, G. and Vecchiocattivi, F., *Coupling schemes for atom-diatom interactions, and an adiabatic decoupling treatment of rotational temperature effects on glory scattering*, J. Chem. Phys., **89** 751-761, 1988.

Aquilanti, V., A. Lombardia, and E. Yurtsever, *Global view of classical clusters: the hyperspherical approach to structure and dynamics*, Phys. Chem. Chem. Phys. **4**, 5040-5051, 2002.

Aquilanti, V. and Grossi, G., *Angular momentum coupling schemes in the quantum mechanical treatment of P-state atom collisions* J. Chem. Phys., **73** 1165-1172, 1980.

Aquilanti, V., Cavalli, S., and Grossi, G., Theor. Chem. Acta, **79** 283, 1991.

Aquilanti, V. and Cavalli, S., Few Body Systems, Suppl. **6** 573, 1992.

Aquilanti, V. and Grossi, G., Lett. Nuovo Cimento, **42** 157, 1985.

Aquilanti, V., Cavalli, S. and De Fazio, D., *Angular and hyperangular momentum coupling coefficients as Hahn polynomials*, J. Phys. Chem. **99** 15694, 1995.

Aquilanti, V., Cavalli, S., Coletti, C. and Grossi, G., *Alternative Sturmian bases and momentum space orbitals; an application to the hydrogen molecular ion*, Chem. Phys. **209** 405, 1996.

Aquilanti, V., Cavalli, S. and Coletti, C., *The d-dimensional hydrogen atom; hyperspherical harmonics as momentum space orbitals and alternative Sturmian basis sets*, Chem. Phys. **214**, 1997.

Aquilanti, V., and Avery, J., *Generalized potential harmonics and contracted Sturmians*, Chem. Phys. Letters, **267** 1-8, 1997.

Aquilanti, V., Cavalli, S., Coletti, C., Di Domenico, D., and Grossi, G., Int. Rev. Phys. Chem., **20** 673, 2001.

Aquilanti, V. and Caligiana, A., Chem. Phys. Letters, **366** 157, 2002.

Aquilanti, V., Caligiana A., and Cavalli, S., Int. J. Quantum Chem., **92** 99, 2003.

Aquilanti, V., Caligiana, A., Cavalli, S., and Coletti, C., Int. J. Quantum Chem., **92** 212, 2003.

Aquilanti, V., and Caligiana, A., in **Fundemental World of Quantum Chemistry: A Tribute to the Memory of P.O. Löwdin, I**, E.J. Brändas and E.S. Kryachko, Eds., Kluwer, Dordrecht, 297, 2003.

Aquilanti, V., et al.: Int. J. Quant. Chem. **89**, 277 (2002); Aquilanti, V., Lombardi, A., Littlejohn, R.G.: Theo. Chem. Acc. **111**, 400 (2004); Kuppermann, A.: J. Phys. Chem. **108**, 8894 (2004); Lepetit, B., Wang, D., Kuppermann, A.: J. Chem. Phys. **125**, 133505 (2006); Kuppermann, A.: Phys. Chem. Chem. Phys. **13**, 8259, 2011.

Arfken, G., **Mathematical Methods for Physicists**. Academic Press, Waltham, 1966.

Atkins, P.W., Child, M.S. and Phillips, C.S.G., *Tables for Group Theory*, Oxford, 1970.

Avery, J., **Creation and Annihilation Operators**, McGraw-Hill, 1976.

Avery, J. and Ørmen, P.J., *Transferable integrals in a deformation density approach to molecular orbital calculations; IV. Evaluation of angular integrals by a vector-pairing method*, Int. J. Quantum Chem. **18** 953-971, 1980.

Avery, J., **Hyperspherical Harmonics; Applications in Quantum Theory**, Kluwer Academic Publishers, Dordrecht, 1989.

Avery, J., *Hyperspherical Sturmian basis functions in reciprocal space*, in **New Methods in Quantum Theory**, Tsipis, C.A., Popov, V.S., Herschbach, D.R., and Avery, J.S., Eds., Kluwer, Dordrecht, 1996.

Avery, J. and Antonsen, F., *A new approach to the quantum mechanics of atoms and small molecules*, Int. J. Quantum Chem., Symp. **23** 159, 1989.

Avery, J. and Antonsen, F., *Iteration of the Schrödinger equation, starting with Hartree-Fock wave functions*, Int. J. Quantum Chem., **42** 87, 1992.

Avery, J. and Antonsen, F., Theor. Chim. Acta, **85** 33, 1993.

Avery, J. and Herschbach, D. R., *Hyperspherical Sturmian basis functions*, Int. J. Quantum Chem., **41** 673, 1992.

Avery, J. and Wen, Z.-Y., *A Formulation of the quantum mechanical many-body in terms of hyperspherical coordinates*, Int. J. Quantum Chem. **25** 1069, 1984.

Avery, J., *Correlation in iterated solutions of the momentum-space Schrödinger equation*, Chem. Phys. Lett., **138** (6) 520-4, 1987.

Avery, J., *Hyperspherical harmonics; Some properties and applications*, in **Conceptual Trends in Quantum Chemistry**, Kryachko, E.S., and Calais, J.L., Eds., Kluwer, Dordrecht, 1994.

Avery, J., Hansen, T.B., Wang, M. and Antonsen, F., *Sturmian basis sets in momentum space*, Int. J. Quantum Chem. **57** 401, 1996.

Avery, J., and Hansen, T.B., *A momentum-space picture of the chemical bond* Int. J. Quantum Chem. **60** 201, 1996.

Avery, J., *Many-particle Sturmians*, J. Math. Chem., **21** 285, 1997.

Avery, J., *A formula for angular and hyperangular integration*, J. Math. Chem., **24** 169-174, 1998.

Avery, J. and Antonsen, F., *Relativistic Sturmian basis functions*, J. Math. Chem. **24** 175, 1998.

Avery, J., *Many-electron Sturmians applied to atoms and ions*, J. Mol. Struct. **458** 1, 1999.

Avery, J., *Many-electron Sturmians as an alternative to the SCF-CI method*, Adv. Quantum Chem., **31** 201, 1999.

Avery, J., **Hyperspherical Harmonics and Generalized Sturmians**, Kluwer Academic Publishers, Dordrecht, Netherlands, 196 pages, 2000.

Avery, J.S.: J. Phys. Chem. **97**, 2406 (1993); Avery, J.S.: J. Comp. Appl. Math. **233**, 1366, 2010.

Avery, J. *Selected applications of hyperspherical harmonics in quantum theory*, J. Phys. Chem. **97**, 2406, 1993.

Avery, J. and Antonsen, F., *Evaluation of angular integrals by harmonic projection*, Theor. Chim. Acta. **85**, 33, 1993.

Avery, J. and Antonsen, F., *Magnetic interactions and 4-currents*, Journal of Molecular Structure (Theochem), **261**, 69, 1992.

Avery, J., Antonsen, F. and Shim, I., *4-Currents in relativistic quantum chemistry*, Int. J. Quantum Chem. **45**, 573, 1993.

Avery, J., *Fock transforms in reciprocal-space quantum theory*, J. Math. Chem. **15** 233, 1994.

Avery, J., *The generalized Sturmian method and inelastic scattering of fast electrons*, J. Math. Chem. **27**, 279-292, 2000.

Avery, J., *Sturmian methods in quantum theory*, Proc. Workshop on Concepts in Chemical Physics, G.D. Billing and N. Henriksen, editors, Danish Technical University, 2001.

Avery, J. and Shim, R., *Molecular Sturmians, Part 1*, Int. J. Quantum Chem., **83**, 1-10, 2001.

Avery, J., *Sturmians*, in **Handbook of Molecular Physics and Quantum Chemistry**, S. Wilson, ed., Wiley, Chichester, 2003.

Avery, J., *Harmonic polynomials, hyperspherical harmonics, and Sturmians*, in **Fundemental World of Quantum Chemistry, A Tribute Volume to the Memory of Per-Olov Löwdin**, E.J. Brändas and E.S. Kryachko, eds., Kluwer Academic Publishers, Dordrecht, 261-296, 2003.

Avery, J. and Avery, J., *The generalized Sturmian method for calculating spectra of atoms and ions*, J. of Math. Chem., **33** 145-162, 2003.

Avery, J. and Avery, J., *Kramers pairs in configuration interaction*, Adv. Quant. Chem., **43** 185-206, 2003.

Avery, J., *Many-center Coulomb Sturmians and Shibuya-Wulfman integrals*, Int. J. Quantum Chem., 2003.

Avery, J., Avery J. and Goscinski, O., *Natural orbitals from generalized Sturmian calculations*, Adv. Quant. Chem., **43** 207-16, 2003.

Avery, J., Avery, J., Aquilanti, V., and Caligiana, A., *Atomic densities, polarizabilities and natural orbitals derived from generalized Sturmian calculations*, Adv. Quant. Chem., **47** 156-173, 2004.

Avery, J. and Avery, J., *Generalized Sturmian solutions for many-particle Schrödinger equations*, J. Phys. Chem. A, **41** 8848, 2004.

Avery, J. and Sauer, S., *Many-electron Sturmians applied to molecules*, in **Quantum Systems in Chemistry and Physics, Volume 1**, A. Hernández-Laguna, J. Maruani, R. McWeeney and S. Wilson editors, Kluwer Academic Publishers, 2000.

Avery, J. and Coletti, C., *Generalized Sturmians applied to atoms in strong external fields*, J. Math. Chem. **27** 43-51, 2000.

Avery, J. and Coletti, C., *Many-electron Sturmians applied to atoms and ions in strong external fields*, in **New Trends in Quantum Systems in Chemistry and Physics**, J. Marauani et al. (eds.), pages 77-93, Kluwer Academic Publishers, 2001.

Avery, J. and Shim, R., *Core ionization energies of atoms and molecules calculated using the generalized Sturmian method*, Int. J. Quantum Chem. **79** 1-7, 2000.

Avery, J.E., *Fast electron repulsion integrals for molecular Coulomb Sturmians*, Adv. Quantum Chem., **67** 129-151, 2013.

Avery, J., *The generalized Sturmian method and inelastic scattering of fast electrons*, J. Math. Chem. **27**, 279-292, 2000.

Avery, J., *Sturmian methods in quantum theory*, Proc. Workshop on Concepts in Chemical Physics, G.D. Billing and N. Henriksen, editors, Danish Technical University, 2001.

Avery, J., *Sturmians*, in **Handbook of Molecular Physics and Quantum Chemistry**, S. Wilson, ed., Wiley, Chichester, 2003.

Avery, J., *Harmonic polynomials, hyperspherical harmonics, and Sturmians*, in **Fundemental World of Quantum Chemistry, A Tribute Volume to the Memory of Per-Olov Löwdin**, E.J. Brändas and E.S. Kryachko, eds., Kluwer Academic Publishers, Dordrecht, 261-296, 2003.

Avery, J. and Avery, J., *The generalized Sturmian method for calculating spectra of atoms and ions*, J. of Math. Chem., **33** 145-162, 2003.

Avery, J. and Avery, J., *Kramers pairs in configuration interaction*, Adv. Quant. Chem., **43** 185-206, 2003.

Avery, J., *Many-center Coulomb Sturmians and Shibuya-Wulfman integrals*, Int. J. Quantum Chem., 2003.

Avery, J., Avery J. and Goscinski, O., *Natural orbitals from generalized Sturmian calculations*, Adv. Quant. Chem., **43** 207-16, 2003.

Avery, J., Avery, J., Aquilanti, V., and Caligiana, A., *Atomic densities, polarizabilities and natural orbitals derived from generalized Sturmian calculations*, Adv. Quant. Chem., **47** 156-173, 2004.

Avery, J. and Avery, J., *Generalized Sturmian solutions for many-particle Schrödinger equations*, J. Phys. Chem. A, **41** 8848, 2004.

Avery, J. and Avery, J., **Generalized Sturmians and Atomic Spectra**, World Scientific, 2006.

Avery, J. and Avery, J., *The Generalized Sturmian Library*, http://sturmian.kvante.org, 2006a.

Avery, J. and Avery, J., *Atomic core-ionization energies; approximately piecewise-linear relationships*, J. Math. Chem. **46**, 164-181, 2009.

Avery, J., *Harmonic polynomials, hyperspherical harmonics and atomic spectra*, J. Comp. Appl. Math., 2009.

Avery, J.S., Rettrup, S., and Avery, J.E., **Symmetry-Adapted Basis Sets**, World Scientific, 2012.

Avery, J., and Avery, J., *Can Coulomb Sturmians be used as a basis for N-electron molecular calculations?*, J. Phys. Chem A, **113**, 2009.

Avery, J.E., *New Computational Methods in the Quantum Theory of Nanostructures*, Ph.D. Thesis, University of Copenhagen, 2011.

Bacca, S., Barnea, N., Schwenk, A.: Phys. Rev. **C 86**, 034321, 2012.

Bach, V., *Approximative theories for large Coulomb systems.* In: Rauch, J., Simon, B. (eds.) **Quasiclassical Methods**. Springer, New York (1997); Heinzi C., et al.: Phys. Rev. **A 76**, 052104, 2007.

Ballot, J.L., Beiner, M., Fabre de la Ripelle, M., In: Calogero, F., Ciofi degli Atti, Y.C. (eds.) **The Nuclear Many Body Problem, vol. 1**. Bologna (1973); Fabre de la Ripelle, M. et al.: Lett. al Nuo. Cim. **1**, 584, 1971.

Ballot, L. and Farbre de la Ripelle, M., *Application of the hyperspherical formalism to trinucleon bound-state problems*, Ann. Phys., **127** 62, 1980.

Ballot, J.L., Fabre de la Ripelle, M.: Ann. Phys. (N.Y.) 127, **62** (1980).

Ballot, J.L., Navarro, J.: J. Phys. **B 8**, 172, 1975.

Bandar, M. and Itzyksen, C., *Group theory and the H atom*, Rev. Mod. Phys. **38** 330, p 346, 1966.

Bang, J.M. and Vaagen, J.S., *The Sturmian expansion: a well-depth-method for orbitals in a deformed potential*, Z. Phys. A, **297** (3) 223-36, 1980.

Bang, J.M., Gareev, F.G., Pinkston, W.T. and Vaagen, J.S., Phys. Rep. **125** 253-399, 1985.

Barletta, P., Kievsky, A.: Phys. Rev. **A 64**, 042514, 2001.

Barnea, N., Liedemann, W., Orlandini, G.: Phys. Rev. **C 81**, 064001 (2010); Orlandini, G., Barnea, N., Leidemann, W.: J. Phys. Conf. **Series 312**, 092049, 2011.

Barnea, N., et al.: Few-Body Syst. **39**, 1, 2006.

Barreto, Patricia R. P. et al., J. Phys. Chem. **A 111.49**, 12754-12762, 2007.

Baretty, Reinaldo, Ishikawa, Yasuyuki, and Nieves, Jose F., *Momentum space approach to relativistic atomic structure calculations*, Int. J. Quantum Chem., Quantum Chem. Symp., **20** 109-17, 1986.

Baryudin, L. E. and Telnov, D. A., *Sturmian expansion of the electron density deformation for 3d-metal ions in electric field*, Vestn. Leningr. Univ., Ser 4: Fiz., Khim. (1), p 83-6, 1991.

Bellazzini, J., Frank, R.L., Lieb, E.H., Seilinger, R.: Rev. Math. Phys. **26**, 1350021, 2014.

Benesch, Robert and Smith, Vedene H. Jr., *Natural orbitals in momentum space and correlated radial momentum distributions. I. The 1S ground state of Li+*, Int. J. Quantum Chem., Symp, **4** 131-8, 1971.

Benguria, R.D., Loss, M., Siedentop, H.: J. Math. Phys. **49**, 012302 (2008); Benguria, R.D., Frank, R.L., Loss, M.: Math. Res. Lett. **15**, 613, 2008.

Berry, R.S., Ceraulo, S.C. and Batka, J., *Valence electrons in atoms: Collective or independent-particle-like?*, in Herschbach, D.R., Avery, J. and Goscinski, O., Eds., **Dimensional Scaling in Chemical Physics**, Kluwer, Dordrecht, 1993.

Berry, R. Stephen and Smirnov, Boris M., *Clusters as tools to link macro and micro approaches*, Computational and Theoretical Chemistry **1021** 2-6, 2013.

Berry, R. Stephen and Smirnov, Boris M., *Configurational transitions in processes involving metal clusters*, Physics Reports **527** 205-250, 2013.

Bethe. H.A. and Jackiw, R., **Intermediate Quantum Mechanics, 3rd edition**, The Benjamin Cummings Publishing Company, 1986.

Bhattacharyya, S., Das, T.K.: Pramana (J. Phys.) **44**, 183 (1995); Bhattacharyya, S., Das, T.K., Kanta, K.P., and Ghosh, A.K.: Phys. Rev. **C 50**, 2228, 1994.

Bhattacharyya, S., Das, T.K.: J. Comput. Phys. (USA) **114**, 308, 1994.

Bhattacharyya, S., Das, T.K., Chakrabarti, B.: Phys. Rev. **A 88**, 053614 (2013); Kling, S., Pelster, A.: Phys. Rev. **A 76**, 023609 (2007); Dorfman, K.E., et al.: Phys. Rev. **A 83**, 033609, 2011.

Biedenharn, L.C. and Louck, J.D., **Angular Momentum in Quantum Physics**, Addison Wesley, Reading, Mass, 1981.

Biedenharn, L.C. and Louck, J.D., **The Racah-Wigner Algebra in Quantum Theory**, Addison Wesley, Reading, Mass, 1981.

Biswas, A., Chakrabarti, B., Das, T.K.: J. Chem. Phys. **133**, 104502, 2010.

Biswas, A., Das, T.K., Salasnich, L., Chakrabarti, B.: Phys. Rev. **A 82**, 043607, 2010.

Biswas, A., Chakrabarti, B., Das, T.K., Salasnich, L.: Phys. Rev. **A 84**, 043631, 2011.

Blinder, S.M., *On Green's functions, propagators, and Sturmians for the nonrelativistic Coulomb problem*, Int. J. Quantum Chem., Quantum Chem. Symp., **18** 293-307, 1984.

Blume, D., Greene, C.H.: J. Chem. Phys. **112**, 8053 2000.

Blume, D., Greene, C.H.: Phys. Rev. **A 63**, 063601 (2001); Bortolotti, D.C.E., et al.: Phys. Rev. Lett. **97**, 160402 (2006); Kalas, R.M., Blume, D.: Phys. Rev. **A 77**, 032703, 2008.

Boroun, G.R., Abdolmaleki, H.: Phys. Scr. **80**, 065003, 2009.

Bose, S.N., Z. Phys. **26**, 178, 1924.

Botero, J. and Greene, C.H., *Positronium negative ion: An adiabatic study using hyperspherical coordinates*, Phys. Rev. **A32**, 1249, 1985.

Bradley, C.C., Sackett, C.A., Tollett, J.J., Hulet, R.G.: Phys. Rev. Lett. **75**, 1687, 1995.

Bransden, B.H., Joachain, C.J.: **Physics of Atoms and Molecules**. Longman, New York, 1983 Lieb, E.H., Simon, B.: Adv. Math. **23**, 22, 1977.

Bransden, B.H., Noble, C.J., and Hewitt, R.N., *On the reduction of momentum space scattering equations to Fredholm form*, J. Phys. B: At., Mol., Opt. Phys., **26** (16) 2487-99, 1993.

Brandsen, B.H., and Joachim, C.J., **Physics of Atoms and Molecules; 2nd Ed.**, Prentace-Hall, 2003.

Brink, D.M. and Satchler, D.R., *Angular Momentum*, Clarendon, Oxford, 1968.

Burden, F.R., *Origins of Molecular Structure. 1. 3-body calculations from* $e_2^+e^-$ *to* H_2^+ *using a hyperspherical basis.*, J. Phys. **B16**, 2289, 1983.

Burov, V.V. et al., *Elastic and inelastic form factors of light nuclei in the method of hyperspherical functions*, J. Phys. **G7**, 137, 1981.

Caligiana. A., **Sturmian Orbitals in Quantum Chemistry**, Ph.D. thesis, University of Perugia, Italy, October, 2003.

Chakrabarti, B., Kundu, A., Das, T.K.: J. Phys. **B 38**, 2457, 2005

Chakrabarti, B., Das, T.K.: Molecular Phys. **107**, 1817, 2009.

Chakrabarti, B., Das, T.K.: Phys. Rev. **A 78**, 063608, 2008.

Chattopadhyay, R., Das, T.K., Mukherjee, P.K.: Phys. Scr. **54**, 601, 1996.

Chattopadhyay, R., Das, T.K.: Fiz. **A 5**, 11, 1996.

Chen, J. Cheng Y., and Ishihara, T., *Hydrogenic- and Sturmian-function expansions in three-body atomic problems*, Phys. Rev., **186** (1) 25-38, 1969.

Chiu, T.W., *Non-relativistic bound-state problems in momentum space*, J. Phys. A: Math. Gen, **19** (13) 2537-47, 1986.

Cinal, M., *Energy functionals in momentum space: exchange energy, quantum corrections, and the Kohn-Sham scheme*, Phys. Rev. A, **48** (3) 1893-902, 1993.

Clark, C.W. and Taylor, K.T., *The quadratic Zeeman effect in hydrogen Rydberg series: application of Sturmian functions*, J. Phys. B, **15** (8) 1175-93, 1982.

Cohen, L., *Generalized phase-space distribution functions*, J. Math. Phys., **7** 781-786, 1966.

Cohen, L. and Lee, C., *Correlation hole and physical properties: a model calculation*, Int. J. Quantum Chem., **29** (3) 407-24, 1986.

Coleman, A.J. and Yukalov, V.I., **Reduced Density Matrices: Coulson's Challenge**, Springer-Verlag, New York, 2000.

Clare, R.B. and Levinger, J.S., *Hypertriton and Hyperspherical Harmonics*, Phys. Rev. **C31**, 2303, 1985.

Chakrabarti, B., Kundu, A., Das, T.K.: J. Phys. **B 38**, 2457, 2005.

Chakrabarti, B., Das, T.K., Debnath, P.K.: Phys. Rev. **A 79** 053629, 2009.

Coleman, A.J., *The Symmetric Group Made Easy*, Adv. Quantum Chem. **4**, 83-108, 1968.

Coletti, C., **Struttura Atomica e Moleculare Come Rottura della Simetria Ipersferica**, PhD thesis. Chemistry Department, University of Perugia, 1998.

Condon, E.U. and Shortley, G.H., **Theory of Atomic Spectra**, Cambridge, 1953.

Cooper, F., Khare, A., Sukhatme, U.: Phys. Rep. **251**, 267, 1995.

Correggi, M., Dell'Antonio, G., Finco, D., Michelangeli, A., Teta, A., *A class of Hamiltonians for a three-particle fermionic system at unitarity* (arXiv:1505.04132v1 [math-ph] 15 May 2015).

Cotton, F.A., **Chemical Applications of Group Theory**, Interscience, 1963.

Coulson, C.A., *Momentum distribution in molecular systems. I. Single bond. III. Bonds of higher order*, Proc. Camb. Phys. Soc., **37** 55, p 74, 1941.

Coulson, C.A. and Duncanson, W.E., *Momentum distribution in molecular systems. II. C and C-H bond*, Proc. Camb. Phys. Soc., **37** 67, 1941.

Coxeter, H.S.M. et al. Editors, **Mathematical Expositions, Volume 21: The Collected Papers of Alfred Young, 1873-1940**, University of Toronto Press, Toronto, 1977.

Cutcosky, R.E., *Fadeev-Popov Zeros and Confinement Color in a Hyperspherical Gauge Model*, Phys. Rev. Letters, **51**, 538, 1983.

Cutcosky, R.E., *Harmonic Functions and Matrix Elements for Hyperspherical Quantum Field Models*, J. Math. Phys., **25**, 939, 1984.

Dalfovo, F., Giorgini, S., Pitaevskii, L.P., Stringari, S.: Rev. Mod. Phys. **71**, 463, 1999.

Danilin, B.V., et al.: Phys. Rev. **C 55**, R577, 1997.

Das, G.P., Ghosh, S.K., and Sahni, V.C., *On the correlation energy density functional in momentum space*, Solid State Commun, **65** (7) 719-21, 1988.

Das, T.K., Bhattacharyya, S.: Pramana (J. Phys.) **40**, 189, 1993.

Das, T.K., De, T.B.: Pramana (J. Phys.) **28**, 645, 1987.

Das, T.K., Chakrabarti, B., Canuto, S.: J. Chem. Phys. **134**, 164106, 2011.

Das, T.K., Chakrabarti, B.: Phys. Rev. **E 62**, 4347, 2000.

Das, T.K., Chattopadhyay, R., Mukherjee, P.K.: Phys. Rev. **A 50**, 3521, 1994.

Das, T.K., Coelho, H.T., Torreao, J.R.A.: Phys. Rev. **C 45**, 2640, 1992.

Das, T.K., De, T.B.: Pramana J. Phys. **28**, 645, 1987.

Das, T.K., Chakrabarti, B.: Phys. Rev. **A 70**, 063601, 2004.

Das, T.K., Canuto, S., Kundu, A., Chakrabarti, B.: Phys. Rev. **A 75**, 042705, 2007.

Das, T.K., Kundu, A., Canuto, S., Chakrabarti, B.: Phys. Lett. **A 373**, 258, 2009.

Das, T.K., et al.: Lett. al Nuo. Cim. **33**, 1, (1982); Coelho, H.T. et al.: Phys. Letts. **B 109**, 255, 1982; Das, T.K., et al.: Phys. Rev. C 26, R754, 1982; Phys. Rev. C 26, 697, 1982; Coelho, H.T., et al.: Phys. Rev. **C 28**, 1812 (1983); Robilotta, M.R., et al.: Phys. Rev. **C 31**, 646 (1985); De, T.B., Das, T.K.: Pramana J. Phys. **30**, 183, 1988; Ghosh, A.K., Das, T. K.: Pramana J. Phys. **36**, 313, 1991; Kievsky, A., et al.: Few-Body Syst. **45**, 115, 2009; Phys. Rev. **C 81**, 044003 (2010); Viviani, M., et al.: Few-Body Syst. 54, 885 2013. Phys. Rev. Lett. 111, 172302 (2013); Das, T.K., Coelho, H.T., Fabre de la Ripelle, M.: Phys. Rev. **C 26**, 2288, 1982; Fujita, J., Miyazawa, H.: Prog. Theor. Phys. **17**, 360, 1957.

Das, T.K., **Hyperspherical Harmonic Expansion Techniques: Applocation to Problems in Physics**, Springer, 2016.

Das, G.P., Ghosh, S.K., and Sahni, V.C., *On the correlation energy density functional in momentum space*, Solid State Commun, **65** (7) 719-21, 1988.

De, T.B., Das, T.K.: Phys. Rev. **C 36**, 402, 1987.

de Diego, R., Garrido, E., Fedorov, D.V., Jensen, A.S.: Nucl. Phys. **A 786**, 71, 2007.

de Tourreil, R., Sprung, D.W.L.: Nucl. Phys. **A 201**, 193, 1973.

Delaripelle, M.F., *A Bridge Between the Hyperspherical and Integro-Differential Approaches to Many-Body Bound States*, Few Body Systems, **1**, 181, 1986.

De Windt, L., Defranceschi, M., and Delhalle, J., *Variation-iteration method in momentum space: determination of Hartree-Fock atomic orbitals*, Int. J. Quantum Chem., **45** (6) 609-18, 1993.

Debnath, P.K., Chakrabarti, B., Das, T.K.: Int. J. Quantum Chem. **111**, 1333, 2011.

Defranceschi, M., Suard, M., and Berthier, G., *Epitome of theoretical chemistry in momentum space*, Folia Chim. Theor. Lat., **18** (2) 65-82, 1990.

Defranceschi, M., *Theoretical investigations of the momentum densities for molecular hydrogen*, Chem. Phys., **115** (3) 349-58, 1987.

Defranceschi, M. and Delhalle, J., *Momentum space calculations on the helium atom*, Eur. J. Phys., **11** (3) 172-8, 1990.

Delande, D. and Gay, J.C., *The hydrogen atom in a magnetic field. Spectrum from the Coulomb dynamical group approach*, J. Phys. B: At. Mol. Phys., **19** (6) L173-L178, 1986.

Deloff, A. and Law, J., *Sturmian expansion method for bound state problems*, Phys. Rev. C, **21** (5) 2048-53, 1980.

Delves, L. M.: Nucl. Phys. **9**, 391 (1959); **20**, 275, 1960.

Denteneer, P.J.H. and Van Haeringen, W., *The pseudopotential-density-functional method in momentum space: details and test cases*, J. Phys. C, **18** (21) 4127-42, 1985.

Dorr, M., Potvliege, R.M., and Shakeshaft, R., *Atomic hydrogen irradiated by a strong laser field: Sturmian basis calculations of rates for high-order multiphoton ionization, Raman scattering, and harmonic generation*, J. Opt. Soc. Am. B: Opt. Phys., **7** (4) 433-48, 1990.

Douglas, M., *Coulomb perturbation calculations in momentum space and application to quantum-electrodynamic hyperfine-structure corrections*, Phys. Rev. A, **11** (5) 1527-38, 1975.

Drake, G.W.F. and Goldman, S.P., *Relativistic Sturmian and finite basis set methods in atomic physics*, Adv. At. Mol. Phys., **25** 393-416, 1988.

Dube, L.J. and Broad, J.T., *Sturmian discretization. II. The off-shelf Coulomb wavefunction*, J. Phys. B: At., Mol., Opt. Phys., **23** (11) 1711-32, 1990.

Dube, Louis J. and Broad, J. T., *Sturmian discretization: the off-shell Coulomb wave function*, J. Phys. B: At., Mol. Opt. Phys., **22** (18) L503, 1989.

Duchon, C., Dumont-Lepage, M.C., and Gazeau, J.P., *On two Sturmian alternatives to the LCAO method for a many-center one-electron system*, J. Chem. Phys., **76** (1) 445-7, 1982.

Duchon, C., Dumont-Lepage, M.C., and Gazeau, J.P., *Sturmian methods for the many-fixed-centers Coulomb potential*, J. Phys. A: Math. Gen., **15** (4) 1227-41, 1982.

Duffy, P., Casida, M.E., Brion, C.E., and Chong, D.P., *Assessment of Gaussian-weighted angular resolution functions in the comparison of quantum-mechanically calculated electron momentum distributions with experiment* Chem. Phys., **159** (3) 347-63, 1992.

Duncanson, W.E., *Momentum distribution in molecular systems. IV. H molecule ion, H_2^+*, Proc. Camb. Phil. Soc., **37** 47, 1941.

Dutta, S.K., Das, T.K., Khan, M.A., Chakrabarti, B.: Few-Body Syst. **35**, 33 (2004); Dutta, S.K., Das, T.K., Khan, M.A., Chakrabarti, B.: Int. J. Mod. Phys. **E 13**, 811, 2004.

Dutta, S.K., Das, T.K., Khan, M.A., Chakrabarti, B.: J. Phys. **G: Nucl. Part. 29**, 2411, 2003.

Dzhibuti, R.I. and Tsiklauri, S.M., *Study of the Complete Breakup of Hypertritium on the Basis of Hyperspherical Functions*, Soviet J. of Nuclear Phys., **34**, 33, 1981.

Dziembowski, Z., Fabre de la Ripelle, M., Miller, G.A.: Phys. Rev. **C 53**, R2038, 1996.

Dzhibuti, R.I. and Tsiklauri, S.M., *Iterative Solutions of the Equations of the Method of Hyperspherical Functions in the Momentum Representation for Hypernuclei and Charmed Supernuclei*, Soviet J. of Nuclear Phys., **39**, 704, 1984.

Dzhibuti, R.I. et al., *Hyperspherical Basis for Continuum Spectrum*, Nuclear Phys., **A276**, 421, 1977.

Edmonds, A.R.: **Angular Momentum in Quantum Mechanics**. Princeton University Press, Princeton, 1957.

Efros, V.D., *Matrix Elements of n-n Forces in a Hyperspherical Basis*, Soviet J. of Nuclear Phys., **27**, 302, 1978.

Einstein, A., Sitzber. Kgl. Preuss. Akad. Wiss. p. 261 (1924); ibid p. 3, 1925.

Elminyawi, I., Levinger, J.S.: Phys. Rev. **C 28**, 82, 1983.

Epstein, P.S., Proc. Natl. Acad. Sci. (USA), **12** 637, 1926.

Englefield, M.J., **Group theory and the Coulomb problem**, Wiley-Interscience, New York, 1972.

Erdös, L., Schlein, B., Yau, H.-T.: Invent. Math. **167**, 515 (2007); Phys. Rev. **A 78**, 053627, 2008.

Erdös, L., Schlein, B., Yau, H.-T.: J. Am. Math. Soc. **22**, 1099 (2009); Ann. Math. **1722**, 291, 2010.

Erens, G., Visschers, J.L., Van Wageningen, R.: Ann. Phys. (N.Y.) **67**, 461, 1971.

Ershov, S.N., Vaagen, J.S., Zhukov, M.V.: Phys. Rev. **C 86**, 034331, 2012.

Eyre, D.and Miller, H.G., *Sturmian projection and an L2 discretization of three-body continuum effects*, Phys. Rev. C: Nucl. Phys., **32** (3) 727-37, 1985.

Eyre, D. and Miller, H.G., *Sturmian approximation of three-body continuum effects*, Phys. Lett. B, **153B** (1-2) 5-7, 1985.

Eyre, D. and Miller, H.G., *Sturmian expansion approximation to three-body scattering*, Phys. Lett. B, **129B** (1-2) 15-17, 1983.

Fabre de la Ripelle, M., Braun, M., Sofianos, S.A.: Prog. Theor. Phys. **98**, 1261, 1997.

Fabre de la Ripelle, M., Sofianos, S.A., Adam, R.M.: Ann. Phys. **316**, 107, 2005.

Fabre de la Ripelle, M.: Ann. Phys. (N.Y.) **147**, 281, 1983.

Fabre de la Ripelle, M.: Few-Body Syst. **1**, 181, 1986.

Fabre de la Ripelle, M., Fiedeldey, H., Sofianos, S.A.: Phys. Rev. **C 38**, 449, 1988.

Fabre de la Ripelle, M., Ballot, J.L., Navarro, J.: Phys. Lett. **B 143**, 19, 1984.

Fabre de la Ripelle, M., Fiedeldey, H., Sofianos, S.A.: Few-Body Syst. **Suppl. 2**, 493, 1987.

Fabre de la Ripelle, M., Fiedeldey, H., Sofianos, S.A.: Phys. Rev. **C 38**, 449, 1988.

Fabre de la Ripelle, M., Fiedeldey, H., Sofianos, S.A.: Few-Body Syst. **6**, 157, 1989.

Fang, K.K., *The Hyperspherical Model Applied to the Quark Model of Nucleons*, Bull. Amer. Phys. Soc. **22**, 69, 1977.

Fang, K.K., Levinger, J.S., Fabre de la Ripelle, M.: Phys. Rev. **C 17**, 24, 1978.

Fang, K.K. et al., *Trinucleon Photoeffect to Isospin 3/2 Using Coupled Hyperspherical Harmonics*, Phys. Rev. **C17**, 24, 1978.

Fang, K.K., *Calculation of* $\alpha + n + p \to \alpha + n + p$ *Elastic Scattering Using Hyperspherical Harmonics*, Phys. Rev. **C19**, 1637, 1979.

Fang, K.K., *Three-Body to Three-Body Elastic Scattering Using Hyperspherical Harmonics*, Phys. Rev. **C15**, 1204, 1977.

Fano, U., *Wave propagation and diffraction on a potential ridge*, Phys. Rev. **A22** 2660, 1980.

Fano, U., *Unified treatment of collisions*, Phys. Rev. **A24** 2402, 1981.

Fano, U., *Correlations of two excited electrons*, Rep. Prog. Phys. **46** 97, 1983.

Fano, U. and Rao, A.R.P., **Atomic Collisions and Spectra**, Academic Press, Orlando, Florida, 1986.

Feizi, H., A.A. Rajabi, and M.R. Shojaei, *A Simple Approach to Study the Isospin Effect in Mass Splitting of Three-Nucleon Systems Using Hyperspherical Functions*, Comm. Theor. Phys. **58**, 141-145, 2012.

Fernández Rico, J., Ramírez, G., López, R., and Fernández Alonso, J.I., Collect. Czech. Chem. Comm., **53** 2250, 1987.

Fernández Rico, J., López, R., Ema, I., and Ramírez, G., preprints, 1997.

Flores, J.C., *Kicked quantum rotator with dynamic disorder: a diffusive behavior in momentum space*, Phys. Rev. A, **44** (6) 3492-5, 1991.

Fock, V.A., Z. Phys., **98** 145, 1935.

Fock, V.A., *Hydrogen atoms and non-Euclidian geometry*, Kgl. Norske Videnskab Forh, **31** 138, 1958.

Fonseca, A.C. and Pena, M.T., *Rotational-invariant Sturmian-Faddeev ansatz for the solution of hydrogen molecular ion (H2+): a general approach to molecular three-body problems*, Phys. Rev. A: Gen. Phys., **38** (10) 4967-84, 1988.

Fonseca, A.C., *Four-body equations in momentum space*, Lect. Notes Phys., **273** (Models Methods Few-Body Phys.), p 161-200, 1987.

Frolov, A.M., Wardlaw, D.M.: Eur. Phys. Jour. **D 63**, 339 2011.

Gadre, S.R. and Bendale, R.D., *Maximization of atomic information-entropy sum in configuration and momentum spaces*, Int. J. Quantum Chem., **28** (2) 311-14, 1985.

Gadre, S.R. and Chakravorty, S., *The self-interaction correction to the local spin density model: effect on atomic momentum space properties*, Chem. Phys. Lett., **120** (1) 101-5, 1985.

Gallaher, D.F. and Wilets, L., *Coupled-state calculations of proton-hydrogen scattering in the Sturmian representation*, Phys. Rev., **169** (1) 139-49, 1968.

Garcilazo, H., Vijande, J., Valcarce, A.: J. Phys. G: Nucl. Part. Phys. **34**, 961 2007.

Gasaneo, G. et al., *Theory of hyperspherical Sturmians for three-body reactions*, J. Phys. Chem. A, **113**, 14573-14582, 2009.

Gasaneo, G. et al., *Multivariable hypergeometric solutions for three charged particles*, J. Phys. B **30**, L265-L271, 1997.

Gasaneo, G., and Mmacek, J.H., *Hyperspherical adiabatic eigenfunctions for zero-range potentials*, J. Phys. B. **35**, 2239-2250, 2002.

Gasaneo, G., and Colavecchia, F.D., *Two-body Coulomb wave functions as a kernel for three alternative integral transformations*, J. Phys. A **36**, 8443-8462, 2003.

Gattobigio, M., Kievsky, A., Viviani, M., Barletta, P.: Phys. Rev. **A 79**, 032513, 2009; Gattobigio, M., Kievsky, A., Viviani, M.: Phys. Rev. **C 83**, 024001, 2011; Deflorian, S., et al., EPJ Web Conf. **66**, 02025, 2014.

Gaunt, J.A., Phil. Trans. Roy. Soc. Lond., **228** 151, 1929; Proc. Roy. Soc., **A222** 513, 1929.

Gazeau, J.P. and Maquet, A., *A new approach to the two-particle Schrödinger bound state problem*, J. Chem. Phys., **73** (10) 5147-54, 1980.

Gazeau, J.P. and Maquet, A., *Bound states in a Yukawa potential: a Sturmian group theoretical approach*, Phys. Rev. A, **20** (3) 727-39, 1979.

Gerry, C.C., *Inner-shell bound-bound transitions from variationally scaled Sturmian functions*, Phys. Rev. A: Gen. Phys., **38** (7) 3764-5, 1988.

Geller, M., *Two-center Coulomb integrals*, J. Chem. Phys., **41** 4006, 1964.

Gerratt, J., *General theory of spin-coupled wave functions for atoms and molecules*, Adv. At. Mol. Phys. **7**, 141-221, 1971.

Ghosh, A.K., Das, T.K.: Phys. Rev. **C 42**, 1249, 1990.

Ghosh, S.K., *Quantum chemistry in phase space: some current trends*, Proc. Indian Acad. Sci., Chem. Sci., **99** (1-2) 21-8, 1987.

Girlanda, L., et al.: Phys. Rev. Lett. **105**, 232502, 2010.

Girlanda, L., et al.: J. Phys. Conf. **Series 527**, 012022, 2014; Piarulli, M., et al., (52 Int. Winter Meeting on Nuclear Physics, Bormio, 2014; Proc. of Sci. **(Bormio 2014)** 040, 2014.

Gloeckle W., *Few-body equations and their solutions in momentum space*, Lect. Notes Phys., **273** (Models Methods Few-Body Phys.), p 3-52, 1987.

Gogny, D., Pires, P., de Tourreil, R.: Phys. Lett. **B 32**, 591, 1970.

Goscinski, O., *Preliminary Research Report No. 217*, Quantum Chemistry Group, Uppsala University, 1968.

Goscinski, O. and Lindner, P., **Natural Spin Orbitals and Generalized Overlap Amplitudes, Preliminary Research Report No. 238**, Quantum Chemistry Group, Uppsala University, 1969.

Goscinski, O., Adv. Quantum Chem., **41** 51-85, 2003.

Gradshteyn, I.S. and Ryshik, I.M., **Tables of Integrals, Series and Products**, Academic Press, New York, 1965.

Grant, I.P., in **Relativistic Effects in Atoms and Molecules**, Wilson, S., Ed., Plenum Press, 1988.

Grant, I.P., in **Atomic, Molecular and Optical Physics Handbook**, Drake, G.W.F. Ed., Chapt 22, p 287, AIP Press, Woodbury New York, 1996.

Greene, C.H., Phys. Rev., **A23** 661, 1981.

Gross, E.K.U. and Dreizler, R.M. (eds.), *Density Functional Theory*, Plenum Press, New York, 1995.

Gruzdev, P.F., Soloveva, G.S. and Sherstyuk, A.I., *Calculation of neon and argon steady-state polarizabilities by the method of Hartree-Fock SCF Sturmian expansion*, Opt. Spektrosk., **63** (6) 1394-7, 1987.

Griffeth, J.S., *The Theory of Transition Metal Ions*, Cammbridge, 1964.

Gruber, B., Editor, **Symmetries in Science, VI: From the Rotation Group to Quantum Algebras**, Plenum, New York, 1993.

Haftel, M.I. and Mandelzweig, V.B., *A fast convergent hyperspherical expansion for the helium ground state*, Phys. Lett., **A120** 232, 1987.

Han, C.S., *Electron-atom scattering in an intense radiation field*, Phys. Rev. A: At., Mol., Opt. Phys., **51** (6) 4818-23, 1995.

Haldar, S.K., Chakrabarti, B., Das, T.K.: Phys. Rev. **A 82**, 043616, 2010.

Haldar, S.K., Chakrabarti, B., Das, T.K., Biswas, A.: Phys. Rev. **A 88**, 033603, 2013.

Hall, G.G., *Applied Group Theory*, Longmans, 1967.

Hammermesh, M., **Group Theory and Its Applications to Physical Problems**, Addison-Wesley, 1964.

Harris, D., and Bertolucci, M.D., **Symmetry and Spectroscopy: An Introduction to Vibrational and Electronic Spectroscopy**

Harris, F.E. and Michels, H.H., Adv. Chem. Phys. **13**, 205, 1967.

Hartt, K. and Yidana, P.V.A., *Analytic Sturmian functions and convergence of separable expansions*, Phys. Rev. C: Nucl. Phys., **36** (2) 475-84, 1987.

Heddle, D. P., Kwon, Y.R. and Tabakin, F., *Coulomb plus strong interaction bound states-momentum space numerical solutions*, Comput. Phys. Commun., **38** (1) 71-82, 1985.

Heine, V., *Group Theory in Quantum Mechanics*, Pergamon, 1960.

Henderson, G. A., *Variational theorems for the single-particle probability density and density matrix in momentum space*, Phys. Rev. A, **23** (1) 19-20, 1981.

Herrick, D.R., *Variable dimensionality in the group-theoretic prediction of configuration mixings for doubly-excited helium*, J. Math. Phys., **16** 1046, 1975.

Herrick, D.R., *New symmetry properties of atoms and molecules*, Adv. Chem. Phys., **52** 1, 1983.

Herschbach, D. R., *Dimensional interpolation for two-electron atoms*, J. Chem. Phys., **84** 838, 1986.

Herschbach, D. R., Avery, J., and Goscinski, O., Eds., **Dimensional Scaling in Chemical Physics**, Kluwer, Dordrecht, 1993.

Hillery, M., O'Connell, R.F., Scully, M.O., and Wigner, E.P., *Distribution functions in physics: Fundementals*, Physics Reports, **106** 121-167, 1984.

Hochstrasser, R.M., *Molecular Aspects of Symmetry*, Benjamin, 1966.

Hollas, J.M., *Symmetry in Molecules*, Chapman, 1972.

Holoeien, E. and Midtdal, J., *Variational nonrelativistic calculations for the 2pnp)1,3Pe states of two-electron atomic systems*, J. Phys. B, **4** (10) 1243-9, 1971.

Horacek, J. and Zejda, L., *Sturmian functions for nonlocal interactions*, Czech J. Phys., **43** (12) 1191-201, 1993.

Hosseinbor, A.P., et al.: *4D hyperspherical harmonic (HyperSPHARM) representation of multiple disconnected brain subcortical structures.* In: Mori, K., et al. (eds.) **MICCAI 2013, Part I**. LNCS, vol. 8149, pp. 598-605. Springer, Berlin, 2013.

Hua, L.K., **Harmonic Analysis of Functions of Several Complex Variables in the Classical Domains**, American Mathematical Society, Providence R.I., 1963.

Ihm, J., Zunger, A. and Cohen, M. L., *Momentum-space formalism for the total energy of solids*, J. Phys. C, **12** (21) 4409-22, 1979.

Ishikawa, Y., Rodriguez, W., and Alexander, S.A., *Solution of the integral Dirac equation in momentum space*, Int. J. Quantum Chem., Quantum Chem. Symp., **21** 417-23, 1987.

Ishikawa, Y., Rodriguez, W., Torres, S. and Alexander S.A., *Solving the Dirac equation in momentum space: a numerical study of hydrogen diatomic monopositive ion*, Chem. Phys. Lett., **143** (3) 289-92, 1988.

Islampour, R., Gharibi, M., and Maralinaghi, M., *The molecular Hamiltonian in Jacobi Coordinates*, Molecular Physics, 1879-1890, 2006.

Jain, A. and Winter, T.G., *Electron transfer, target excitation, and ionization in H+ + Na(3s) and H+ + Na(3p) collisions in the coupled-Sturmian-pseudostate approach*, Phys. Rev. A: At., Mol., Opt. Phys., **51** (4) 2963-73, 1995.

Jain, B.L., *A numerical study on the choice of basis sets used for translating ETOs in multi-center LCAO calculations*, ETO Multicent Mol Integr, Proc Int Conf, 1st, Reidel, Dordrecht, Neth, p 129-33, 81, Ed. Weatherford, Charles A., Jones, Herbert W., 1982.

Jaffé, H.H. and Ochrin M., *Symmetry in Chemistry*, Wily, 1965.

Jastrow, R.: Phys. Rev. **98**, 1479, 1955.

Johnson, B.R.: J. Chem. Phys. **69**, 4678, 1978.

Jolicard, G., and Billing, G.D., *Energy dependence study of vibrational inelastic collisions using the wave operator theory and an analysis of quantum flows in momentum space*, Chem. Phys., **149** (3) 261-73, 1991.

Judd, B.R., **Angular Momentum Theory for Diatomic Molecules**, Academic Press, New York, 1975.

Judd, B.R., **Operator Techniques in Atomic Spectroscopy**, McGraw-Hill, 1963.

Judd, B.R., **Second Quantization and Atomic Spectroscopy**, Johns Hopkins Press, 1967.

Kaijser, P. and Lindner, P., *Momentum distribution of diatomic molecules*, Philos. Mag., **31** (4) 871-82, 1975.

Kaijser, P. and Sabin, J.R., *A comparison between the LCAO-X.alpha. and Hartree-Fock wave functions for momentum space properties of ammonia*, J. Chem. Phys., **74** (1) 559-63, 1981.

Kanta, K.P., Das, T.K.: Pramana J. Phys. **39**, L405 (1992); **43**, 313, 1994.
Kaplan, I.G., **Symmetry of Many-Electron Systems**, Academic Press, 1975.
Karule, E. and Pratt, R.H., *Transformed Coulomb Green function Sturmian expansion*, J. Phys. B: At., Mol. Opt. Phys., **24** (7) 1585-91, 1991.
Kaschiev, M. and Shitikova, K.V., *Compressability of Nuclei in the Method of Hyperspherical Functions*, Soviet J. Nuclear Phys. **30**, 766, 1979.
Kellman, M.E. and Herrick, D.R., *Ro-vibrational collective interpretation of supermultiplet classifications of intrashell levels of two-electron atoms*, Phys. Rev. A, **22** 1536, 1980.
Kezerashvili, R.Y., Sov. J. Nucl. Phys. **38**, 293, 1983.
Khan, Md.A, Das, T.K.: Fiz. **B 9**, 55, 2000.
Khan, Md.A., Das, T.K.: Pramana (J. Phys.) **56**, 57, 2001; Abdul Khan, Md., Das, T.K.: Fiz. **B 10**, 83, 2001.
Khan, MdA, Das, T.K., Chakrabarti, B.: Int. J. Mod. Phys. **E 10**, 107, 2001.
Khan, Md.A., Dutta, S.K., Das, T.K.: Fizika **B 8**, 469, 1999.
Khan, M.A., Dutta, S.K., Das, T.K., Pal, M.K.: J. Phys. **G 24**, 1519, 1998; Abdul Khan, Md., Das, T.K.: Pramana (J. Phys.) **57**, 701, 2001.
Klar, H., J. Phys. B, **7** L436, 1974.
Klar, H. and Klar, M., *An accurate treatment of two-elecctron systems*, J. Phys. B, **13** 1057, 1980.
Klar, H., *Exact atomic wave functions - a generalized power-series expansion using hyperspherical coordinates*, J. Phys. A, **18** 1561, 1985.
Klarsfeld, S. and Maquet, A., *Analytic continuation of sturmian expansions for two-photon ionization*, Phys. Lett. A, **73A** (2) 100-2, 1979.
Klarsfeld, S. and Maquet, A., *Pade-Sturmian approach to multiphoton ionization in hydrogenlike atoms*, Phys. Lett. A, **78A** (1) 40-2, 1980.
Knox, R.S. and Gold, A., *Symmetry in the Solid State*, Benjamon, 1964.
Kievsky, A., et al.: Few-Body Syst. **22**, 1, 1997.
Kievsky, A., et al.: J. Phys. **G 35**, 063101 (2008); Marcucci, L.E., et al., Phys. Rev. **C 83**, 014002, 2011.
Kievsky, A., et al.: Phys. Rev. **C 81**, 044003 (2010); J. Phys. **G 35**, 063101, 2008.
Kil'dyushov, M.S., Sov. J .Nucl. Phys., **15** 113, 1972.
Kil'dyushov, M.S., and Kuznetsov, G.I., Sov. J. Nucl. Phys., **17** 1330, 1973.
King, H.F., Stanton, R.E., Kim, H., Wyatt, R.E., and Parr, R.G., J. Chem. Phys., **47** 1936, 1967.
Klepikov, N.P., Sov. J. Nucl. Phys. **19** 462, 1974.
Kievsky, A. et al., (in English), Few-Body Systems **22.1**, 1-10, 1997
Koga, T. and Murai, Takeshi, *Energy-density relations in momentum space. III. Variational aspect*, Theor. Chim. Acta, **65** (4) 311-16, 1984.
Koga, T., *Direct solution of the* $H1s) - H^+$ *long-range interactionproblem in momentum space*, J. Chem. Phys., **82** 2022, 1985.
Koga, T. and Matsumoto, S., *An exact solution of the interaction problem between two ground-state hydrogen atoms*, J. Chem. Phys., **82** 5127, 1985.
Koga, T. and Kawaai, R., *One-electron diatomics in momentum space. II. Second and third iterated LCAO solutions* J. Chem. Phys., **84** (10) 5651-4, 1986.

Koga, T. and Matsuhashi, T., *One-electron diatomics in momentum space. III. Nonvariational method for single-center expansion*, J. Chem. Phys., **87** (3) 1677-80, 1987.

Koga, T. and Matsuhashi, T., *Sum rules for nuclear attraction integrals over hydrogenic orbitals*, J. Chem. Phys., **87** (8) 4696-9, 1987.

Koga, T. and Matsuhashi, T., *One-electron diatomics in momentum space. V. Nonvariational LCAO approaaach*, J. Chem. Phys., **89** 983, 1988.

Koga, T.; Yamamoto, Y., and Matsuhashi, T., *One-electron diatomics in momentum space. IV. Floating single-center expansion*, J. Chem. Phys., **88** (10) 6675-6, 1988.

Koga, T. and Ougihara, T., *One-electron diatomics in momentum space. VI. Nonvariational approach to excited states*, J. Chem. Phys., **91** (2) 1092-5, 1989.

Koga, T., Horiguchi, T. and Ishikawa, Y., *One-electron diatomics in momentum space. VII. Nonvariational approach to ground and excited states of heteronuclear systems*, J. Chem. Phys., **95** (2) 1086-9, 1991.

Kolos, W. and Wolniewicz, L., J. Chem. Phys., **41** 3663, 1964.

Kolos, W. and Wolniewicz, L., J. Chem. Phys., **49** 404, 1968.

Kramer, P. J. and Chen, J.C.Y., *Faddeev equations for atomic problems. IV. Convergence of the separable-expansion method for low-energy positron-hydrogen problems*, Phys. Rev. A, **3** (2) 568-73, 1971.

Krause, J.L. and Berry, R.S., *Electron correlation in alkaline earth atoms*, Phys. Rev. A, **31** (5) 3502-4, 1985.

Kristoffel, N., *Statistics with arbitrary maximal allowed number of particles in the cell of the momentum space (methodical note)*, Eesti. Tead. Akad. Toim., Fuus., Mat., **41** (3) 207-10, 1992.

Krivec, R.: Few-Body Syst. **25**, 199, 1998.

Krivac, R., Haftel, M.I., Mandelzwig, V.B.: Phys. Rev. **A 47**, 911, 1993.

Kundu, A., Chakrabarti, B., Das, T.K., Canuto, S.: J. Phys. **B 40**, 2225, 2007.

Kupperman, A. and Hypes, P.G., *3-dimensional quantum mechanical reactive scattering using symmetrized hyperspherical coordinates*, J. Chem. Phys., **84** 5962, 1986.

Kuppermann, Aron, Israel Journal of Chemistry **43.3-4**, 229-241, 2003.

Kuznetsov, G.I. and Ya. Smorodinskii, Sov. J. Nucl. Phys. 25, p. 447, 1976.

Lakshmanan, M. and Hasegawa, H., *On the canonical equivalence of the Kepler problem in coordinate and momentum spaces*, J. Phys. A: Math Gen, **17** (16), 1984.

Lanczos, C., *An iteration method for the solution of the eigenvalue problem of linear differential and integral operators*, J. Res. Natl. Bur. Stand., **45** 255-82, 1950.

Landau, L.D., and Lifshitz, E.M., **Quantum Mechanics; Non-Relativistic Theory**, Pergamon Press, London, 1959.

Lassettre, E.N., *Momentum eigenfunctions in the complex momentum plane. V. Analytic behavior of the Schrödinger equation in the complex momentum plane. The Yukawa potential*, J. Chem. Phys., **82** (2) 827-40, 1985.

Lassettre, E.N., *Momentum eigenfunctions in the complex momentum plane. VI. A local potential function*, J. Chem. Phys., **83** (4) 1709-21, 1985.

Lee, Y.T., R.J. Gordon, and D.R. Herschbach, *Molecular Beam Kinetics: Reactions of H and D Atoms with Diatomic Alkali Molecules.* J. Chem. Phys. **54**, 2410, 1971.

Leech, J.W. and Newman, D.J., **How to Use Groups**, Methuen, 1969.

Lekala, M.L., et al.: Phys. Rev. **A 89**, 023624, 2014.

Lepetit, B., D. Wang, and A. Kuppermann, J. Chem. Phys. **125.13**, 133505, 2006.

Lewin, M., Nam, P.T., Rougerie, N.: Adv. Math. **254**, 570 (2014); Trans. Am. Math. Soc. in press. arXiv:1405.3220, 2014.

Lewin, M., Seiringer, R.: J. Stat. Phys. **137**, 1040, 2009.

Lieb, E.H.: *Many particle Coulomb systems*, presented at the 1976 session on statistical mechanics of the International Mathematical Summer Center (C.I.M.E.). Bressanone, Italy 21-27 June, 1976.

Lieb, E.H., Loss, M.: Adv. Theor. Math. Phys. **7**, 667 2003. arXiv:math-ph/0408001v1 31 July, 2004.

Lieb, E.H., Seiringer, R., Solovej, J.P., Yngvason, J.: *The mathematics of Bose gas and its condensation.* arXiv:0610117v1 [cond-mat.stat-mech] 4 Oct 2006. (Originally published by Birkhüser Verlag (Basel-Boston-Berlin) in 2005 under the title **The Mathematics of the Bose Gas and its Condensation** as number 34 of its Oberwolfach Seminar series) Lieb, E.H., Seiringer, R.: Phys. Rev. Lett. **88**, 170409, 2002.

Lieb, E.H., Seiringer, R., Yngvason, J.: Phys. Rev. **A 61**, 043602, 2000.

Lieb, E.H., Yngvason, J.: Phys. Rev. Lett. **80**, 2504, 1998.

Lieb, E.H., Seiringer, R., Yngvason, J.: Phys. Rev. Lett. **94**, 080401, 2005.

Lieb, E.H., Seiringer, R.: Commun. Math. Phys. **264**, 505, 2006.

Lieb, E.H., Seiringer, R., Yngvason, J.: Phys. Rev. **A 79**, 063626, 2009.

Liebig, Susanna, **Antisymmetrisation in a Jacobi coordinate based no-core shell model approach**, Dissertation zur Erlangung des Doktorgrades (Dr. rer. nat.) der Mathematisch-Naturwissenschaftlichen Fakultät der Rheinischen Friedrich-Wilhelms-Universität Bonn, 102 pages, Bonn, 3013.

Lin, C.D., *Analytical channel functions for 2-electron atoms in hyperspherical coordinates*, Phys. Rev. A, **23** 1585, 1981.

Lin, C.D., Phys. Rev. Lett. **35** 1150, 1975.

Lin, C.D., *Analytical channel functions for 2-electron atoms in hyperspherical coordinates*, Phys. Rev. A, **23** 1585, 1981.

Lin, C.D.: Phys. Rep. **257**, 1, 1995.

Lin, C.D.: Phys. Rev. **A 10**, 1986, 1974.

Liu, F.Q., Hou, X.J. and Lim, T.K., *Faddeev-Yakubovsky theory for four-body systems with three-body forces and its one-dimensional integral equations from the hyperspherical-harmonics expansion in momentum space*, Few-Body Syst., **4** (2) 89-101, 1988.

Liu, F.Q. and Lim, T.K., *The hyperspherical-harmonics expansion method and the integral-equation approach to solving the few-body problem in momentum space*, Few-Body Syst., **5** (1) 31-43, 1988.

Linderberg, J. and Öhrn, Y., *Kinetic energy functional in hyperspherical coordinates*, Int. J. Quantum Chem., **27** 273, 1985.

Lizengevich, A.I., *Momentum correlations in a system of interacting particles*, Ukr Fiz Zh (Russ Ed), **33** (10) 1588-91, 1988.

Longuit-Higgins, H.C., *The symmetry groups of non-rigid molecules*, Molecular Physics **6**, 445-460, 1963.

López, R., Ramírez, G., García de la Vega, J.M., and Fernández Rico, J., J. Chim. Phys., **84** 695, 1987.

Louck, J.D., *Generalized orbital angular momentum and the n-fold degenerate quantum mechanical oscillator*, J. Mol. Spectr., **4** 298, 1960.

Louck, J.D. and Galbraith, H.W., Rev. Mod. Phys., **44** 540, 1972.

Loebl, E.M., Editor, *Group Theory and Its Applications*, Academic Press, 1968.

Lomont, J.S., **Applications of Finite Group Theory**, Academic Press, 1959.

Loss, M., Weidl, T.: J. Eur. Math. Soc. **11**, 1365 (2009); Frank, R.L., Loss, M.: J. Math. Pure Appl. **97**, 39, 2012.

Löwdin, P.O., Phys. Rev. **97** 1474, 1955.

Löwdin, P.O., Appl. Phys. Suppl., **33** 251, 1962.

Macek, J.: J. Phys. **B 1**, 831, 1968.

Mahapatra, S., Das, T.K., Dutta, S.K.: Int. J. Mod. Phys. **E 18**, 1741, 2009.

Manakov, N.L., Rapoport, L.P. and Zapryagaev, S.A., *Sturmian expansions of the relativistic Coulomb Green function*, Phys. Lett. A, **43** (2) 139-40, 1973.

Mandelzweig, Victor B., Nuclear Physics **A 508**, 63-72, 1990.

Marcucci, L. E. et al., Phys. Rev. **C 80** (3), 034003. 2009.

Mason, J.K. and C.A, Schuh, Acta Materialia **56.20**, 6141-6155, 2008.

Maquet, Alfred, M., Philippe and Veniard, Valerie, *On the Coulomb Sturmian basis*, NATO ASI Ser, Ser C, **271** (Numer. Determ. Electron. Struct. At., Diat. Polyat. Mol.), 295-9, 1989.

Marcucci, L.E., et al.: Phys. Rev. **C 80**, 034003, 2009.

McCarthy, I.E. and Rossi, A.M., *Momentum-space calculation of electron-molecule scattering*, Phys. Rev. A: At., Mol., Opt. Phys., **49** (6) 4645-52, 1994.

McCarthy, I.E. and Stelbovics, A.T., *The momentum-space coupled-channels-optical method for electron-atom scattering*, Flinders Univ. South Aust., Inst. At. Stud., (Tech Rep) FIAS-R, (FIAS- R-111,) 51 pp., 1983.

McWeeny, R., and Coulson, C.A., *The computation of wave functions in momentum space. I. The helium atom*, Proc. Phys. Soc. (London) A, **62** 509, 1949.

McWeeny, R., *The computation of wave functions in momentum space. II. The hydrogen molecule ion*, Proc. Phys. Soc. (London) A, **62** 509, 1949.

McWeeny, R., **Methods of Molecular Quantum Mechanics**, (second edition), 1989.

McWeeny, R., **Symmetry**, Pergamon, 1963.

Meijer, P.H.E. and Bauer, E., **Group Theory**, North Holland, 1962.

Meremianin, A.V.: J. Math. Phys. **50**, 013526, 2009.

Michels, M.A.J., *Evaluation of angular integrals in rotational invariants*, Int. J. Quantum Chem. **20**, 951-952, 1981.

Mitnik D.M., et al.: Comp. Phys. Comm. **182**, 1145, 2011.

Monkhorst, H.J. and Harris, F.E., *Accurate calculation of Fourier transform of two-center Slater orbital products*, Int. J. Quantum Chem., **6** 601, 1972.

Monkhorst, H.J. and Jeziorski, B., *No linear dependence or many-center integral problems in momentum space quantum chemistry*, J. Chem. Phys., **71** (12) 5268-9, 1979.

Moore, C.E., **Atomic Energy Levels; Circular of the National Bureau of Standards 467**, Superintendent of Documents, U.S. Government Printing Office, Washington 25 D.C., 1949.

Morse P.M. and Feshbach, H., **Methods of Theoretical Physics**, McGraw Hill, 1953.

Motenberg, M., Bivens, R., Metropolis, N., and Wooten, J.K. Jr. *The 3-j and 6-j Symbols*, Technology Press, MIT, Cambridge Mass., 1959.

Murnagham, F.D. **The Theory og Group Representations**, Johns Hopkins Press, Baltimore, 1938.

Nag, R., Sanyal, S., Mukherjee, S.N.: Phys. Rev. **D 36**, 2788, 1987; Nag, R., Sanyal, S., Mukherjee, S.N.: Prog. Theor. Phys. **83**, 51, 1990; Nag, R., Mukherjee, S.N., Sanyal, S.: Phys. Rev. **C 44**, 1709, 1991; Nag, R., Sanyal, S.: J. Phys. **G 20**, L125, 1994; Sanyal, S., Nag, R.: J. Phys. **G 22**, L7, 1996; Sanyal, S., Nag, R.: Phys. Rev. **D 59**, 014027, 1998.

Nam, P.T., Rougerie, N., Seiringer, R.: *Ground states of large Bose systems: the Gross-Pitaevskii limit revisited.* arXiv:1503.07061v2 [math-ph] 7 April, 2015.

National Institute for Fusion Science (NIFS), *Atomic and Molecular Databases: Data for Autoionizing States*, http://dprose.nifs.ac.jp/DB/Auto/

National Institute of Standards and Technology (NIST), *NIST Atomic Spectra Database*, http://physics.nist.gov/asd.

National Institute of Standards and Technology (NIST), *CODATA Recommended Values of the Fundamental Physical Constants: 2002*, http://physics.nist.gov/cuu/Constants/

Navasa, J. and Tsoucaris, G., *Molecular wave functions in momentum space*, Phys. Rev. A, **24** 683, 1981.

Nguyen, N.B., Nunes, F.M., Thompson, I.J., Brown, E.F.: Phys. Rev. Lett. **109**, 141101, 2012.

Nikiforov, A.F., Suslov, S.K., and Uvarov, V.B., **Classical Orthogonal Polynomials of a Discrete Variable**, Springer-Verlag, Berlin, 1991.

Norbury, J.W., Maung, K.M. and Kahana, D.E., *Exact numerical solution of the spinless Salpeter equation for the Coulomb potential in momentum space*, Phys. Rev. A: At., Mol., Opt. Phys., **50** (5) 3609-13, 1994.

Novosadov, B.K., Opt. Spectrosc., **41** 490, 1976.

Novosadov, B.K., Int. J. Quantum Chem., **24** 1, 1983.

Oehm, W., Sofianos, S.A., Fiedeldey, H., Fabre de la Ripelle, M.: Phys. Rev. **C 43**, 25, 1991.

Oehm, W., Sofianos, S.A., Fiedeldey, H., Fabre de la Ripelle, M.: Phys. Rev. **C** **44**, 81, 1991.

Ojha, P.C., *The Jacobi-matrix method in parabolic coordinates: expansion of Coulomb functions in parabolic Sturmians*, J. Math. Phys. (N Y), **28** (2) 392-6, 1987.

Okhrimenko, I.P. and A.I. Steshenko, *Calculation of Collective Excitations of the Ne-20 Nucleus by the Method of Generalized Hyperspherical Functions with Use of the sp (2,R) Basis*, Sov. J. Nucl. Phys. **34**, 488, 1981.

Orchin, M. and Jaffé, H.H., *Symmetry, Orbitals and Spectra*, Wily, 1971.

Orgel, L.E., *An Introduction to Transition Metal Chemistry Ligand Field Theory*, Methuen, 1960.

Ovcharenko, V.I., I.P. Okhrimenko, and A.I. Steshenko, *Study of the Properties of Collective Excitations in O-18 and Ne-18 Using the Method of Generalized Hyperspherical Functions*. Sov. J. Nucl. Phys. **35**, 372, 1982.

Paldus, J., *Coupled Cluster Methods*, in Wilson, S., Editor, **Handbook of Molecular Physics and Quantum Chemistry**, Wiley, 2003.

Paldus, J., *Dynamical Groups*, in Drake, G.W.F. Editor, **Atomic, Molecular and Optical Physics Handbook**, American Institute of Physics, New York, 1996.

Paldus, J., and Gould, M.D., *Unitary Group Approach to Reduced Density Matrices, II*, Theor. Chim. Acta **86**, 83.96, 1993.

Park, I.H., Kim, H.J. and Kang, J.S., *Computer simulation of quantum mechanical scattering in coordinate and momentum space* Sae. Mulli., **26** (4) 155-67, 1986.

Parr, R.G., Yang, W.: **Density-Functional Theory of Atoms and Molecules**. Oxford University Press, Oxford (1989); Jones, R.O., Gunnarsson, O.: Rev. Mod. Phys. **61**, 689 (1989); Argaman N., Makov, G.: Am. J. Phys. **68**, 69 (2000); Frank, R.L., Lieb, E.H., Seiringer, R., Siedentop, H.: Phys. Rev. **A 76**, 052517, 2007.

Pathak, R.K., Kulkarni, S.A. and Gadre, S.R., *Momentum space atomic first-order density matrixes and "exchange-only" correlation factors*, Phys. Rev. A, **42** (5) 2622-6, 1990.

Pathak, Rajeev K., Panat, Padmakar V. and Gadre, S.R., *Local-density-functional model for atoms in momentum space*, Phys. Rev. A, **26** (6) 3073-7, 1982.

Pauncz, R., **The Symmetric Group in Quantum Chemistry**, CRC Press, Boca Raton, 1995.

Pekeris, C.L.: Phys. Rev. **112**, 1649, 1958.

Pekeris, C.L.: Phys. Rev. **126**, 1470, 1962.

Pethick, C.J., Smith, H.: **Bose-Einstein Condensation in Dilute Gases**. Cambridge University Press, Cambridge, 2002.

Petrochen, M.I. and Trifanov, E.D., *Group Theory in Quantum Mechanics*, Iliffe, 1969.

Pickl, P.: Rev. Math. Phys. **27**, 1550003, 2015.

Pinchon, D. and Hoggan, P.E., *Rotation matrices for real spherical harmonics: general rotations of atomic orbitals in space-fixed axes*, J. Phys. A. **40**, 1597-16100, 2007.

Pinchon, D. and Hoggan, P.E., *Gaussian approximation of exponential type orobitals based on B functions*, Int. J. Quantum Chem. **109**, 135-148, 2009.

Pitaevskii, L., Stringari, S.: **Bose-Einstein Condensation**. Oxford University Press, Oxford (2003); Dalfovo, F., Pitaevskii, L., Stringari, S.: Encyclopedia of Mathematical Physics, **vol. 1**, p. 312. Elsevier, Amsterdam, 2006.

Plante, D.R., Johnson, W.R., and Sapirstein, J., Phys. Rev. **A49** 3519, 1994.

Potvliege, R.M. and Shakeshaft, R., *Determination of the scattering matrix by use of the Sturmian representation of the wave function: choice of basis wave number*, J. Phys. B: At., Mol. Opt. Phys., **21** (21) L645, 1988.

Potvliege, R.M. and Smith, Philip H.G., *Stabilization of excited states and harmonic generation: Recent theoretical results in the Sturmian-Floquet approach*, NATO ASI Ser, Ser B, **316** (Super-Intense Laser-Atom Physics) 173-84, 1993.

Purcell, J.E., et al.: Nucl. Phys. **A 848**, 1, 2010.

Purwanto, W., Zhang, S.: Phys. Rev. **A 72**, 053610, 2005; Zezyulin, D.A., et al.: Phys. Rev. **A 78**, 013606, 2008; Song, J.L., Zhou, F.: Phys. Rev. Lett. **103**, 025302, 2009; Chakrabarti, B., Das, T.K., Debnath, P.K.: J. Low Temp. Phys. **157**, 527, 2009; Goswami, S., Das, T.K., Biswas, A.: Phys. Rev. **A 84**, 053617, 2011; Goswami, S., Das, T.K., Biswas, A.: J. Low Temp. Phys. **172**, 184, 2013; Haldar, S.K., et al.: Eur. Phys. J. **D 68**, 262, 2014.

Pyykkö, P., **Relativistic Theory of Atoms and Molecules. A Bibliography, 1916-1985**, Lecture Notes in Chemistry, **41**, 1986.

Pyykkö, P., Chem. Rev., **88** 563, 1988.

Rahman, N.K., *On the Sturmian representation of the Coulomb Green's function in perturbation calculation*, J. Chem. Phys., **67** (4) 1684-5, 1977.

Rawitscher, G.H. and Delic, G., *Sturmian representation of the optical model potential due to coupling to inelastic channels*, Phys. Rev. C, **29** (4) 1153-62, 1984.

Rawitscher, G. H. and Delic, G., *Solution of the scattering T matrix equation in discrete complex momentum space*, Phys. Rev. C, **29** (3) 747-54, 1984.

Raynal, J., Revai, J.: Nuovo Cimenti **A 68**, 612, 1970.

Read, F.H., *A new class of atomic states: The 'Wannier-ridge' resonances*, Aus. J. Phys. **35** 475-99, 1982.

Rettrup, S., **Lecture Notes on Quantum Chemistry**, Department of Chemistry, University of Copenhagen, 2003.

Rezaei, B., Boroun, G.R., Abdolmaleki, H.: Int. J. Math. Phys. Quan. Eng. **7**, 446, 2013.

Richards, J.-L., **The Non-Relativistic Three-Body Problem for Baryons** (99 pages), Physics Reports, **212**, 1992.

Richard, J.-M.: Phys. Rep. **212**, 1 (1992); Vassallo, A.: **Baryon structure with hypercentral constituent quark model**. Ph.D. Thesis, Universita degli Studi di Genova, 2004.

Roberts, J.L., et al.: Phys. Rev. Lett. **86**, 4211, 2001.

Robinson, G. de B., **Representation Theory of Symmetric Groups**, University of Toronto Press, Toronto, 1961.

Rodriguez, W. and Ishikawa, Y., *Fully numerical solutions of the Hartree-Fock equation in momentum space: a numerical study of the helium atom and the hydrogen diatomic monopositive ion*, Int. J. Quantum Chem., Quantum Chem. Symp., **22** 445-56, 1988.

Rodriguez, W. and Ishikawa, Y., *Fully numerical solutions of the molecular Schrödinger equation in momentum space*, Chem. Phys. Lett., **146** (6) 515-17, 1988.

Rohwedder, B. and Englert, B.G., *Semiclassical quantization in momentum space*, Phys. Rev. A: At., Mol., Opt. Phys., **49** (4) 2340-6, 1994.

Rose, M.E., *Elementary Theory of Angular Momentum*, Wiley, 1957.

Rotenberg, M., Ann. Phys. (New York), **19** 262, 1962.

Rotenberg, M., *Theory and application of Sturmian functions*, Adv. At. Mol. Phys., **6** 233-68, 1970.

Roy, R.R., Nigam, B.P.: **Nuclear Physics: Thory and Experiment**. John Wiley and Sons Inc., New York 1967.

Royer, A., *Wigner function as the expectation value of a parity operator*, Phys. Rev. A, **15** 449-450, 1977.

Rudin, W., **Fourier Analysis on Groups**, Interscience, New York, 1962.

Rutherford, D.E., **Substitutional Analysis**, Edinburgh University Press, Edinburgh, 1948.

Sadovoi, A.A., I.K. Averyanov, and A.I. Golubev, Sov. J. Nucl. Phys. **24**, p. 581, 1982.

Sanyal, S., Mukherjee, S.N.: Phys. Rev. **C 29**, 665 (1984); ibid **31**, 33 (1985); ibid **36**, 67 1987.

Schmider, H., Smith, V.H. Jr. and Weyrich, W., *On the inference of the one-particle density matrix from position and momentum-space form factors*, Z. Naturforsch., A: Phys. Sci., **48** (1-2) 211-20, 1993.

Schmider, H., Smith, V.H. Jr. and Weyrich, W., *Reconstruction of the one-particle density matrix from expectation values in position and momentum space*, J. Chem. Phys., **96** (12) 8986-94, 1992.

Schneider, T.R.: Phys. Lett. **B 30**, 439 1972.

Schönland D.S., *Molecular Symmetry*, Van Norstrand, 1965.

Schuch, D., *On a form of nonlinear dissipative wave mechanics valid in position- and momentum-space*, Int. J. Quantum Chem., Quantum Chem. Symp., **28** (Proceedings of the International Symposium on Atomic, Molecular, and Condensed Matter Theory and Computational Methods, 1994) 251-9, 1994.

Seiringer, R.: *Cold quantum gases and Bose-Einstein condensation.* In: Guiliani, A., Mastropietro, V., Yngvason, J. (eds.) **Quantum Many Body Systems. Lecture Notes in Mathematics, vol. 2051**, p. 5592. Springer, New York, 2012.

Seiringer, R.: Commun. Math. Phys. **229**, 491, 2002.

Seiringer, R.: J. Phys. **A 36**, 9755, 2003.

Seiringer, R.: *Dilute trapped Bose gases and Bose-Einstein condensation.* In: Derezinski, J., Siedentop, H. (eds.) **Large Coulomb Systems. Lecture Notes in Physics, vol. 695**, pp. 251-276. Springer, New York, 2006.

Shabaev, V.M., *Relativistic Coulomb Green function with regard to finite size of the nucleus*, Vestn. Leningr. Univ., Fiz., Khim. (2), p 92-6, 1984.

Shakeshaft, R. and Tang, X., *Determination of the scattering matrix by use of the Sturmian representation of the wave function*, Phys. Rev. A: Gen. Phys., **35** (9) 3945-8, 1987.

Shakeshaft, R., *A note on the Sturmian expansion of the Coulomb Green's function*. J. Phys. B: At. Mol. Phys., **18** (17) L611-L615, 1985.

Shakeshaft, R., *Application of the Sturmian expansion to multiphoton absorption: hydrogen above the ionization threshold*, Phys. Rev. A: Gen. Phys., **34** (6) 5119-22, 1986.

Shakeshaft, R., *Coupled-state calculations of proton-hydrogen-atom scattering with a Sturmian expansion*, Phys. Rev. A, **14** (5) 1626-33, 1976.

Shakeshaft, R., *Sturmian expansion of Green's function and its application to multiphoton ionization of hydrogen*, Phys. Rev. A: Gen. Phys., **34** (1) 244-52, 1986.

Shakeshaft, R., *Sturmian basis functions in the coupled state impact parameter method for hydrogen(+) + atomic hydrogen scattering*, J. Phys. B, **8** (7) 1114-28, 1975.

Shibuya, T. and Wulfman, C.E., *Molecular orbitals in momentum space, Proc. Roy. Soc. A*, **286**, 376, 1965.

Shulgina, N.B., Jonson, B., Zhukov, M.V.: Nucl. Phys. **A 825**, 175, 2009.

Shulgina, N.B., Danilin, B.V., Grigorenko, L.V., Zhukov, M.V., Bang, J.M.: Phys. Rev. **C 62**, 014312, 2000.

Shull, H. and Löwdin, P.-O., *Superposition of configurations and natural spin-orbitals. Applications to the He problem*, J. Chem. Phys., **30** 617, 1959

Siedentop, H. (ed.): **Complex Quantum Systems: Analysis of Large Coulomb Systems**. Lecture Notes Series, **vol. 24**, Institute of Mathematical Sciences, National University of Singapore. World Scientific Publishing, Singapore, 2013.

Simas, A.M., Thakkar, A.J. and Smith, V.H. Jr., *Momentum space properties of various orbital basis sets used in quantum chemical calculations*, Int. J. Quantum Chem., **21** (2) 419-29, 1982.

Simonov, Yu.A.: Sov. J. Nucl. Phys. **3**, 461, 1966; Simonov, Yu. A.: Sov. J. Nucl. Phys. **7**, 722, 1968.

Sinanoglu, O., and Bruekner, K., *Three Approaches to Electron Correlation in Atoms*, Yale University Press, 1970.

Slater, J.C., **Quantum Theory of Atomic Structure**, McGraw-Hill, 1960.

Sloan, I.H. and Gray, J.D., *Separable expansions of the t-matrix*, Phys. Lett. B, **44** (4) 354-6, 1973.

Sloan, Ian H., *Sturmian expansion of the Coulomb t matrix*, Phys. Rev. A, **7** (3) 1016-23, 1973.

Smirnov, Yu. F. and Shitikova, K.V., Sov. J. Part. Nucl., **8** 344, 1976.

Smith, F.T., *Generalized angular momentum in many-body collisions*, Phys. Rev., **120** 1058, 1960.

Smith, F.T., *A symmetric representation for three-body problems. I. Motion in a plane*, J. Math. Phys., **3** 735, 1962.

Smith, F.T., *Participation of vibration in exchange reactions*, J. Chem. Phys., **31** 1352-9, 1959.

Smith, F.T.: Phys. Rev. **120**, 1058, 1960.

Smith, V.H. Jr., *Density functional theory and local potential approximations from momentum space considerations*, in **Local Density Approximations Quantum Chemostry and Solid State Physics**, (Proc. Symp.), Plenum, New York, N. Y, p 1-19, 82, Eds. Dahl, J.P., Avery, J., 1984.

Smith, F.T., *Generalized angular momentum in many-body collisions*, Phys. Rev., **120** 1058, 1960.

Smith, F.T., *A symmetric representation for three-body problems. I. Motion in a plane*, J. Math. Phys., **3** 735, 1962.

Smith, F.T., *Participation of vibration in exchange reactions*, J. Chem. Phys., **31** 1352-9, 1959.

Smorodinskii, Ya., and Efros, V.D., Sov. J. Nucl. Phys. **17** 210, 1973.

Sofianos, S.A., et al.: Phys. Rev. **A 87**, 013608, 2013.

Sofianos, S.A., et al.: Few-Body Syst. **54**, 1529, 2013.

Sofianos, S.A., Adam, R.M., Belyaev, V.B.: Phys. Rev. **C 84**, 064304, 2011.

Solovej, J.P.: In: Francoise, J.-P., Naber, G.L., Oxford, T.S.T. (eds.) **Encyclopedia of Mathematical Physics, vol. 5**, pp. 8-14. Elsevier 2006. Seiringer, R., Solovej, J.P.: Bull. AMS, **50**, 169, 2013.

Sorensen, O., et al.: Phys. Rev. **A 65**, 051601 (2002); Sorensen, O., Federov, D.F., Jensen, A.S.: Phys. Rev. **66**, 032507 (2002); Phys. Rev. Lett. **89**, 173002 (2002); Biswas, A., Das, T.K.: J. Phys. **B 41**, 231001 2008.

Springborg, M. and Dahl, J.P., *Wigner's phase-space function and atomic structure*, Phys. Rev. A, **36** 1050-62, 1987.

Stringari, S.: Phys. Lett. **A 347**, 150 2005.

Sugano, S., Tanabe, Y. and Kamimura, H., **Multiplets of Transition Metal Ions in Crystals**, Academic Press, 1960.

Sultanov, R.A., Guster, D.: J. Comput. Phys. **192**, 231, 2003.

Suno, H., Esry, B.D.: Phys. Rev. **A 78**, 062701, 2008.

Suzuki, H., **Electronic Absorbtion Spectra and Geometry of Organic Molecules**, Academiic Press, 1967.

Szmytkowski, R., *The Dirac-Coulomb Sturmians and the series expansion of the Dirac-Coulomb Green function; Application to the relativistic polarizability of the hydrogenlike atom*, J. Phys. A, **31** 4963, 1998.

Szmytkowski, R., *The continuum Schrödinger-Coulomb and Dirac-Coulomb Sturmian functions*, J. Phys. A, **31** 4963, 1998.

Szmytkowski, R., *The continuum Schrödinger-Coulomb and Dirac-Coulomb Sturmian functions*, J. Phys. A, **31** 4963, 1998.

Taieb, R., Veniard, V., Maquet, A., Vucic S. and Potvliege R.M., *Light polarization effects in laser-assisted electron-impact-ionization (e,2e) collisions: a Sturmian approach*, J. Phys. B: At., Mol. Opt. Phys., **24** (14) 3229-40, 1991.

Tang, J.Z., Watanabe, S., Matsuzawa, M.: Phys. Rev. **A 46**, 2437, 1992.

Tang, X. and Shakeshaft, R., *A note on the solution of the Schrödinger equation in momentum space*, Z. Phys. D: At., Mol. Clusters, **6** (2) 113-17, 1987.

Tarter, C.B., J. Math. Phys., **11** 3192, 1970.

Tel-nov, D.A., *The d.c. Stark effect in a hydrogen atom via Sturmian expansions*, J. Phys. B: At., Mol. Opt. Phys., **22** (14) L399-L404, 1989.

Thakkar, A.J. and Koga, T., *Analytic approximations to the momentum moments of neutral atoms*, Int. J. Quantum Chem., Quantum Chem. Symp., **26** (Proc. Int. Symp. At., Mol., Condens. Matter Theory Comput. Methods, 1992) 291-8, 1992.

Thakkar, A.J. and Tatewaki, H., *Momentum-space properties of nitrogen: improved configuration-interaction calculations*, Phys. Rev. A, **42** (3) 1336-45, 1990.

Timofeyuk, N.K.: Phys. Rev. **A 86**, 032507 2012.

Tinkham, M., **Group Theory and Quantum Mechanics**, McGraw-Hill, 1964.

Tiwari, R.P., Shukla, A.: Comput. Phys. Commun. **174**, 966, 2006.

Tzara, C., *A study of the relativistic Coulomb problem in momentum space*, Phys. Lett. A, **111A** (7) 343-8, 1985.

Ugalde, J.M., *Exchange-correlation effects in momentum space for atoms: an analysis of the isoelectronic series of lithium 2S and beryllium 1S*, J. Phys. B: At. Mol. Phys., **20** (10) 2153-63, 1987.

Urch, D.S., **Orbitals and Symmetry**, Penguin, 1970.

Vallieres, M., Coelho, H.T., Das, T.K.: Nucl. Phys. **A 271**, 95 (1976); Coelho, H. T., Das, T.K., Vallieres, M.: Rev. Bras. Fis. **7**, 237, 1977.

Valliéres, M., Das, T.K., and Coelho, H.T., *Generalized N-body Harmonic Oscillator Basis Functions for the Hyperspherical Approach*, Nuclear Physics **A257**, 389-396, 1976.

Vallieres, M., Coelho, H.T., Das, T.K.: Nucl. Phys. A 271, 95 (1976); Consoni, L., Das, T.K., Coelho, H.T.: Rev. Bras. Fis. 13, 137, 1983.

Van Haeringen, H. and Kok, L.P., *Inequalities for and zeros of the Coulomb T matrix in momentum space*, Few Body Probl Phys, Proc Int IUPAP Conf, 10th, North-Holland, Amsterdam, Neth, p 361-2, 83, Ed. Zeitnitz, Bernhard, 1984.

Verma, S.P., Lall, N.R.: Indian J. Pure Appl. Phys. **48**, 851, 2010.

Vijande, J., Barnea, N., Valearce, A.: Nucl. Phys. **A 790**, 542c, 2007.

Vilenkin, N.K., **Special Functions and the Theory of Group Representations**, American Mathematical Society, Providence, R.I., 1968.

Vilenkin, N. Ya., Kuznetsov, G.I., and Smorodinskii, Ya.A., Sov. J. Nucl. Phys., **2** 645, 1966.

Viviani, M., et al.: **arXiv:1109**.3625 [nucl-th]. 2011; Kovalchuk, V.I., Kozlovskii, I.V.: Phys. Part. Nucl. 43, 294, 2012; Marcucci, L.E., et al.: Phys. Rev. **C 80**, 034003, 2009; Kievsky, A., et al.: Few-Body Syst. **54**, 2395, 2013.

Vladimirov, Yu.S. and Kislov, V.V., *Charge of the nucleus of a hydrogen-like atom as an eigenvalue of a 6-dimensional wave equation in momentum space*, Izv. Vyssh. Uchebn. Zaved., Fiz., **28** (4) 66-9, 1985.

Wang, D., Kuppermann, A.: J. Phys. Chem. **A 113**, 15384 (2009); Kuppermann, A.: Phys. Chem. Chem. Phys. **13**, 8259, 2011.

Watanabe, H., **Operator Methods in Ligand Field Theory**, Prentace Hall, 1966.

Weatherford, C.A., *Scaled hydrogenic Sturmians as ETOs*, ETO Multicent. Mol. Integr., Proc. Int. Conf., 1st, Reidel, Dordrecht, Neth., p 29-34, 81, Ed. Weatherford, Charles A., Jones, Herbert W., 1982.

Weatherford, C.A., and Jones, H.W., Editors, **International Conference on ETO Multicenter Integrals**, Reidel, Dordrecht, 1982.

Wen, Z.-Y. and Avery, J., *Some properties of hyperspherical harmonics*, J. Math. Phys., **26** 396, 1985.

Weniger, E.J., *Weakly convergent expansions of a plane wave and their use in Fourier integrals*, J. Math. Phys., **26** 276, 1985.

Weniger, E.J. and Steinborn, E.O., *The Fourier transforms of some exponential-type basis functions and their relevance for multicenter problems*, J. Chem. Phys., **78** 6121, 1983.

Weniger, E.J.; Grotendorst, J., and Steinborn, E.O., *Unified analytical treatment of overlap, two-center nuclear attraction, and Coulomb integrals of B functions via the Fourier transform method*, Phys. Rev. A, **33** 3688, 1986.

Weyl, H., **The Theory of Groups in Quantum Mechanics**, Dover, 1950.

Weyl, H., **The Classical Groups, Their Invariants and Representations**, Princeton University Press, 1939.

Whitten, R.C. and Sims, J.S., Phys. Rev. A, **9** 1586, 1974.

Wigner, E.P., **Group Theory and Its Application to the Quantum Mechanics of Atomic Spectra**, Academic Press, 1959.

Wilson, E.B., **Molecular Vibrations: The Theory of Infrared and Raman Spectroscopy**.

Wilson, S., Editor, **Handbook of Molecular Physics and Quantum Chemistry**, Wiley, 2003.

Windt, L. de, Fischer, P., Defranceschi, M., Delhalle, J. and Fripiat, J. G., *A combined analytical and numerical strategy to solve the atomic Hartree-Fock equations in momentum space*, J. Comput. Phys., **111** (2) 266-74, 1994.

Winter, T.G. and Alston, S.G., *Coupled-Sturmian and perturbative treatments of electron transfer and ionization in high-energy helium p-He+ collisions*, Phys. Rev. A, **45** (3) 1562-8, 1992.

Winter, T.G., *Electron transfer and ionization in collisions between protons and the ions lithium(2+) and helium(1+) studied with the use of a Sturmian basis*, Phys. Rev. A: Gen. Phys., **33** (6) 3842-52, 1986.

Winter, T.G., *Coupled-Sturmian treatment of electron transfer and ionization in proton-neon collisions*, Phys. Rev. A, **48** (5) 3706-13, 1993.

Winter, T.G., *Electron transfer and ionization in collisions between protons and the ions helium(1+), lithium(2+), beryllium(3+), boron(4+), and carbon(5+) studied with the use of a Sturmian basis*, Phys. Rev. A: Gen. Phys., **35** (9) 3799-809, 1987.

Winter, T.G., *Electron transfer in p-helium(1+) ion and helium(2+) ion-atomic helium collisions using a Sturmian basis*, Phys. Rev. A, **25** (2) 697-712, 1982.

Winter, T.G., *Sturmian treatment of excitation and ionization in high-energy proton-helium collisions*, Phys. Rev. A, **43** (9) 4727-35, 1991.

Winter, T.G., *Coupled-Sturmian treatment of electron transfer and ionization in proton-carbon collisions*, Phys. Rev. A, **47** (1) 264-72, 1993.

Winter, T.G., *Electron transfer and ionization in proton-helium collisions studied using a Sturmian basis*, Phys. Rev. A, **44** (7) 4353-67, 1991.

Wolfram Research, *Mathematica 11.1*, Wolfram Research Inc., Champaign, Illinois, 1991–2017.

Wulfman, C.E., *Semiquantitative united-atom treatment and the shape of triatomic molecules*, J. Chem. Phys. **31** 381, 1959.

Wulfman, C.E., *Dynamical groups in atomic and molecular physics*, in **Group Theory and its Applications**, Loebel, E.M. Ed., Academic Press, 1971.

Wulfman, C.E., *Approximate dynamical symmetry of two-electron atoms*, Chem. Phys. Letters **23** (3), 1973.

Wulfman, C.E., *On the space of eigenvectors in molecular quantum mechanics*, Int. J. Quantum Chem. **49** 185, 1994.

Wulfman, C.E., **Dynamical Symmetry**, World Scientific, 2011.

Yalcin, Z., Simsek, M.: Int. J. Quan. Chem. **88**, 735 2002.

Yngvason, J.: *Topics in the mathematical physics of cold Bose gases*, presented at the 5th Warsaw school of statistical physics in June 2013. arXiv:1402.0706v2 [cond-mat.quant-gas] 14 Feb 2014.

Yurtsever, E., Yilmaz, O., and Shillady, D.D., *Sturmian basis matrix solution of vibrational potentials.* Chem. Phys. Lett., **85** (1) 111-16, 1982.

Yurtsever, E., *Franck-Condon integrals over a sturmian basis. An application to photoelectron spectra of molecular hydrogen and molecular nitrogen*, Chem. Phys. Lett., **91** (1) 21-6, 1982.

Zernike, F., Brinkman, H.C.: Proc. Kon. Ned. Akad. Wettensch. **38**, 161, 1935.

Zhang, R., Deng, C.: Phys. Rev. **A 47**, 71, 1993.

Zhukov, M.V., et al.: Phys. Rep. **231**, 151, 1993.

Zofka, J. and M. Sotana, *Hyperspherical Model for Hyper-Nuclei Including Lambda-nn Interactions*, Nukleonika **25**, 609, 1980.

Index

d-dimensional Laplacian operator, 49

Action principle, 155
Alternative Coulomb Sturmians, 106
Alternative hyperspherical harmonics, 103
Alternative plane wave expansion, 50
Alternative Sturmians, 110
Angular coordinates, 12
Angular integration, 142
Antisymmetric functions, 85
Approximate Schrödinger equation, 200
Approximate separability, 199
Aquilanti, V., 103
Associated Legendre polynomials, 30, 214, 227
Atomic calculations, 91
Atomic mass, 203
Atomic units, 81, 83, 101, 203
Autoionizing states, 98
Automatic selection of basis sets, 88

Basis functions, 95
Beltrami, Eugenio, 178
Bessel transformed radial functions, 229
Binding energies, 216
Binding energy, 98
Binomial coefficients, 98
Boundary conditions, 170
Box normalization, 170

Calculus of variations, 153
Canonical decomposition, 3, 5, 12
Canonically conjugate momentum, 164
Carbonlike isoelectronic series, 98
Cartesian coordinates, 1, 156, 174
Center of mass coordinates, 157
Chain of point masses, 165
Chain rule, 2
Chains of subgroups, 17, 18, 34
Christoffel symbol, 183
Clusters, melting, 196
Coletti, C., 103
Collection of harmonic oscillators, 172, 225
Commutation relations, 223
Complete set of orthonormal functions, 50
Completeness, 85, 87, 88
Completeness of Coulomb Sturmians, 73
Completeness relation, 52
Confluent hypergeometric function, 68, 91
Conjugate function, 95
Conjugate momenta, 157
Conservation laws, 157
Conservative system, 161
Continuous media, 167

Continuous systems, 167
Contraction of tensors, 175
Contravariant metric tensor, 174, 179, 198, 200
Contravariant tensors, 174
Convergence, 73
Coulomb interaction operator, 140
Coulomb Sturmian basis sets, 216
Coulomb Sturmians, 82, 103, 104, 128
Coulomb Sturmians in d dimensions, 71
Coulomb Sturmians in direct space, 76
Coulomb Sturmians in reciprocal space, 76
Coupling coefficients, 131, 230
Coupling coefficients for hyperspherical harmonics, 130
Covariant metric tensor, 174, 178, 198, 200
Covariant tensors, 174
Creation and annihilation operators, 172, 222, 225
Curvature, intrinsic, 182
Curved spaces, 183
Curvilinear coordinate transformation, 177
Cyclic coordinates, 156

Das, T.K., 216
Degeneracy, 96
Degeneracy of hyperspherical harmonics, 18, 29
Degenerate configurations, 96
Densities, 1-center, 137, 141, 143
Derivation of the secular equations, 93
Determination of K_λ, 29
Disjoint sets of coordinates, 8, 9
Distributions, 88
Dominant nuclear attraction, 96
Doubling the frequency, 230
Doubly-excited states, 98

Effective charge, 92
Einstein convention, 174, 178
Electromagnetic fields, 167
Electromagnetic potentials, 169
Electron exchange, 85
Elliptic Coulomb Sturmians, 112
Energy-independent matrix, 95
Equilibrium position, 161, 196
Equilibrium positions, 196
Euclidean space, 1, 173
Euler, Leonhard, 155
Euler-Lagrange equations, 155, 159, 177, 183
Excited states, 96
Expansions of a plane wave, 46
Experimental values, 98, 100
Exponential-type orbitals, 90

Factorable potential energy, 200
Fermat's principle, 153, 158
Fermat, Pierre de, 153
Few-electron atoms, 96
Fields, 153, 167
First-iterated solution, 87
Fluorinelike isoelectronic series, 99
Fock's original treatment, 55
Fock's relationship, 129
Fock's relationships, 131, 132, 140
Fock's transformation, 57
Fock's treatment of hydrogenlike atoms, 18, 55
Fock's treatment, generalized, 60
Force constant, 196
Force constant of a spring, 166, 196
Four-dimensional hyperspherical harmonics, 130, 140
Fourier coefficients, 170
Fourier convolution theorem, 52, 84, 86
Fourier transformed harmonic oscillator, 229
Fourier transformed hydrogenlike orbitals, 58
Fourier transformed orbitals, 18
Fourier transforms, 84, 87, 106, 140
Fourier transforms in d dimensions, 45

Fourier transforms of Coulomb Sturmians, 76, 137
Fourier transforms, preservation of symmetry, 50
Free electromagnetic field, 172
Frequency spectrum, 165

Gaussian basis functions, 90
Gegenbauer polynomials, 25, 48
Gegenbauer polynomials, properties, 27
Gegenbauer polynomials, table, 26
General canonical decomposition, 8
General hyperangular integration theorem, 15
General relativity, 183
Generalized angular integral, 17
Generalized angular momentum operator, 11
Generalized coordinates, 158
Generalized Laplace equation, 3
Generalized Laplacian operator, 1, 9, 11, 31, 49, 60, 67
Generalized solid angle element, 13
Generalized solid angle, total, 13
Generalized Sturmian method, 96
Generalized Sturmian secular equations, 93, 95
Generalized Sturmians, 81, 83, 84, 88, 89
Generating function, 25, 29
Geodesics, 183
Geometrical optics, 153, 158, 184
Goscinski, O., 81, 91
Goscinskian configurations, 82, 92, 93, 216
Green's first identity, 15
Green's function, 87
Green's function iteration, 88
Green's function of Δ, 48
Green's function, alternative form, 87
Ground states, 100

Hamilton's characteristic function, 159, 183
Hamilton's eikonal equation, 159
Hamilton's eikonal function, 158, 183
Hamilton's equations of motion, 160, 161
Hamilton, Sir William Rowan, 154
Hamilton-Jacobi equation, 161
Hamiltonian, 95, 154, 160
Hamiltonian density, 169
Harmonic functions, 3, 8
Harmonic normal modes, 170
Harmonic oscillator in d dimensions, 221, 226, 227
Harmonic oscillator Lagrangians, 166, 171
Harmonic oscillator wave functions, 198
Harmonic oscillator, d-dimensional, 226
Harmonic polynomial of highest order, 7, 31
Harmonic polynomials, 3, 11
Harmonic projection, 7
Hartrees, 101
Heliumlike isoelectronic series, 98, 100, 216
Hermite polynomials, 221
Hermite, Charles, 221
Hermitian matrix, 110
Hermitian operator, 14
Hermiticity, 15, 75, 82
Highly-excited states, 98
Hilbert space, 13, 29
Hilbert subspace, 105
Homogeneous functions, 2–4, 8
Homogeneous functions, properties, 3
Homogeneous polynomial, 1, 12, 14, 31
Homogeneous polynomials, 5
Homogeneous wave equation, 170
Hua, 26
Huygens' principle, 183
Huygens, Christian, 183
Hybridization, 96
Hydrogenlike orbitals in d dimensions, 67, 70
Hydrogenlike spin-orbitals, 91, 93
Hydrogenlike wave equation, 92, 104

Hyperangles, 12
Hyperangular integration, 14, 130
Hyperangular integration theorem, general, 15
Hyperangular overlap integral, 122
Hypergeometric functions, 91
Hyperradius, 1, 11, 12, 32, 67
Hyperspheres, 183
Hyperspherical Bessel functions, 47, 230
Hyperspherical Bessel transforms, 49, 229
Hyperspherical coordinates, 178
Hyperspherical harmonics, 12, 17, 18, 48

Index of refraction, 184
Infinitesimal element of length, 182
Infinitesimal length, 174
Inner product, 84
Integral equation, 84, 86
Integration by parts, 168
Interelectron repulsion, 90, 93
Interelectron repulsion integrals, 90, 140, 141, 143, 150
Interelectron repulsion matrix, 95, 97
Intrinsic curvature, 182
Invariance, 167
Invariance under Fourier transforms, 129
Invariance under rotation, 27
Irreducible representations, 51, 88
Isoelectronic series, 98
Isoenergetic basis sets, 81, 88, 91
Isoenergetic solutions, 74
Isotropic harmonic oscillator, 226
Iteration, 86
Iteration using a basis, 88

Jacobi, Carl Gustav, 176
Jacobian, 176
Jacobians, 178

Kepler's second law, 157
Kinetic energy density, 167
Kinetic energy matrix, 134
Kinetic energy operator, 92
Kinetic energy term, 83
Koga, T., 121
Kronecker delta, 174

Löwdin, P.O., 71
Lagrange, Joseph-Louis, 154
Lagrangian densities, 167
Lagrangian density, 167
Lagrangians, 153
Laguerre polynomials, 203
Laplace-Beltrami operator, 177, 198, 200
Large-Z approximation, 96, 98
Least action, 155
Legendre polynomials, 25
Levi-Civita tensor, 176
Levi-Civita, Tulio, 176
Light rays, 184
Linear combination of Coulomb Sturmians, 128
Liquids, 196
Lithium-like isoelectronic series, 100
Local minima, 199, 200
Lorentz invariance, 167, 169
Lorentz transformation, 169

Major quantum numbers, 81
Many center Sturmians, 128
Many-particle problems, 81
Mass-weighted coordinates, 163, 196, 226
Mass-weighted potential energy matrix, 163
Matsuhashi, T., 121
Melting of a cluster, 196
Melting of a molecule, 196
Method of Trees, 43
Metric tensors, 173
Minimal geodesics, 183
Minor indices, 81
Mixed tensors, 174
Molecular integrals, 137
Molecular ions, 89
Molecular orbitals, 123, 128
Molecular spectra, 89, 90

Molecular vibrations and rotations, 166
Molecules, 89
Molecules, melting, 196
Momentum-space, 57
Momentum-space orthonormality, 84, 106
Momentum-space Schrödinger equation, 86
Momentum-space wave equation, 122
Monomials, 1
Morse oscillator bound states, 203
Morse oscillator. exact solutions, 202
Morse potential, 201
Motion of the nucleus, 93

N-electron Schrödinger equation, 95
Neon-like isoelectronic series, 99
Newtonian mechanics, 154
NIST tables, 100
Nitrogen-like isoelectronic series, 99
Non-melted state, 196
Non-relativistic case, 100
Non-relativistic wave equation, 93
Non-standard trees, 37
Non-uniform medium, 158
Non-weighted orthonormality, 87
Non-zero frequencies, 201
Nonhomogeneous polynomials, 17
Normal coordinates, 164, 166
Normal mode coordinates, 200, 201
Normal modes, 161, 166, 167, 170, 196
Normalization, 64, 82, 89, 93, 94, 227
Normalization constant, 221
Normalization of hyperspherical harmonics, 35
Normalized lowering operator, 224
Normalized raising operator, 224
Normalizing constant, 63
Nuclear attraction, 90, 93
Nuclear attraction integrals, 135
Nuclear attraction matrix, 95, 134
Nuclear attraction matrix diagonal and energy-independent, 95
Nuclear attraction potential, 91
Nuclear charge, 98
Nuclear charge, weighted, 91
Nuclear positions, 128, 138
Number operator, 226

Optimal basis set, 95
Orthogonality, 14
Orthonormal set of harmonics, 33
Orthonormality, 28, 70, 92, 130
Orthonormality conditions, 17
Orthonormality relation, 28, 228–230
Orthonormality relations, 172
Overcompleteness, 147
Overlap integrals, 129
Overlap matrix, 147
Oxygen-like isoelectronic series, 99

Particles, 153
Path integrals, 155
Pauli principle, 98
Perturbation method, 110
Perturbing potential, 118
Photon annihilation operators, 173
Photon creation operators, 173
Plane wave, 230
Plane wave, alternative expansion, 50
Plane wave, Sturmian expansion, 85
Plane waves in d dimensions, 46
Point groups, 114
Point masses, 196
Potential, 93
Potential energy, 211
Potential energy density, 167
Potential energy matrix, 88, 196
Potential-weighted orthogonality, 81
Potential-weighted orthonormality, 74, 75, 82, 83, 89, 93, 94, 105, 136, 138
Primitive configurations, 98
Principal quantum number, 93, 97
Product of Coulomb Sturmians, 138
Products of basis functions, 213
Projection, 13, 29
Projection operator, 13, 85
Properties of homogeneous functions, 3

Properties of homogeneous polynomials, 1, 10
Pure angular function, 12
Pythagorean theorem, 153

Quantum numbers, 93

R-block, definition, 96
R-blocks, 98, 99
Radial functions in d dimensions, 69
Raising and lowering operators, 223
Rank of a tensor, 174
Real spherical harmonics, 110
Relativistic correction, 216
Relativistic effects, 98, 100
Roots of the interelectron repulsion matrix, 98, 99
Rotations, 166, 197

Scalar density, 177
Scalar product, 129
Scalars, 175
Scaling parameter, 83, 93, 95
Schrödinger equation, 83, 199
Schrödinger equation in reciprocal space, 56
Schrödinger equation, integral form, 86
Second-iterated solution, 87
Secular equations, 90, 93, 95, 116, 163
Separability, 91, 201
Shibuya, T., 120, 132
Shibuya-Wulfman integrals, 90, 132, 135
Shibuya-Wulfman orbitals, 120
Shifted roots, 97
Shull, H., 71
Simple harmonic oscillators, 165
Simultaneous equations, 5
Slater determinant, 89, 91
Slater exponents, 96
Small displacements from equilibrium, 162, 199, 201
Small displacements from local minima, 200
Small internuclear separations, 147
Snell-Descartes laws of refraction, 184
Sobolev spaces, 82, 84, 85
Solid angle element, 57, 62, 130
Solution in direct space, 67
Space-time continuum, 183
Spectroscopic tables, 100
Spectrum of energies, 95
Spherical Bessel functions, 47
Spherical harmonics, 91
Spherical polar coordinates, 173
Spin-orbit coupling, 93
Spin-spin coupling, 93
Spring-like bonds, 196
Square-planar array, 112
Standard tree, 35, 178
Sturmian basis sets, 84
Sturmian expansion of a plane wave, 85
Sturmian secular equations, 83, 88, 93, 95
Sturmians, 2k, 137
Subspaces, 8
Sum rule, 48
Sum rule for spherical harmonics, 28
Sum rule, generalized, 28
Symmetry, 88, 99
Symmetry under iteration, 88
Symmetry-adapted basis functions, 116
Symmetry-adapted basis sets, 88
Synthesis of the wave function, 91
System of N electrons, 85
Systems of rays, 158
Systems of trajectories, 159

Table of hyperspherical harmonics, d=4, 18
Table of hyperspherical harmonics, d=6, 37
Taylor series expansion, 161, 196
Tensor capacities, 177
Tensor densities, 177
Tensors, 174
Theory of hyperspherical harmonics, 25
Total solid angle, 30, 230

Transformation law, 28
Transformation laws, 174
Translational wave function, 201
Translations, 166, 197
Trial Lagrangian density, 170
Truncated basis set, 88
Turning points, 91

Unit matrix, 97
Unitarity requirement, 28
Unitary transformation matrix, 196
Unitary transformations, 103, 110

Variational calculus, 153
Variational optimization, 96
Variational principles, 167
Velocity-dependent forces, 161
Vibrations, 166
Virial theorem, 92
Volume element, 175
Volume element in p-space, 62

Wave equation, 170
Wave equations, 167
Wave fronts, 158, 183
Wave lengths, 167
Wavelets, 183
Weighted nuclear charge, 91, 92
Weighted potential, 75, 81, 84, 89
Weighting factors, 90
Wulfman, C.E., 120, 132

Zero-frequency modes, 166, 197